TEACHER'S RESOURCE BOOK

# BSCS Biology
## *A Molecular Approach*

BLUE VERSION

8th *edition*



BSCS
5415 Mark Dabling Blvd.
Colorado Springs, CO 80918-3842

Jon Greenberg, Revision Editor

EVERYDAY LEARNING
Chicago, Illinois

**BSCS Biology: A Molecular Approach**

**BSCS Administrative Staff**
Chair, Board of Directors: Timothy H. Goldsmith
Executive Director: Rodger W. Bybee
Associate Director: Susan Loucks-Horsley
Financial Officer: Marcia Mitchell

**BSCS Project Staff**
Principal Investigator, 1998–1999: Michael Dougherty, Ph.D.
Project Director, 1999–2000: Jon Greenberg, Ph.D.
Production Director: Wendy Haggren, Ph.D.
Investigations Editor: Mark Bloom, Ph.D.
Editors: Barbara C. Resch and John Pesta
Art Coordinator: Jean P. Milani
Production Assistants: Rose Johnson and Linda Ward
Safety Reviewers: Geof Smith and Brendon Taylor
Development/Marketing Manager: Laura Engleman

**Everyday Learning Staff**
Editorial: Anna Belluomini, Kathleen Ludwig
Design: Jess Schaal
Production: Elizabeth Gabbard

**Additional Credits**
Project Manager: Karen Roberts
Editors: Fred Middendorf, Marianne Miller
Editorial, Design, and Production Services:
Laurel Technical Services
Design: Kim Buening Torbeck, Imbue Design
Photo Research: Fred Middendorf

**Photo Acknowledgments**
Cover image: Dr. Dennis Kunkel/Phototake
Photo credits: **15.1a** ©Alain Dex/Publiphoto/Photo Researchers, Inc.; **15.1b** CORBIS/Royalty Free; **15.1c, d** ©Dan Kline/Visuals Unlimited; **15.1e** ©Tom Stack

ISBN 0-538-69409-2



Everyday Learning Corporation
P.O. Box 812960
Chicago, IL 60681
www.everydaylearning.com

1 2 3 4 5 6 7 8 9 QW 05 04 03 02 01 00

# INTRODUCTION

The *Teacher's Resource Book* is a collection of copymasters designed to supplement and extend the text material in a variety of ways. Each item is keyed to the chapter or the chapter and section to which it most closely relates.

**Supplementary Topics** are in-depth essays that augment the content of the text. The essays expand on material in the text or introduce new topics and ideas to students. Each topic includes questions to stimulate classroom discussion; you may wish to develop additional questions to expand the discussions. The Supplementary Topics also include additional genetics problems to help students increase their understanding of the concepts.

The **Extended Laboratory Program** includes a variety of laboratory investigations that can be used either instead of or in addition to the investigations in the student textbook. This program includes 3 investigations that use gel electrophoresis, 2 investigations that may be conducted as microcomputer-based labs, 15 traditional investigations, a proposed laboratory practical examination, a materials list, and instructions for preparing solutions used in the investigations. The sequence of the Extended Laboratory Program follows the sequence of chapters in the text. Safety information appears in the teacher's notes for the investigations; refer to these notes before doing any of the investigations. See page 25 for more information.

**Current Literature** provides reprints of articles from *Science* magazine that describe current developments in biology; the articles relate to the material in the text but offer a different approach. These timely articles can be used as enrichment, as student readings, or as springboards for class or small-group discussion.

**Research Problems in Biology** suggest topics for independent laboratory research. The problems are designed to promote higher-level thinking skills associated with scientific investigation. Such skills involve problem recognition and clarification, hypothesis formation, and various aspects of experimental design. Working individually or in small groups, students have an opportunity to become engaged in the design aspects (including the design of safety plans) of a variety of proposed lab and field investigations and to develop their technical laboratory skills. In addition, students may gain practical experience in data organization, analysis, and interpretation. See page 187 for more information.

The Research Problems in Biology cover 14 topic areas related to eight of the chapters in the *BSCS Biology, eighth edition.* Approximately three specific research problems are suggested for each topic area. Your *Teacher's Resource Book* also offers suggestions on how students should approach the procedures, and, whenever necessary, it provides more specific procedures for them to follow. The book also provides information about materials and safety for both teachers and students.

**Blackline Masters** enable you to create teaching transparencies with reproductions of textual art and diagrams. You may choose to use many of these as student worksheets. The reproductions have been chosen for their usefulness in discussion and for their clarification of difficult concepts. Some blackline masters are required for particular investigations.

# CONTENTS

# CONTENTS

Chapter/Section Reference

## ◆ Extended Laboratory Program, continued



## ◆ Current Literature

# CONTENTS

Chapter/Section Reference

## ◆ Current Literature, continued



## ◆ Research Problems In Biology

# CONTENTS

# CONTENTS

# Supplementary Topics

# SUPPLEMENTARY TOPICS

# SUPPLEMENTARY TOPIC 1

◆ Chapter 6 Cell Structures and Their Functions ◆

**Technology for Studying Cells**

The possibility that some parts of the cell become lost or distorted during fixation and staining is a worry to researchers who study cells. However, if they use a variety of techniques to examine the cells and if all of these techniques produce the same image, then the researchers can feel confident that the image represents a real structure.

The most convincing way to establish that a structure actually exists is to examine living cells under a microscope without fixation or staining. This requires special optical systems designed to take advantage of the light-refracting properties of cells. When light passes through a living cell, the light waves bend. Light waves that pass through a relatively dense part of a cell, such as the nucleus, bend more than those that pass through the cytoplasm, which is less dense.

Regular light microscopes depend on a light source that is mounted underneath the platform, or stage. The specimen is illuminated from below, and little light refraction occurs between the light source and the microscope slide. The specimen is viewed from above through a series of glass lenses that magnify the image. Two types of microscopes, the phase-contrast microscope and the differential-interference-contrast microscope, exploit the different densities of different regions of a cell. Each region bends light waves to a different extent, and as a result, transparent structures become visible. The two types of microscopes differ in that the differential-interference-contrast microscope uses polarized light. With either microscope, many details of the living cell are clearly defined.

A simpler method uses the light scattered by the various cell components to reveal the details of an unstained cell. The method known as dark-field microscopy uses illumination from the side, rather than from underneath. As a result, only scattered light enters the microscope lenses, and the cell appears as a light object against a dark background.

Light microscopy enables us to observe living cells and to study processes such as mitosis. Because many cell processes occur rather slowly, scientists often use time-lapse photos or video recordings to study them.

The clarity of cells viewed with various light-microscope techniques is somewhat limited since structures cannot be resolved if they are smaller than the wavelength of the light being used for illumination. This limitation can be overcome by using electrons rather than light. The faster an electron moves, the shorter its wavelength and, therefore, the greater the resolution of the sample. Although special sample preparation problems and radiation damage limit the resolution of objects, electron microscopes still offer a resolution about 100 times greater than that of light microscopes. At best, this resolving power, or ability to render an object distinctly visible, is about 0.1 nm (1 Å).

The transmission electron microscope (TEM) is similar to the light microscope, but it is much larger. It uses bulky magnets in place of glass lenses to focus the electron beam. The electron microscope also must operate in a vacuum since air molecules will scatter the electron beam. Relative to a light microscope, the electron microscope is oriented in an upside down configuration (see Figure 1.1). "Illumination" comes from an electron-emitting filament at the top of a 2-meter-high cylinder. Electrons are accelerated and passed through a tiny hole, forming an electron beam that passes down the cylinder. The electron beam is focused by magnetic coils placed along the length of the cylinder. The sample is placed in the vacuum and exposed to the beam of electrons. Because different cell components have different densities, some of the electrons are scattered. The remainder are focused to form an image on a phosphorescent screen.

Samples viewed with electron microscopes require special preparation. First, the sample must be dried. To avoid distortion during drying, the wet tissue must be fixed with substances that stabilize lipids and that covalently bond protein molecules to one another. Next, the sample must be injected with a resin that infiltrates the cells and forms a solid block. The sample can then be cut into extremely thin sections with an instrument called a microtome.

Contrast in electron-microscope images depends on the atomic number of the atoms in the specimen. The higher the atomic number, the more electrons become scattered and, therefore, the greater the contrast. Molecules made up of atoms with low atomic numbers are stained through exposure to heavy metal salts, such as uranium or lead. The different components of the cell show varying degrees of contrast depending on the amount of salts absorbed. For example, lipids stain darkly, showing the location of cell membranes.

Electron microscopes can produce three-dimensional images. The scanning electron microscope (SEM) uses an electron beam reflected from the surface of the sample rather than passed through it. First, the sample is fixed and dried. Then it is coated with a thin layer of heavy metal evaporated onto it—a process called shadowing. In the SEM, the sample is scanned with the focused electron beam. The reflected electrons are detected and converted to an image on a television screen. The greatest resolution is obtained by an electron microscope that combines aspects of TEM and SEM. This type of instrument, called the scanning-tunneling electron microscope, or STEM, is described in Biological Challenges, Chapter 2.

Two other methods of preparing samples—freeze-fracturing and freeze-etching—are particularly useful in allowing three-dimensional images to be obtained using an electron microscope. The sample is frozen at a very cold temperature, and a cold knife is used to fracture the specimen. The knife does not cut cleanly through the frozen cell but splits it open, rather like breaking apart a

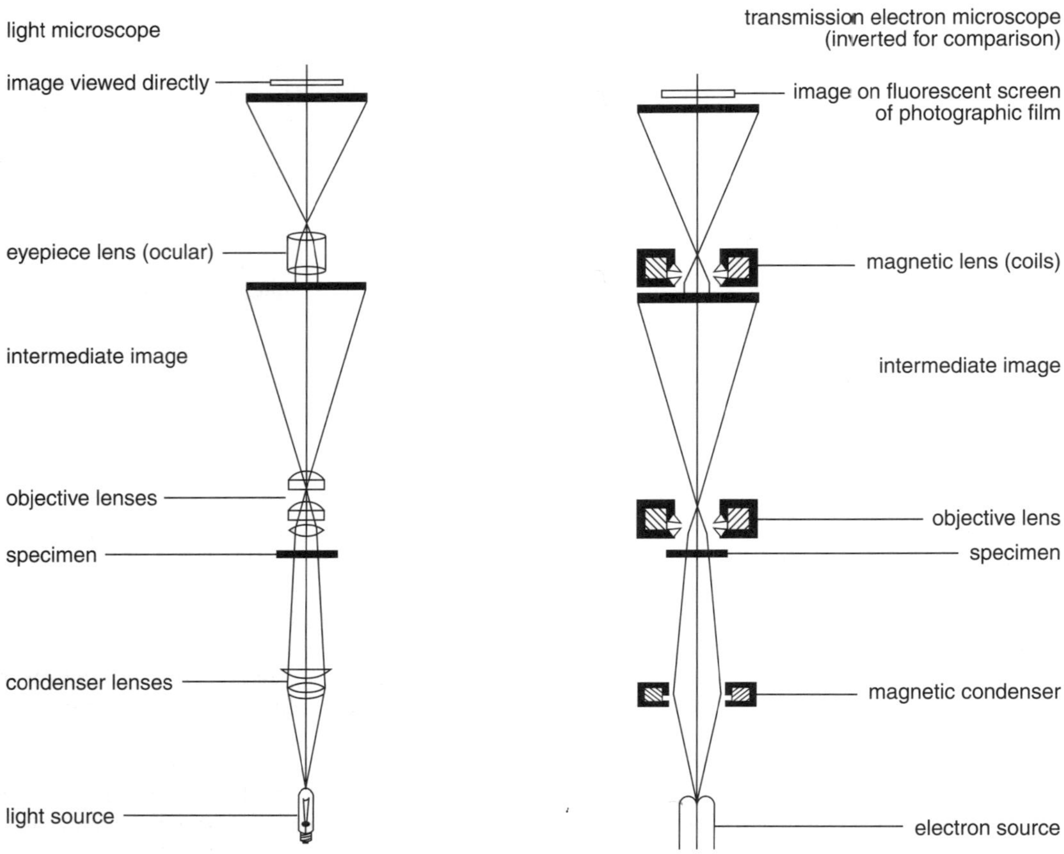


**FIGURE 1.1**
A comparison of a light microscope and a transmission electron microscope.

muffin. The surface of the fractured cell is enhanced by etching—the removal of the water in a vacuum. This process leaves membrane proteins intact with one of the lipid bilayers.

A fine mist of platinum is sprayed at an angle onto the fractured cell surface. Shadows occur where the platinum mist is blocked by irregularities and bumps of the cell. A film of carbon is added to strengthen the platinum coating. The original cell is dissolved with acids and enzymes, leaving only the platinum-coated replica. The three-dimensional replica can then be examined under the electron microscope.

**Discussion Questions**

1. What limits the resolution of the light microscope?
2. In microscopy, how can one distinguish between a real structure and a structure that is created as a consequence of the sample preparation?
3. Why can't X rays be used to increase the resolution of the light microscope?
4. What are the advantages and disadvantages of light and electron microscopy?

# SUPPLEMENTARY TOPIC 2

◆ Chapter 7 Transport Systems ◆

**EKGs**

At some time, possibly while giving you a physical examination, a doctor may have taken an electrocardiogram (EKG) of your heart. An electrocardiogram is a recording of the electrical changes that occur in the muscle fibers within the heart as it beats. The heartbeat is controlled by specialized cells in a region of the heart called the pacemaker. Cells in the pacemaker generate rhythmic electrical impulses by changing the charge distributions across their cell membranes. This process is known as depolarization of the cell membrane. These electrical changes can be detected on the surface of the body because body fluids can conduct electrical currents.

To record an EKG, metal electrodes are placed in certain locations on the skin. These electrodes are connected by wires to a recording instrument that responds to very weak electrical changes, causing a pen to mark a moving strip of paper. When the machine is operating, the up-and-down movements of the pen correspond to electrical changes occurring within the body because of heart muscle activity. The paper moves at a known rate, and the distance between the deflections of the pen can be used to measure the time that elapses between various phases of the cardiac cycle. The procedure is painless.

The EKG pattern includes several deflections, or waves, during each cardiac cycle (see Figure 2.1). Between the cycles, the muscle fibers remain polarized, and no detectable electrical changes occur. Consequently, the pen simply marks along the baseline as the paper moves through the instrument. However, when the pacemaker triggers a cardiac impulse, muscle cells in the atrial fibers are stimulated to change their electrical charge. As a result, the pen is deflected. When this electrical change is completed, the pen returns to the base position. This deflection is called a P wave, and it corresponds to the depolarization of the atrial fibers just before they contract.

When the cardiac impulse reaches the muscle fibers of the ventricles, they are stimulated to rapidly depolarize. The ventricular walls of the heart have more surface area than the atrial walls; and the amount of electrical change is, therefore, much greater. Subsequently, the pen is deflected to a greater degree than before. Once again, when the change is completed, the pen returns to the baseline, leaving a mark called the QRS complex. Near the end of the EKG pattern, the pen is deflected for a third time, recording a T wave. This wave is caused by the ventricular muscle fibers' slowly regaining their resting potential, or repolarizing. The record of the atrial repolarization does not appear in the pattern because the atrial fibers repolarize at the same time the ventricular fibers depolarize.

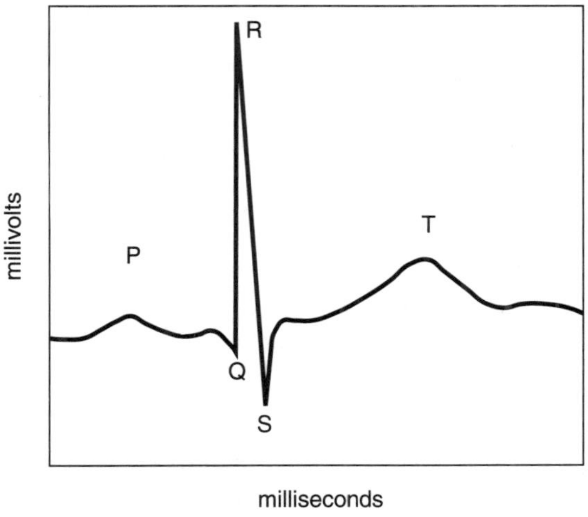


**FIGURE 2.1** An EKG pattern showing the P wave, QRS complex, and T wave.

EKG patterns reveal how well a heart conducts impulses; therefore, a physician can study an EKG to judge the condition of the heart. Although slight variations in heart actions sometimes occur normally, marked changes in the usual rate, or rhythm, may indicate cardiovascular disease. Such abnormal action is called arrhythmia. Sometimes a patient who has this condition can be fitted with an artificial pacemaker. Such a device contains a small battery- or nuclear-powered electrical stimulator. To install an artificial pacemaker, the surgeon threads electrodes through the veins into the right ventricle and implants the stimulator beneath skin in the shoulder or abdomen. The artificial pacemaker then transmits rhythmic electrical impulses to the heart, and the muscle fibers respond by contracting. Although some artificial pacemakers operate at a fixed rate, others respond to changing needs of body cells for increased or decreased blood supply.

**Discussion Questions**

1. Why are EKGs important tools?
2. Suppose a teacher were setting up a classroom investigation using EKGs of students. What bioethical issues would this raise? Would such an investigation be a good idea? Explain.
3. Research the effects of nitroglycerine tablets on patients with angina. How does the drug work? What other drugs are used to treat angina? How do they work?

# SUPPLEMENTARY TOPIC 3

◆ Chapter 13 Patterns of Inheritance ◆

**Gene Interactions**

When a trait is controlled by a single gene and is expressed in a dominant or recessive form, the predicted $F_2$ phenotypic ratio is 3:1. In the case of codominance, the $F_2$ phenotypic ratio is 1:2:1. Unexpected ratios among $F_2$ offspring may mean that the inheritance of a trait is controlled by more than one gene. When purple-flowered sweet peas are crossed with certain white-flowered sweet peas, all the $F_1$ plants have purple flowers. Purple seems to be dominant over white. The $F_2$ generation, however, shows a 9:7 ratio of purple to white plants. This ratio differs from the 3:1 expected in a monohybrid cross or even the 9:3:3:1 ratio of a standard dihybrid cross.

Figure 3.1 illustrates the interaction of genes in determination of sweet-pea color. Two different genes in the pea affect petal color. Alleles of these genes are represented by *P*, *p*, *C*, and *c*. Purpleness is produced by the allele *P*, but it cannot be expressed unless at least one *C* allele is present. Any combination of *P* and *p* with *cc* produces white flowers. The genotype *pp* with any combination of *C* and *c* also produces white flowers. This phenomenon is called epistasis, a term that describes how one gene interferes with the expression of another independently inherited gene.

A cross between a black guinea pig and an albino guinea pig produces only black $F_1$ offspring. When these heterozygotes are crossed, the $F_2$ offspring occur in a ratio of 9 black: 3 brown: 4 albino. What might this ratio mean? Two genes appear to be involved. One gene may be present as the *B* allele that causes the deposition of pigment molecules in the coat. The homozygous *bb* genotype results in less pigment deposition; thus, the animal is brown. Another gene controls the production of the skin pigment melanin. A *cc* genotype results in no pigment production. An animal with this genotype is an albino. Even when a *B* allele is present, the guinea pig will be albino if it has the *cc* genotype. In this example of epistasis, an allele of one gene *(cc)* obliterates the phenotypic expression of the allelic alternatives of the other gene *(B)*.

In any complex metabolic pathway, a nonfunctional enzyme (coded for by a mutant allele) may prevent the successful actions of enzymes later in the pathway. If a chemical precursor is not produced, it cannot serve as a substrate for another enzyme. For example, a precursor chemical, H, is necessary for the production of the A and B red blood cell surface proteins. An allele labeled *H* is responsible for the formation of this precursor. Most type O persons form H and have no antibodies in their sera for H, but they have antibodies for A and B blood proteins. Researchers also have identified the Bombay blood type, so named because it was discovered in that city. This blood type is homozygous for the mutant allele *h* and forms no H. An individual with this blood type has antibodies for the H precursor but cannot produce A or B blood proteins, even with a genotype for Type A or Type B type blood. The *h* gene is said to be epistatic to the *A* and *B* genes.

**Discussion Questions**

1. Given the information about the guinea pigs, determine how the 9:3:4 ratio results from a cross of two black *CcBb* animals.
2. Review the information in the situation involving the *H* allele. A woman carrying the gene for blood group B (she gave birth to an AB child by a father with Type A blood) had no B protein on her red cells. Explain her blood genotype and phenotype.

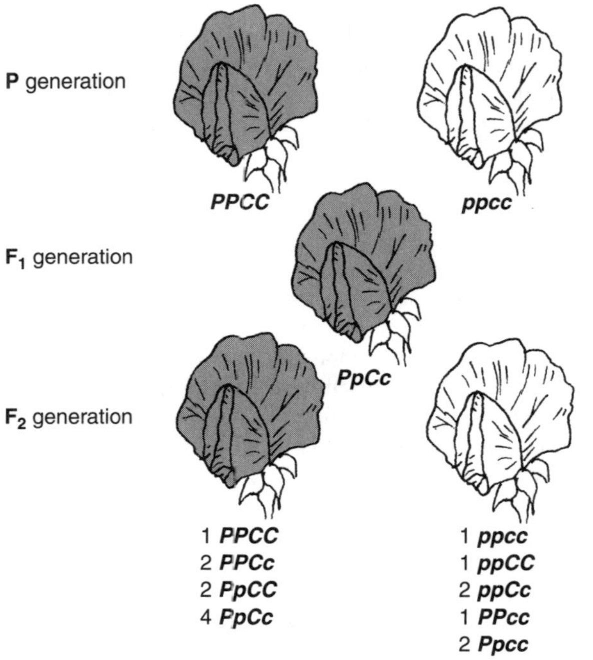


**FIGURE 3.1** Interaction of genes in the determination of color in sweet peas. The P allele is needed for the production of purple pigment, but this allele is not expressed in plants homozygous for allele *c*. The *C* allele is necessary if the pigment coded by the *P* gene is to be produced.

# SUPPLEMENTARY TOPIC 4

◆ Chapter 13 Patterns of Inheritance ◆

**Genetics Problems**

1. The polled (hornless) trait in cattle is dominant. The horned trait is recessive. A certain polled bull is mated to three cows. Cow A, which is horned, gives birth to a polled calf. Cow B, also horned, produces a horned calf. Cow C, which is polled, produces a horned calf. Name the genotypes of all parents.
2. In shorthorn cattle, when a red bull *(RR)* is crossed with a white cow *(rr)*, the offspring are roan (intermingled red and white hairs). How could a rancher establish a herd of roan cattle?
3. In sheep, white coat is dominant. Black is recessive. Occasionally, a black sheep appears in a flock. Black wool is worthless. How could a farmer eliminate the genes for black coat from the flock?
4. In summer squash, white fruit color is dominant. Yellow is recessive. A squash plant that is homozygous for white is crossed with a homozygous yellow one. Predict the appearance of (a) the $F_1$ generation, (b) the $F_2$, and (c) the offspring of a cross between an $F_1$ individual and a homozygous white individual.
5. Identify the possible blood types of children in the following families:
   (a) Type A mother, Type A father
   (b) Type A mother, Type O father
   (c) Type B mother, Type AB father
   (d) Type AB mother, Type AB father
   (e) Type A mother, Type B father
6. Before Mendel, the popular concept of heredity was the theory of bloodline inheritance. According to this theory, the parents' traits are blended in the offspring, just as two liquids blend when mixed together. Mendel's theory rested on the idea that traits are transmitted by particles (genes) and do not blend. Give evidence in support of the older theory. Then show how the results of Mendel's experiments fail to fit that theory.
7. How is variation introduced into populations that reproduce asexually?
8. At the present time, there is no such thing as an all-blue tulip. The first one found will be quite valuable. How might a tulip breeder increase the chances of finding a blue tulip?
9. A die (singular of "dice") has six sides. What is the probability that an even number will come up on one throw of a die?
10. If base pairs in DNA molecules consist only of adenine-thymine or cytosine-guanine, what must be the ratio of the amount of adenine to thymine? of the amount of cytosine to guanine? of the amount of adenine to cytosine?
11. In garden peas, tall vine is dominant and short vine is recessive. If a homozygous tall plant is crossed with a homozygous short plant, what genotypes are possible in the $F_1$ generation?
12. In garden peas, inflated pod is dominant and constricted pod is recessive.
    (a) If a plant homozygous for inflated pod is crossed with a plant homozygous for constricted pod, what ratio of phenotypes would you expect to find in the $F_2$ generation?
    (b) What ratio of genotypes would you expect?
13. In guinea pigs, short hair is dominant and long hair is recessive. A shorthaired male and a shorthaired female produced mostly shorthaired offspring, but a few were longhaired. Show how you can determine the genotypes of the parents.
14. In laboratory mice, the normal gray color is dominant over the albino color. Starting with purebred albino and purebred gray as parents, what is the ratio of phenotypes in the $F_2$ generation?
15. A pea plant that was homozygous for axial flowers (the dominant strain) was crossed with a plant that was homozygous for terminal flowers.
    (a) What ratio of genotypes would you expect in the $F_2$ generation?
    (b) What ratio of phenotypes would you expect?
16. In the following cases, $Z$ stands for a certain dominant trait and $z$ stands for the recessive trait. Identify the ratios of genotypes that you would expect from the following crosses: (a) $ZZ \times zz$, (b) $Zz \times Zz$; (c) $Zz \times zz$, (d) $Zz \times ZZ$.
17. In a certain species of plant, one purebred variety has hairy leaves and another purebred variety has smooth leaves. A cross of the two varieties produces only offspring with smooth leaves. Predict the ratio of phenotypes in the $F_2$ generation.
18. In corn, yellow seed color is dominant and white is recessive. A certain ear of corn had a mixture of yellow seeds and white seeds. From what color seeds could the parents have grown?
19. In a certain animal, one breed always has a hairy tail and another breed always has a naked tail. How would you determine which trait is dominant?
20. In garden peas, axial flower position is dominant and terminal flower position is recessive; also, tall vine is dominant and short vine is recessive. A plant that is known to be purebred for tall vine and axial flowers is crossed with a plant having short vines bearing terminal flowers.
    (a) What is the phenotype of the offspring?
    (b) What is the genotype of the offspring?
    (c) Predict the types of offspring (phenotypes) that would appear in the $F_2$ generation and their ratios.
21. In snapdragons, red flower color ($R$) is codominant with white ($r$), the heterozygotes being pink; the normal (broad) leaves ($B$) are codominant with narrow (grasslike) leaves ($b$), the heterozygotes having leaves of medium breadth. If a red-flowered, broad-leaved plant is crossed with a white-flowered, narrow-leaved one, what will be the phenotypes and their expected ratio in the $F_2$ generation?
22. Several different genetic situations are described below. In each case, decide which of the following

types of inheritance are involved:

(a) codominance
(b) linkage with little or no crossing over
(c) two genes on different chromosomes
(d) multiple genes on different chromosomes
(e) multiple alleles
(f) linkage with a high percentage of crossing over

Situation 1: You have a number of plants belonging to an $F_2$ generation. About 25% have no thorns, about 25% have long thorns, and about 50% have short thorns. What type of inheritance is involved?

Situation 2: In another set of plants, also an $F_2$ generation, you find the following phenotype ratios: 9 red flowers and broad leaves, 3 red flowers and narrow leaves, 3 white flowers and broad leaves, 1 white flower and narrow leaves. What type of inheritance is involved?

Situation 3: You discover a species of bird. In the population, some have long crests on their heads, some have scarcely any crests, and many have crests showing all possible intermediate lengths. What type of inheritance is involved?

Situation 4: In a certain species of insect, three eye colors are known: red, orange, and yellow. If purebred red-eyed individuals are crossed with purebred orange-eyed, all the offspring are red-eyed. If purebred orange-eyed are crossed with purebred orange-eyed, all the offspring are orange-eyed. If purebred red-eyed are crossed with purebred yellow-eyed, all the offspring are red-eyed. What is the most likely method of inheritance?

Situation 5: In another case in which you study two traits of a plant at the same time, you find in the $F_2$ generation that 75% have hairy stems and green seeds while the other 25% have smooth stems and brown seeds. What type of inheritance is involved?

Situation 6: Among many varieties of popcorn, one has an average ear length of 6 cm (ranges from 4 to 9) and another has an average ear length of 16 cm (ranges from 12 to 19). When the two varieties are crossed, the $F_2$ offspring have ears with an average length of 11 cm (ranges from 5 to 19). What is the most likely method of inheritance?

Situation 7: The $F_2$ generation of a dihybrid cross in a plant showed the following ratio of phenotypes: 66% had hairy stems and round fruit, 9% had hairy stems and pear-shaped fruit, 9% had smooth stems and round fruit, 16% had smooth stems and pear-shaped fruit. What type of inheritance is involved?

23. In poultry, rose comb is a dominant trait (*R*) and single comb is recessive (*r*). A rose-combed male is mated with two rose-combed females. Female A produces 14 chicks, all rose-combed. Female B produces nine chicks, seven rose-combed and two single-combed. What are the genotypes of the three parent birds?

24. The Jones family has eight children, all of whom are girls. What is the chance that the next child will be a boy? Explain.

25. In humans, normal pigmentation is a dominant trait (*C*) and albinism is a recessive trait (*c*).
(a) A normally pigmented man marries an albino woman. Their first child is an albino. What are the genotypes of these three individuals?
(b) If the couple in (a) has more children, what would they probably be like?
(c) An albino man marries a normally pigmented woman. They have nine children, all normally pigmented. What are the genotypes of the parents and the children?

26. In dogs, wire hair is a dominant trait and smooth hair is recessive. Two wirehaired dogs produce a male puppy that is wirehaired. To find out most quickly whether the puppy carries the factor for smooth hair, to what type of female should you mate him?

27. In guinea pigs, the coat may be rough or smooth. Certain rough-coated guinea pigs, when crossed with smooth-coated ones, produce all rough-coated offspring. Other rough-coated guinea pigs that are crossed with smooth-coated ones produce equal numbers of rough-coated and smooth-coated offspring. Smooth-coated guinea pigs crossed together always produce smooth-coated offspring. Explain these results. Write the genotypes for all the animals involved in the crosses.

28. Identify the gametes that would be produced by organisms having the following genotypes: (a) *AaBB*, (b) *aaBB*, (c) *AAbb*, (d) *AaBBCc*.

29. In horses, black is a dominant trait (*B*) and chestnut is recessive (*b*); also, the trotting gait is dominant (*T*) and pacing is recessive (*t*).
(a) If a homozygous black pacer is mated to a homozygous chestnut trotter, what will be the genotype(s) and phenotype(s) of the $F_1$ generation?
(b) If two $F_1$ individuals were mated, what types of offspring could they have and in what proportions?
(c) If an $F_1$ male were mated to a homozygous female black pacer, what types of offspring could be produced and in what proportions?

30. In cocker spaniels, black is a dominant trait (*B*) and red is recessive (*b*). Solid color is dominant (*S*), and white spotting is recessive (*s*). A red male was mated to a black-and-white female. They had five puppies: one black, one red, one black-and-white, and two red-and-white. What were the genotypes of the parents?

# SUPPLEMENTARY TOPIC 5

◆ Chapter 13 Patterns of Inheritance ◆

**Heredity and the Environment**

Scientists have long debated whether genetics or lifestyle plays the greater role in determining how long we live, and the answer has been elusive. In March 1988, however, a group of Danish scientists reported that both factors exert a strong influence on life expectancy.

University of Copenhagen geneticist Thorklid I. A. Sorensen and his colleagues studied adoptees from 960 families. Studies of people who grow up in adoptive homes are common in genetic research. By comparing adoptees with both their biological and adoptive parents, geneticists can investigate the genetic and environmental factors contributing to various traits and diseases.

For example, if children brought up in adoptive homes are found to die at a relatively early age of the same diseases or disorders as their natural parents, then genetics was probably the larger factor in their deaths. If, on the other hand, adoptees die prematurely of the same causes as their adoptive parents, one can conclude that the home environment—including eating, drinking, smoking, and exercise habits—played a greater role in their deaths.

The adoptees studied by Sorensen's research team were born between 1924 and 1926, and all were adopted by people unrelated to them. The scientists compared the causes of premature death—between the ages of 16 and 58—among the adoptees with the causes of death of their biological and adoptive parents.

The survey revealed that the death of a biological parent before age 50 almost doubled the general overall risk of premature death for the adoptee. The death of an adoptive parent before age 50 did not increase the general risk of death for the adoptee.

When Sorensen's group examined specific causes of premature death, however, the picture changed considerably. The scientists found a fivefold risk for adoptees if a biological parent had died of an infectious disease, a fourfold risk if a biological parent had died of heart disease or stroke, but only a slightly increased risk if a biological parent had died of cancer. If the adoptive parent had died of heart disease or stroke, the adoptee's risk roughly tripled, and—the most surprising finding—there was a fivefold increase in risk to the adoptee if the adoptive parent had died of cancer.

Some of these findings seemed to contradict prevailing views about the role of genetics in disease. For example, scientists had expected a shared family environment to be more important than genetic factors in infectious disease. On the other hand, recent research has shown a strong genetic basis in most cancers. How, then, does one explain Sorensen's data?

With respect to infectious diseases, Sorensen explained that the high death rate from infections among adoptees and their biological parents might mean that some people's immune systems are genetically weaker than other people's. Sorensen said the high incidence of cancer among adoptees whose adoptive parents had cancer could be due to such environmental factors as cigarette smoking.

Sorensen's study suggests that it is probably fruitless to try to determine whether genetics or environment is the more important cause of premature death. The research indicates that most major chronic diseases result from the cumulative effect of environmental factors in people who are genetically susceptible to those diseases.

**Discussion Questions**

1. Why did Sorensen choose to study the factors contributing to the premature deaths of parents and adoptees?
2. Would these same findings be likely to occur if the study focused on deaths that were not considered premature? Explain.
3. How might you go about conducting further studies of whether genetics or the environment plays a larger role in longevity? Be sure to include provisions for controls.

**Source**

# SUPPLEMENTARY TOPIC 6

◆ Chapter 15 Advances in Molecular Genetics ◆

**The Molecular Biology of Cancer**

President Nixon declared war on cancer in 1971. He provided an army of scientists with the research dollars they needed to study the genetic influences on the disease. The timing was fortunate since the tools of molecular biology were just being developed. Although scientists could study cancer in animals and even in cultured cells growing in petri dishes, they could not hope to isolate specific genes associated with cancer (oncogenes). Recombinant-DNA technology gave scientists the ability to dissect the human genome, setting the stage for the dramatic successes of the 1980s.

After initial success isolating oncogenes from animal tumors and cultured cells, the first human oncogene, called *ras*, was cloned from a human bladder carcinoma in 1982. When scientists compared the DNA sequence of the *ras* oncogene to that of the same gene isolated from normal cells, they were surprised to find that the cancer gene differed from the normal gene in just one nucleotide position. The presence of the altered *ras* protein helps direct the cell to keep dividing nonstop. This loss of control over cell division leads to the formation of a tumor and the disease we call cancer. The *ras* oncogene functions as a dominant mutation. If a single copy of the normal gene becomes mutated, then cancer will result. In the years immediately following 1982, several other human oncogenes were isolated, and each functioned in a dominant fashion.

In 1987, scientists studying a cancer of the eye called retinoblastoma expanded our view of oncogenes. This type of cancer, which affects young children, occurs in hereditary and nonhereditary forms. The disease can be successfully treated if it is detected at an early stage. The researchers noticed that a number of children with retinoblastoma were missing a piece from chromosome 13. Using this clue, they tested a number of DNA markers until they found one that pinpointed the location of the retinoblastoma gene. Unlike the previously discovered oncogenes, the mutated retinoblastoma (Rb) gene behaved as a recessive mutation. Both copies of the Rb gene had to be mutated for cancer to occur. This new type of oncogene was called a tumor-suppressor gene.

Humans (and other organisms) possess intricate programs that direct specific cells to divide, other cells to stop dividing, and even some cells to kill themselves. During development, this program ensures that our arms grow to the same length and that cells between our toes die so we won't have webbed feet like a duck. Even when we are adults, some of our cells, such as skin cells, are programed to divide rapidly, whereas others, including many brain cells, stop dividing altogether. There are probably several hundred genes that have important roles to play in regulating cell growth. These genes, in their unmutated state, are called proto-oncogenes. Each one is essential for our good health, but at any time they can mutate and contribute to cancer. It is the mutated gene that is referred to as an oncogene. Fortunately, cancer is a multihit process. Several specific genes must be mutated, all within the same cell, before a tumor will develop. A recent study demonstrated that normal human cells could be made cancerous by introducing mutations in three specific proto–oncogenes.

Dominant acting oncogenes, such as *ras*, behave like an accelerator in a car. They direct the cell to keep dividing. Conversely, recessive acting oncogenes, such as Rb, behave like car brakes. They stop the cell from dividing. They are recessive because a single working copy of the gene makes enough protein to function as a brake. Only when both gene copies are mutated does the cell start to divide out of control. These are called tumor-suppressor genes because their normal function is to suppress cell growth.

It is important to remember that cancer is ultimately a genetic disease in which good genes have gone astray. Some people are born with mutations that predispose them to cancer. Usually, however, mutations leading to cancer are collected during the individual's lifetime. This explains why the environment has such a profound effect on the incidence of cancer. If an individual leads a less healthy lifestyle—for example, by smoking—or works in a less healthy environment, such as an asbestos factory, then that individual is more likely to collect the mutations necessary for cancer to develop.

By discovering the molecular pathways involved in the regulation of cell growth, scientists can gain a deeper understanding of the different types of cancer and the genes that drive them. This new insight is leading to development of diagnostic tests, new anticancer medicines, and even the promise of treatment through gene therapy. In the future, doctors will be able to identify the specific mutations associated with a patient's cancer. This knowledge will permit them to target therapy, not only to the type of cancer but also to the exact mutations found in the patient's tumor.

**Discussion Questions**

1. Why is cancer more prevalent among the elderly than among the young?
2. Suppose research leads to a cure for cancer. How might this affect the medical profession and the health of the general population?
3. What bioethical issues may develop as a result of new cancer tests and treatments?

**References**

Hahn, W. C. et. al. "Creation of Human Tumor Cells with Defined Genetic Elements." *Nature* 400, 1999.

Weinberg, R. A. *One Renegade Cell: How Cancer Begins.* Basic Books, 1998.

# SUPPLEMENTARY TOPIC 7

◆ Chapter 15 Advances in Molecular Genetics ◆

**Dominance and Recessiveness**

Gregor Mendel studied seven traits in pea plants. Each trait came in two forms, such as wrinkled or smooth seeds. Mendel interpreted his results to mean that there was a factor (today, we would say an allele) for each form of a given trait (gene). When a plant received two copies of the same allele, its corresponding phenotype was observed. Surprisingly, however, when a single copy of each type of allele was present, only one phenotype—always the same one—was observed. In Mendel's day, this went against the prevailing attitude, which expected a blending of the two possible phenotypes. Mendel reasoned that one allele was dominant relative to the other, which he termed recessive. The recessive phenotype is seen only when the plant contains two copies of the recessive allele.

With our present understanding of molecular genetics, we can reflect on the concept of dominance at a much deeper level. We can ask whether the DNA sequences for the dominant and recessive alleles are related to each other. Returning to the example of wrinkled and smooth seeds, we can ask whether one gene product produces wrinkled seeds and a different gene product produces smooth seeds. The gene associated with Mendel's wrinkled and smooth seeds was cloned in 1990. The gene codes for a protein called starch branching enzyme. When a functional form of the enzyme is present, normal starch grains are produced, resulting in the smooth phenotype. The recessive allele results from the insertion of an extra DNA sequence, which is called a transposable element, into the gene for starch branching enzyme. The insertion of this extra DNA into the gene disrupts its expression and prevents a functional enzyme from being produced. Without a functional starch branching enzyme, the seed cannot form normal starch grains and takes on a wrinkled appearance. Mutations resulting in loss of a functional gene product are a common mechanism for creating recessive alleles. As long as one unmutated copy of the gene is present, enough functional protein is made to produce the dominant phenotype.

More rarely, mutations can give rise to dominant alleles. Mutations producing dominant alleles are associated with genes for structural or regulatory proteins. For example, the blood disease hereditary spherocytosis is caused by a dominant mutation. The mutation is located in a gene that encodes a protein called ankyrin. Ankyrin is a protein found in the membranes of red blood cells. It helps the blood cell to maintain its characteristic doughnut shape. Mutations in one copy of the ankyrin gene lead to production of defective ankyrin protein. Even though one allele is directing the synthesis of normal ankyrin protein, the presence of the defective ankyrin from the mutated allele causes the cell to lose its doughnut shape and take on a spherical form. The spherical blood cells are more fragile and don't live as long as normal blood cells. This leads to anemia, which is a symptom of the disorder.

**Discussion Questions**

1. How can gene mutations result in what looks like blended inheritance or incomplete dominance?
2. What is the reason that dominant mutations are associated with genes for structural or regulatory proteins?
3. In gene therapy, doctors introduce healthy copies of a gene into the cells of a patient with a genetic disorder. Would this approach work equally well for dominantly and recessively inherited diseases? Explain.

**References**

Bhattacharyya, M. K., A. M. Smith, T. H. Ellis, C. Hedley, and C. Martin. "The Wrinkled-Seed Character of Pea Described by Mendel Is Caused by a Transposon-like Insertion in a Gene Encoding Starch-Branching Enzyme." *Cell*, January 12, 1990.

Costa, F. F., P. Agre, P. C. Watkins, J. C. Winkelmann, T. K. Tang, K. M. John, S. E. Lux, and B. G. Forget. "Linkage of Dominant Hereditary Spherocytosis to the Gene for the Erythrocyte Membrane-Skeleton Protein Ankyrin." *New England Journal of Medicine* 323, 1990.

# SUPPLEMENTARY TOPIC 8

◆ Chapter 15 Advances in Molecular Genetics ◆

**Mothers and Fetuses: Whose Rights Are More Important?**

Should a pregnant woman be forced to change a behavior that might harm the fetus that is developing in her uterus? Should a pregnant woman be allowed to refuse to change her behavior on the grounds that she has the right to do as she pleases with her own body? Whose rights are more important?

These questions are not merely intellectual exercises for specialists in bioethics. They are real questions that have arisen because of progress in medical genetics. One of the most troubling problems involves a genetic disorder called phenylketonuria (PKU).

PKU is inherited as an autosomal recessive disorder. Affected individuals lack an enzyme called phenylalanine hydroxylase. As a result, they cannot break down phenylalanine, an amino acid that is present in many common foods. Instead of breaking down phenylalanine to tyrosine, another amino acid, the phenylalanine is broken down into a substance that damages the developing brain. Affected children who eat foods that contain large amounts of phenylalanine—breads, eggs, cheese, fruits, vegetables, and some meats—will suffer severe mental retardation.

A simple blood test can detect PKU shortly after birth, and the test is mandatory in almost all states. Children diagnosed with PKU are placed on a special diet, which prevents development of mental retardation. Neurologists used to recommend that children stay on the diet until they were at least 12 years old, but there is accumulating evidence that lifelong adherence to the diet is beneficial.

Unfortunately, the special medical diet for children with PKU is not very appetizing, and it is very expensive—about $5,000 to $7,000 per year. Nonetheless, mandatory screening and the special diet have saved many children from profound metal retardation. So what is the problem?

Many of the young girls who were first treated with the special diet are now young women who are beginning to have children of their own. These young women are no longer on the special diet, but they still have PKU. In other words, they still cannot break down phenylalanine properly. The toxic substances that the mother produces cross the placenta and enter the fetus, causing retardation, microcephaly (small head), and other problems, including heart malformations.

A mother with PKU can prevent such damage to the fetus by going back on the diet during pregnancy. Because this diet is unpleasant, some mothers are reluctant to return to it. It is ironic that the problems caused by maternal PKU result from our success in treating the disorder in children. This demonstrates how unintended consequences can occur when we use science and technology to solve a problem or otherwise improve our quality of life.

**Discussion Questions**

1. Suppose a mother with PKU does not want to go back on the diet. What rights does she have to control what happens to her own body? Do these rights include the right to jeopardize the health of a developing fetus?
2. Behaviors such as drinking alcohol and smoking can damage a developing fetus. Should a pregnant woman be required to give up these behaviors? Are her civil liberties more important than the rights of the fetus?
3. Identify other situations in which medical and technological knowledge has resulted in bioethical difficulties.

# SUPPLEMENTARY TOPIC 9

◆ Chapter 19 Changes in Species ◆

**Continental Drift and Evolution**

In 1915, Alfred Wegner, a German scientist, proposed the idea that South America and Africa had once been joined in a supercontinent he called Pangaea. His theory was not well accepted by most of the geologists and biologists of the time, but during the last 20 years, researchers have gathered data that support his idea. We now know that about 200 million years ago, all the continents were together in one huge land mass (see Figure 9.1).

The earth's upper mantle and crust are divided into plates. According to the theory of plate tectonics, new material from the mantle is added to the plates along ridges of the ocean bottom. The Mid-Atlantic Ridge is one example; other ridges exist in the Pacific Ocean. The addition of new material to the earth's crust forces the plates to move slowly away from the ridges. At other locations, the plates are grinding into one another and forcing crustal materials into the mantle through deep ocean trenches.

The continents are carried along on the surface of these plates as they move over the liquid interior of the earth. Unless two landmasses are on the same plate, their positions relative to one another change. For example, North America and Europe are currently drifting apart at a rate of about 2 cm per year. Many important geological phenomena (such as mountain building, volcanic activity, and earthquakes) happen at the boundaries of the plates. The San Andreas Fault in California is part of a border where two plates slide past each other. Mount Saint Helens in Washington State is near another plate boundary. The Himalayas are the result of a collision of the Indian subcontinent with Asia, a process that is still continuing today.

The history of the earth helps explain the current geographic distribution of species. For example, the emergence of islands, such as the Galapagos, creates new environments for founding populations, and adaptive radiation fills many of the available niches with new species. On a global scale, the drifting of the continents is the major factor in the spatial distribution of life.

Plate movements rearrange geography all the time, but two past events have been especially significant in the distribution of organisms. About 250 million years ago, plate movements brought all the landmasses together, forming Pangaea. Species that had been evolving in isolation now were competing against one another. The total amount of shoreline was reduced, and there is evidence that the ocean basins increased in depth, draining most of the shallow coastal seas. Just as they do today, most marine species inhabited the shallow waters. The formation of Pangaea destroyed a considerable amount of that habitat. Changing ocean currents also affect life on land and sea.

The other event occurred about 180 million years ago. The breakup of Pangaea served as an isolating mechanism of gigantic proportions. As the continents drifted apart, each became a distinct evolutionary theater, and the organisms diverged. The pattern of continental separation solves the puzzles of geographical distribution, such as the discontinuous distribution of placental and marsupial mammals.

Australia provides a good example. Until about 53 million years ago, Australia was attached to a much warmer Antarctica, as was South America. Marsupials, which are best represented in Australia and South America today, appear to have migrated overland between these continents via Antarctica. As the two continents moved farther away from Antarctica, the land bridges were lost. From then on, as Australia moved steadily northward, its peculiar flora and fauna evolved in isolation.

If mechanisms such as continental drift isolate breeding groups of organisms, why do many of these

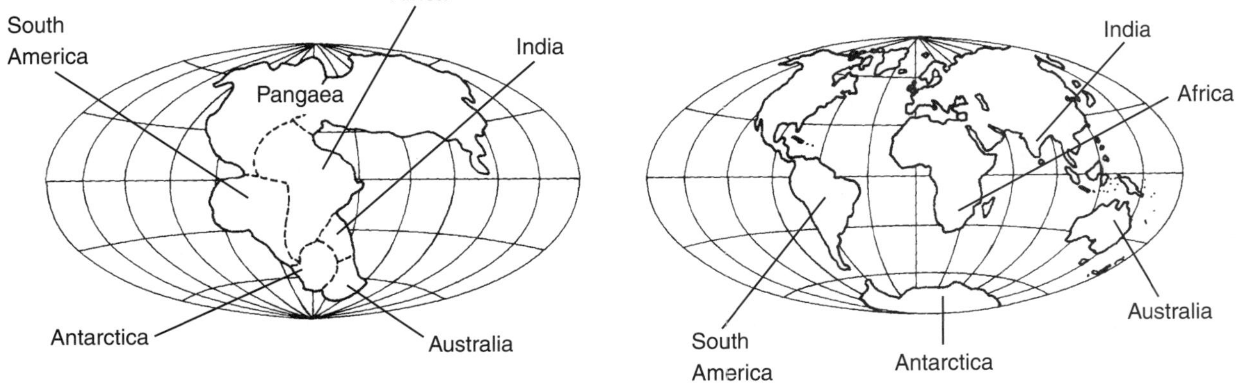


**FIGURE 9.1** All of the earth's continents were part of a supercontinent now known as Pangaea. As the continents separated, forms of life became isolated and evolved independently of one another.

groups appear similar to one another although they are distantly related? For example, the jackrabbit of North America is very similar to the African springhaas, and the pronghorn is similar to the springbok. The answer may lie in the process of natural selection. Organisms that face similar evolutionary challenges may adapt to them in similar ways, and over time, these organisms may come to resemble one another.

**Discussion Questions**

1. How does continental drift help explain the geographic distribution of species?
2. Does the idea that all the continents were once part of a single continent help explain why humans are found in all continents except Antarctica?

**References**

Dewey, J. F. "Plate Tectonics." *Scientific American*, May 1972.

Flessa, K. W. "Biological Effects of Plate Tectonics and Continental Drift." *BioScience* 30, August 1980.

Hallam, A. "Alfred Wegener and the Hypothesis of Continental Drift." *Scientific American*, February 1975.

Hallam, A. "Continental Drift and Fossil Record." *Scientific American*, November 1972.

# SUPPLEMENTARY TOPIC 10

◆ Chapter 20 Human Evolution ◆

**Neanderthals**

In 1856, workers mining for limestone in a cave near Düsseldorf, Germany, in the Neander Valley discovered an unusual humanoid skull and various other bones. Earlier, similar fossils had been found in Belgium (1830) and Gibraltar (1848). Still, it was the discovery in the Neander Valley that generated excitement and led to the name Neanderthal.

The Neander fossils were examined by a number of scientists who came to different conclusions as to their relationship with modern humans. Rudolf Virchow, a German anatomist, was among the first to examine the bones. He concluded that they came from a *Homo sapiens* (modern human), albeit an unfortunate one. He explained that some of the skull's unusual features resulted from the disease rickets. He ascribed the flattened head to some type of injury. The biologist Thomas Huxley concluded that Neanderthals were distant ancestors of modern humans, while the paleontologist Marcellin disagreed. Despite the controversy, Neanderthals were usually portrayed as muscle-bound brutes that lacked the intellectual capacity of modern humans.

This image began to change during the 1960s when anthropologists realized that the first Neanderthal skeleton came from an individual who had suffered from arthritis. It was later determined that Neanderthals did, in fact, walk upright, use tools, and bury their dead. In fact, their average brain capacity was larger than that of some modern humans. A few scientists even went so far as to suggest that if you saw a "cleaned-up" Neanderthal walking down the street wearing a suit, you wouldn't even look twice.

By the 1980s, the pendulum had swung back, and Neanderthals were once again regarded as a blind alley in human evolution. To many, this view was confirmed by a scientific study published in 1997. Scientists extracted mitochondrial DNA from the original Neanderthal skeleton and compared its sequence with that of modern humans. They found about three times as many nucleotide differences between Neanderthal DNA and modern human DNA as they found between pairs of modern humans. The Neanderthal DNA sequence also was found to be equally distant from all modern humans, regardless of their ethnic group. These data are consistent with the idea that Neanderthals do not represent a missing link in human evolution.

Mitochondrial DNA is inherited only from the mother, does not undergo recombination, and mutates more rapidly than nuclear DNA. These characteristics permit analysis of mitochondrial DNA sequences to be used as a molecular "clock." Because mitochondrial DNA sequences change only by mutation, it follows that the more different two sequences are, the farther back in time one must go to reach his or her common ancestor. The accuracy of the clock depends on there being a constant rate of mutation and on the mutations themselves being adaptively neutral. The sequence differences between the Neanderthal sample and samples from modern humans suggest that Neanderthals went their separate way about 550,000–690,000 years ago. Modern humans are believed to have evolved about 200,000 years ago.

Just as Neanderthals were finally being put in their place, a new discovery was announced that muddied the waters again. Paleoarcheologists reported finding in Portugal a 24,500-year-old skeleton that shared characteristics of Neanderthals and early modern humans. This hybrid skeleton of a young boy is the first evidence that humans and Neanderthals may have interbred. If true, this would contradict the idea that humans came out of Africa and displaced the Neanderthals.

Molecular biologists fired back with a report published in 2000. In this study, a second sample of mitochondrial DNA was sequenced from a more recent Neanderthal skeleton. Once again, the sequence variation of the Neanderthal DNA was outside the expected range for humans. This study seems to confirm the earlier DNA analysis.

Scientists remain divided about whether Neanderthals interacted with humans and share our ancestry. The debate over our distant cousin the Neanderthal is likely to continue for some time.

**Discussion Questions**

1. What are some possible explanations for the disappearance of the Neanderthals?
2. How does the accuracy of the molecular clock affect the DNA analysis?
3. Can you reconcile the DNA data that suggest no relationship between Neanderthal and humans with the new fossil discovery that suggests possible interbreeding between the two? Explain.

**Reference**


Krings, M., A. Stone, R. W. Schmitz, H. Krinitzki, M. Stoneking, and S. Paabo. "Neanderthal DNA Sequences and the Origin of Modern Humans." *Cell*, July 7, 1997.

Ovchinnkov, I. V., A. Gotherstrom, G. Romanova, V. M. Kharitonov, K. Liden, W. Goodwin. "Molecular Analysis of Neanderthal DNA from the Northern Caucasus." *Nature* 440, 2000.

# SUPPLEMENTARY TOPIC 11

◆ Chapter 23 Immune Systems ◆

**Cancer and the Immune System**

We live in a sea of germs. There is a bewildering variety of microorganisms that seek to gain access to our bodies. Some of these germs are innocuous. Some are even helpful, aiding in food digestion. Others can make us sick or even kill us.

The immune system has evolved as our shield against germs. Its job is to detect foreign invaders and destroy them before they harm us. The immune system must be able to distinguish self from nonself and destroy invaders with the appropriate response. If the immune system is too lax, then invaders can slip through and make us ill. On the other hand, if the immune system is too aggressive and fails to distinguish self from nonself, it may attack its own body. So-called autoimmune diseases, such as multiple sclerosis and systemic lupus erythematosus, result from the immune system's attacking the body's own healthy tissues. These diseases can be every bit as devastating as diseases caused by pathogens. Scientists hope that by understanding the immune system, they will be able to control autoimmune disease and to fight other diseases such as cancer.

Scientists have wondered about the relationship between cancer and the immune system for more than a hundred years. One hypothesis that has been in and out of fashion is the notion that cancer is always present in the body and that the immune system keeps it in check. This idea is supported by the observation that tumors sometimes shrink in cancer patients who contract bacterial infections. The infection is thought to stimulate an immune response that spills over and attacks the cancer cells. A number of doctors and scientists have tried to develop immune-based therapies for cancer patients. Among these was Dr. William Coley, who worked at Memorial Hospital in New York City from 1892 to 1936. Hoping to stimulate the immune system to attack the tumors, he purposefully infected cancer patients with bacteria and killed-bacterial vaccines. Although his treatments helped some patients, the results were decidedly mixed, and his approach was not generally accepted.

In spite of the mixed results, this "immunosurveillance" model of cancer persisted. By the 1970s, it became clear that although the immune system can attack tumors, it seems to restrict its attack to tumors caused by viral infections. Because virus-induced cancers represent only a small fraction of the total incidence of the disease, the immunosurveillance model began to lose favor. Subsequent advances in molecular biology have given scientists a much deeper understanding of the immune system. Today they are trying to manipulate specific components of the immune system to fight cancer.

Cytokines are molecules produced by the immune system in response to infection. Their job is to help coordinate the body's immune response. Scientists have isolated many cytokines (such as interferons, interleukins, and tumor necrosis factor) and have used them as nonspecific immunotherapies. However, the results have been discouraging, and today cytokines are being tested in combination with other treatments.

The idea of using antibodies as "magic bullets" to specifically destroy cancer cells has been referred to as the Holy Grail of cancer research. The problem with this approach is that for antibodies to be useful in attacking cancer cells, they must "see" the cancer cells as foreign. The surfaces of all cells have proteins and polysaccharides that the immune system can recognize. The immune system uses these target molecules, or antigens, to distinguish self from nonself. Over the years, many researchers claimed to have identified antigens specific to cancer cells. Unfortunately, these claims have not stood the test of time. Recently, the hope of developing "magic bullets" received a boost with the invention of monoclonal antibodies.

Monoclonal antibodies were first developed by Cesar Milstein and Georges J. F. Köhler of the University of Cambridge in 1975. By fusing an antibody-producing cell with a cancer cell, they created an immortal cell line that could make an unlimited supply of a specific antibody molecule. Investors and biotechnology companies were quick to capitalize on this new approach. Researchers created monoclonal antibodies that were attached to toxins or made radioactive to increase their lethality. Unfortunately, when turned loose against cancer cells, monoclonal antibodies failed to meet the investors' high expectations, and their interest quickly faded.

Research using monoclonal antibodies has demonstrated that normal cells and cancer cells display essentially the same set of antigens on their surfaces. This explains why most cancer treatments using monoclonals destroy both normal and cancerous cells. The existence of this problem does not mean that monoclonal antibodies have no role to play in the treatment of cancer. They can be very effective in certain situations. For example, some antigens on normal cells may not be accessible to the blood-borne antibodies or the cancer cells may express higher levels of certain antigens than normal cells or perhaps the damage inflicted on normal tissue may be reversible. Furthermore, monoclonals may be used to target other cells needed for cancer progression, such as those responsible for supplying blood to the growing tumor.

Another problem associated with monoclonal treatments stems from the fact that the antibodies are produced in mouse cells. Once injected into human subjects, the monoclonals are recognized as foreign and neutralized. This problem is being overcome by disguising the monoclonal antibodies so that they can evade the patient's immune system. The process, which is called humanization, involves replacing nonessential parts of the mouse antibody with their corresponding human parts. Because there are now so many antigens to target and so many antibodies to test, clinical studies cannot possibly test them all.

Cancer treatments that use antibodies are termed passive because they provide patients with something that their own immune systems do not produce. In contrast, vaccines offer an active approach because they stimulate the patient's own immune system to fight the disease. Cancer vaccines have been tried for many years, but they have had only limited success, partly because scientists had no way to monitor the effect of the vaccine on the immune system. There was no way to tell why one vaccine showed promise and another did not. Discoveries of more and more oncogenes and tumor-suppressor genes provide many potential antigen targets for vaccine development. As with monoclonal treatments, a current problem is that there are too many vaccine constructs for the clinical setting to test.

Still another form of treatment is called adoptive immunotherapy. This approach was pioneered by Steven Rosenberg of the National Cancer Institute. Dr. Rosenberg was intrigued by rare cases in which patients with terminal cancer seemingly cured themselves. He speculated that somehow the patient's own immune system was roused to defeat the cancer. Rosenberg's approach is to stimulate the patient's immune system, though not by using a vaccine. He extracts immune cells from the patient, treats them with cytokines to enhance their cell-killing ability, allows them to multiply in cell culture, and then injects the modified cells back into the patient. More recently, he has refined his method to include gene therapy. For example, he inserts a copy of a cytokine gene such as TNF (tumor necrosis factor) into the patient's immune cells and then coinjects the gene-modified cells together with purified IL2 (another cytokine) back into the patient. This approach has worked well for some individuals but not for others.

**Discussion Questions**

1. Why do some cancer patients who contract a bacterial infection see their tumors shrink?
2. What is the difference between immunity and autoimmunity?
3. When using gene therapy to treat cancer, why is it important to use the patient's own cells for modification?

**Reference**

Old, L. J. "Immunotherapy for Cancer." *Scientific American*, September 1996.

# SUPPLEMENTARY TOPIC 12

◆ Chapter 24 Ecosystem Structure and Function ◆

**Population Dynamics**

The size of a population changes over time. Suppose a biologist counted 800 Douglas fir trees on a hill in Colorado in 1979, yet in 1989, only 600 trees stood on the hill. That change is expressed as a rate—the amount of change divided by the amount of time for the change to take place. In this example, the rate is the change in the number of trees divided by the change in time:
$-200$ trees $\div 10$ years $= -20$ trees per year.
In other words, there was an average of 20 fewer trees in the population each year.

What could explain the decrease of 200 fir trees in that 10-year period? Fir trees cannot emigrate, so they must have died. That decrease represents the death rate, or mortality, of the fir-tree population. However, mortality is not the only change that can affect a population. While some fir trees have died, some young trees may have started to grow from seed. Death decreases a population size, and reproduction increases it. The rate at which reproduction increases the population is called the birthrate, or natality.

Every organism needs living space. In general, organisms that move from place to place—animals, for instance—require more space than organisms that do not move, such as corn plants. In addition, large organisms need more space than small ones. You might predict that the amount of space an organism needs is related to the availability of nutrients. That is true to some degree, but another factor that affects organisms in a population is their density—the number of individuals in relation to the space the population occupies. For example, in one study, mice in cages were given more than enough food every day. As the mice reproduced, the density of the population increased and the cages became overcrowded. Some of the female mice stopped taking care of their nests and young. Mice continued to be born, but many newborn mice died from neglect. Eventually, the mortality rate of the young mice reached nearly 100 percent, preventing the population density from increasing further.

Populations in nature vary between upper and lower limits. The lowest limit is zero. If that limit is reached, the population no longer exists. It can be restored only through immigration.

What limits a population on the upper side? Many factors may cause a population to stop growing. In the long run, however, the size of a population is limited by the amount of available resources. If a population is so large that it uses all the water, nutrients, or food in its habitat, it can no longer grow. The greatest number of individuals that a space, along with its available resources, can support is called the carrying capacity.

Do populations affect one another? When different species in a community have some requirement or activity in common, their niches potentially overlap. Some resources, such as oxygen, are unlimited; therefore, the niche overlaps present no problem. In other cases, supplies are limited, and the different species must compete for them. The greater the potential for niche overlap, the greater the potential for competition. When two species compete for resources in the same habitat, the simplest outcomes are that they will continue to coexist or that one will exclude the other from the habitat. The more similar two species are in their use of scarce resources in the same habitat, the less likely they are to coexist.

Of all community interactions, predation is perhaps the most riveting. Stable coexistence for both predator and prey populations is likely when predation prevents the prey population from overshooting its carrying capacity. Predators can do this when they reproduce quickly relative to the prey and are capable of eating more when more prey organisms are available.

Occasionally, humans have neglected to take into consideration that predators can keep a prey population in check. Coyotes have been killed off without regard for the fact that they tend to keep prairie dog populations under control. Similarly, the dingo, a wild Australian dog, was killed off because it attacked sheep. The rabbit and wallaby populations then increased, creating new problems for ranchers. The consequences of interfering with natural cycles and balances can be profound and difficult to correct.

**Discussion Questions**

1. In the coyote and dingo examples, what effects would increased prey populations have on cattle and sheep ranching?
2. In some cases, use of natural predators has brought pest populations under control. One example is the control of prickly pear cactus through the introduction of moth borers. What are some other examples?
3. Sometimes the introduction of nonnative species can have unintended and disastrous consequences. Cite some examples.

 

# SUPPLEMENTARY TOPICS

◆ Answers to Discussion Questions ◆

**Topic 1: Technology for Studying Cells**

1. What limits the resolution of the light microscope?

   Resolution of the light microscope is limited by the wavelength of the light used for illumination.
2. In microscopy, how can one distinguish between a real structure and a structure that is created as a consequence of the sample preparation?

   If a cell structure is seen by using different techniques of sample preparation, then it is likely to be real and not created by the sample preparation.
3. Why can't X rays be used to increase the resolution of the light microscope?

   X rays cannot be used in light microscopy because they cannot be focused by using glass lenses.
4. What are the advantages and disadvantages of light and electron microscopy?

   Light microscopy has the advantage of allowing observation of living cells. Its main disadvantage is its relatively low resolution. Electron microscopy has the advantage of much higher resolution but must use killed and dried samples.

**Topic 2: EKGs**

1. Why are EKGs important tools?

   EKGs provide doctors with a simple and fast method for determining the condition of the heart by assessing its ability to conduct electrical impulses. The EKG pattern can provide evidence of cardiovascular disease.
2. Suppose a teacher were setting up a classroom investigation using EKGs of students. What bioethical issues would this raise? Would such an investigation be a good idea? Explain.

   The teacher cannot know whether all of the students have healthy hearts. Diagnosis of medical problems must be made in a medical setting by a trained professional, not in the inappropriate setting of the classroom. Even if the teacher does not attempt to diagnose, a student may become needlessly frightened over a slight variation in the EKG that a physician would judge normal.
3. Research the effects of nitroglycerine tablets on patients with angina. How does the drug work? What other drugs are used to treat angina? How do they work?

   Students should report that angina is caused by insufficient blood supply to the heart. The resulting lack of oxygen in the heart tissue results in the pain associated with angina. Nitrogycerin and other nitrates relieve the pain by opening blood vessels and increasing blood flow to the heart. Another important class of drugs used to treat angina is the beta blockers. They function by decreasing the amount of work the heart has to do and, therefore, the amount of oxygen the heart needs.

**Topic 3: Gene Interactions**

1. Given the information about the guinea pigs, determine how the 9:3:4 ratio results from a cross of two black *CcBb* animals.

   When the *CcBb* heterozygotes are crossed, 16 possible genotypes are created. Animals having *BB* or *Bb* are black, while *bb* animals are brown. Animals having *CC* or *Cc* will display the color corresponding to their genotype at the *B* locus. Animals with a *cc* genotype will be albino regardless of the genotype at the *B* locus. The 16 possible genotypes are as follows:

| BLACK | BROWN | ALBINO |
|---|---|---|
| *CCBB, CCBb, CCbB*<br>*CcBB, CcBb, CcbB*<br>*cCBB, cCBb, cCbB* | *CCbb, Ccbb, cCbb* | *ccBB, ccBb, ccbB,*<br>*ccbb* |

2. Review the information in the situation involving the *H* allele. A woman carrying the gene for blood group B (she gave birth to an AB child by a father with Type A blood) had no B protein on her red cells. Explain her blood genotype and phenotype.

   The woman's genotype is B, *hh.* She has the Bombay blood type (*hh*) and makes antibodies for the H precursor but cannot make the B protein despite her B genotype.

**Topic 4: Genetics Problems**

1. Bull = heterozygous for polled; Cow A = homozygous for horned; Cow B = homozygous for horned; Cow C = heterozygous for polled.
2. A herd of only roan cattle could not be independently maintained. If any breeding were allowed, both red and white offspring would soon occur. Of course, the rancher could maintain a herd by adding only the offspring of red and white cows (not considered a part of the herd) and preventing the roan animals from interbreeding.
3. The farmer should eliminate from the flock both parents and all offspring whenever a black sheep appears.
4. (a) $F_1$ = all white, (b) $F_2$ = 3/4 white and 1/4 yellow, (c) A cross between an $F_1$ individual and a homozygous white individual would produce only white offspring.
5. (a) A or O children; (b) A or O children; (c) A, B, or AB children; (d) A, B, or AB children; (e) A, B, AB, or O children
6. Codominance could be interpreted by a bloodline theory of inheritance. Also, phenotypic characteristics that are the result of the interaction of several genes could be so interpreted. However, Mendel's results, all of which fortuitously involved simple traits exhibiting the dominance phenomenon, were impossible to interpret by any kind of blending theory.
7. Variation is reduced in species that are confined to asexual reproduction. In such species, variation depends on somatic mutations and/or lateral transmission of genetic material (such as the process of transformation in bacteria).

8. The problem involves the production of mutations, one of which, by chance, might be the desired characteristic. Anything that might increase the mutation rate would be desirable.
9. Because there are three even numbers on the die (2, 4, and 6), there are three ways in which the conditions of the question can be met; therefore, the probability is three-sixths, or 50%.
10. Adenine to thymine, 1:1; cytosine to guanine, 1:1; adenine to cytosine (or any other mismatched pair), indeterminate
11. Only heterozygous dominant, *Tt*
12. (a) 3:1 inflated to constricted; (b) 1:2:1, *PP:Pp:pp*
13. Short hair is dominant, so the parents could have been *SS* × *SS*, *SS* × *Ss*, or *Ss* × *Ss*. A few offspring had long hair, the recessive trait. Because they must have obtained a gene for that trait from each parent, both parents had to be *Ss*.
14. 3:1, gray to albino
15. (a) 1:2:1, *AA:Aa:aa;* (b) 3:1, axial to terminal
16. (a) all *Zz*, (b) 1:2:1, (c) 1:1, (d) 1:1
17. 3:1, smooth to hairy
18. The seeds from which the parents grew could have been both yellow (*Yy* × *Yy*) or one yellow and one white (*Yy* × *yy*).
19. Cross the two breeds.
20. (a) tall offspring, with axial flowers; (b) *TtAa;* (c) 9 tall with axial flowers, to 3 tall with terminal flowers, to 3 short with axial flowers, to 1 short with terminal flowers (9:3:3:1)
21. It will be a mix of red flowers with broad, medium, and narrow leaves and the same for white and pink flowers. Red-, white-, and pink-flowered plants will have a ratio of 1:2:1 broad to medium to narrow leaves; however, there will be twice as many pink-flowered plants as red- or white-flowered plants.
22. Situation 1: (a), Situation 2: (c), Situation 3: (d), Situation 4: (e), Situation 5: (b), Situation 6: (d), Situation 7: (f)
23. Male = *Rr*, female A = *RR*, female B = *Rr*
24. 50%. There are only two possibilities.
25. (a) *Cc*, *cc*, *cc;* (b) 50% normal and 50% albino; (c) The male is *cc*, the female is *CC*, and the children are *Cc*.
26. A female with smooth hair
27. The explanation is simple Mendelian inheritance with rough coat dominant and smooth coat recessive. Cross 1: *RR* × *rr* = *Rr*, cross 2: *Rr* × *rr* = 1*Rr* to 1*rr*, cross 3: *rr* × *rr* = *rr*
28. (a) *AB*, *aB;* (b) *aB;* (c) *Ab;* (d) *ABC*, *ABc*, *aBC*, and *aBc*
29. (a) *BbTt*, black trotter; (b) nine black trotters, to three black pacers, to three chestnut trotters, to one chestnut pacer; (c) one black trotter to one black pacer
30. The male is *bbSs*, and the female is *Bbss*.

**Topic 5: Heredity and the Environment**

1. Why did Sorensen choose to study the factors contributing to the premature deaths of parents and adoptees?

   Premature deaths can have either a genetic or lifestyle association that can be detected through the use of appropriate survey data.
2. Would these same findings be likely to occur if the study focused on deaths that were not considered premature? Explain.

   Since most people do not die prematurely, association between nonpremature death and genetics or lifestyle will be difficult to detect.
3. How might you go about conducting further studies of whether genetics or the environment plays a larger role in longevity? Be sure to include provisions for controls.

   Another approach would examine causes of premature deaths among a population that has undergone a change in lifestyle, such as emigrating to the United States from another country with a different lifestyle. A population of people who did not emigrate would serve as controls. The best comparison, though difficult to conduct, is to compare identical twins who were separated shortly after birth and reared in different environments.

**Topic 6: The Molecular Biology of Cancer**

1. Why is cancer more prevalent among the elderly than among the young?

   Cancer results as a consequence of several oncogenes being mutated within the same cell. Although an individual may be born with mutated oncogenes that predispose toward cancer, most mutations occur during the lifetime of the individual. Elderly people have had more time to collect mutations in their DNA than younger people and, therefore, are more likely to contract cancer.
2. Suppose research leads to a cure for cancer. How might this affect the medical profession and the health of the general population?

   A cure for cancer would allow the medical profession to concentrate resources on the other major killer: heart disease. The public would be relieved of the one of the most feared human diseases and would experience an increase in life expectancy.
3. What bioethical issues may develop as a result of new cancer tests and treatments?

   Bioethical issues arise if predictive tests become available without effective treatments. Furthermore, employers and insurance companies could use the results of such tests to discriminate against people with cancer or even those with a predisposition to cancer. Assuming that effective treatments for cancer are developed, they may be expensive and unavailable to those without medical insurance.

**Topic 7: Dominance and Recessiveness**

1. How can gene mutations result in what looks like blended inheritance or incomplete dominance?

   Typically, when one allele in a gene for an enzyme is mutated, the remaining allele produces enough of the enzyme to maintain the normal phenotype. However, it is possible for the loss of 50% of an enzyme to have a visible effect that can resemble blended inheritance.

2. What is the reason that dominant mutations are associated with genes for structural or regulatory proteins?

   A single mutated allele in a gene for a structural or regulatory protein is sometimes enough to produce a dominant phenotype since the altered protein can produce an aberrant structure or lead to an error in gene regulation. In contrast, a single mutated allele in a gene coding for an enzyme simply produces an inactive enzyme and is usually compensated for by the healthy allele.

3. In gene therapy, doctors introduce healthy copies of a gene into the cells of a patient with a genetic disorder. Would this approach work equally well for dominantly and recessively inherited diseases? Explain.

   No, in most recessive disorders adding a healthy gene will produce the missing protein and restore health, while in most dominant disorders, simply adding the healthy gene is not enough. The mutated gene (or its effect) must be removed.

**Topic 8: Mothers and Fetuses: Whose Rights Are More Important?**

1. Suppose a mother with PKU does not want to go back on the diet. What rights does she have to control what happens to her own body? Do these rights include the right to jeopardize the health of a developing fetus?

   In addressing an ethical issue such as that presented here, it is important that you not imply that there are simple right and wrong answers. Students should use the best information available to evaluate the question. They must become comfortable with the idea that uncertainty—which even the best science cannot eliminate completely—can be managed in an intellectually serious manner. In this example, a student having the opinion that a mother with PKU should not be compelled to go back on the special diet may justify that view on the grounds that the affected infant can immediately be put on the diet and avoid mental retardation—as she herself did.

2. Behaviors such as drinking alcohol and smoking can damage a developing fetus. Should a pregnant woman be required to give up these behaviors? Are her civil liberties more important than the rights of the fetus?

   Again, students will likely hold differing views on this question. Insist that students provide justification for their opinions. For instance, students having the opinion that women should be allowed to smoke and drink during pregnancy can point out that most babies born to such women are, in fact, healthy. Therefore, the relatively small risk incurred by those behaviors do not justify limiting the mother's civil liberties.

3. Identify other situations in which medical and technological knowledge has resulted in bioethical difficulties.

   Students may cite examples such as when to declare a person clinically dead, who gets custody over frozen embryos in a divorce, who should have access to genetic information about a person, and how society should apportion rare organs for donation.

**Topic 9: Continental Drift and Evolution**

1. How does continental drift help explain the geographic distribution of species?

   As continental plates move over the liquid interior of the earth, they promote mountain building, volcanic activity, and the emergence of islands. These processes create new habitats for species to occupy. As islands separate from the mainland or continents move apart, species become geographically isolated and free to pursue their own unique evolution.

2. Does the idea that all the continents were once part of a single continent help explain why humans are found in all continents except Antarctica?

   No, the continents separated from each other about 180 million years ago, whereas the common ancestor of humans and apes lived about 5–6 million years ago.

**Topic 10: Neanderthals**

1. What are some possible explanations for the disappearance of the Neanderthals?

   Neanderthals may have been eliminated through competition with *Homo sapiens*, or they may have interbred with humans and become incorporated into *Homo sapiens*.

2. How does the accuracy of the molecular clock affect the DNA analysis?

   If the molecular clock is unreliable, it calls into question the conclusion that Neanderthals separated from our last common ancestor 550,000–690,000 years ago.

3. Can you reconcile the DNA data that suggest no relationship between Neanderthal and humans with the new fossil discovery that suggests possible interbreeding between the two? Explain.

   Since both the DNA data and the fossil discovery came from just one or two specimens, it is possible that they are not in conflict with each other. These samples may turn out to be unrepresentative of Neanderthals.

**Topic 11: Cancer and the Immune System**

1. Why do some cancer patients who contract a bacterial infection see their tumors shrink?

   The immune system mounts a response to the bacterial infection that spills over to the cancer cells. Unfortunately, this is not a common occurrence.

2. What is the difference between immunity and autoimmunity?

   Immunity refers to the body's protection against foreign invaders such as bacteria and viruses. Autoimmunity refers to a pathological condition in which the immune system fails to accurately distinguish between self and nonself and attacks the body's own tissues.

3. When using gene therapy to treat cancer, why is it important to use the patient's own cells for modification?

   Cells from another individual, even a relative, will be recognized by the patient's immune system as foreign and attacked, as happens with organ transplants. Since it is often easy to obtain cells from the patient, this method can be used to eliminate the need to treat with immunosuppressive drugs.

**Topic 12: Population Dynamics**

1. In the coyote and dingo examples, what effects would increased prey populations have on cattle and sheep ranching?

   Increasing numbers of prey such as rabbits can destroy graze lands needed by sheep and cattle.
2. In some cases, use of natural predators has brought pest populations under control. One example is the control of prickly pear cactus through the introduction of moth borers. What are some other examples?

   The introduction of house cats can help keep rodent populations under control. Less well known is the example of the bacteria *Bacillus popilliae* being used to control Japanese beetles in the eastern United States.
3. Sometimes the introduction of nonnative species can have unintended and disastrous consequences. Cite some examples.

   Some examples include the following: The Indian mongoose was introduced in Hawaii to control rat populations; unfortunately, the mongoose also contributed to the decline of native birds, especially those that were ground-nesting species. The African honeybee was introduced to South America to help boost honey production but escaped, and the aggressive bee has now reached the southwestern United States. The plant known as kudzu was introduced to help combat soil erosion and is now growing out of control in the southeastern United States.

# Extended Laboratory Program

# EXTENDED LABORATORY PROGRAM



 

# ◆ Extended Laboratory Program

*Introduction to the Teacher*

The Extended Laboratory Program consists of 20 investigations related to 17 chapters in the student textbook. It also includes a suggested Laboratory Practical Examination. The investigations can be used to complement those in the textbook, whose table of contents indicates the appropriate section in each chapter for the use of each lab. In addition, the investigations are keyed in the Chapter-at-a-Glance notes in the Teacher's Annotated Edition.

In accordance with the emphasis on technology in the *BSCS Blue Version*, the Extended Laboratory Program includes three electrophoresis investigations as well as several labs that can be microcomputer-based. Electrophoresis plays an increasingly important role in many areas of biological research. These investigations give students an opportunity to become acquainted with this important technology, to gain an understanding of its underlying principles, and to experience some of its applications. A number of investigations involve data gathering that can be best accomplished by using a microcomputer such as the Trekker-T-4™ system. These investigations may be performed without the microcomputer, but its use is recommended where appropriate. Several investigations in the Extended Laboratory Program are favorites from the *BSCS Blue Version* that could not be included in the student textbook because of space limitations. For the most part, these investigations are traditional wet labs that encourage students to refine their laboratory skills even further. The investigations emphasize laboratory technique, measurement, and analytical and observational skills. The Laboratory Practical Examination focuses on experimental design and interpretation of data and helps you assess each student's inquiry skills.

The Master Materials List includes all materials required for the Extended Laboratory Program. The Preparation of Solutions section provides directions for preparing all solutions and reagents needed for the investigations. It also provides information about chemical safety and storage. *Before using any of the investigations in the Extended Laboratory Program, review the Guidelines for Laboratory Safety in the Teacher's Annotated Edition.* Make certain students are familiar with the Laboratory Safety material in their textbooks. Insist that students and their parents or guardians sign a safety contract. For your convenience, the sample laboratory safety agreement from the Teacher's Annotated Edition is reproduced on the next page.

## Laboratory Safety Agreement

I, ______________________, agree to abide by the following laboratory safety regulations whenever performing a biology investigation. I will

1. use the science laboratory for authorized work only.
2. remove contact lenses and wear safety goggles when instructed to do so.
3. know the four hazard classes of chemicals and control measures.
4. study the laboratory investigation before coming to the lab. If in doubt of any procedure, I will ask my teacher.
5. know how to use the safety equipment provided and know the location of the fire extinguisher, eyewash station, safety shower, and fire blanket.
6. in case of fire, alert the teacher and leave the laboratory.
7. carefully check for the presence of any ignition source—for example, open flames and electric heating coils—before using flammable materials such as alcohol or acetone.
8. place broken glass and disposable materials in their designated containers.
9. immediately report any incident, accident, injury, or unsafe procedure to the teacher.
10. never taste, touch, or smell any substance unless the teacher specifically directs me to do so.
11. handle chemicals carefully, check the label of every bottle or jar before removing the contents, and never return unused chemicals to reagent containers.
12. when heating a substance in a test tube, make sure that the mouth of the test tube points away from other people and away from myself and always use proper equipment to handle hot glassware.
13. tie back long hair, remove dangling jewelry, roll up loose sleeves, and tuck in loose clothing.
14. at the end of the lab, clean the work area—wash and store all materials and equipment and turn off all water, gas, and electrical appliances.
15. wash my hands thoroughly with soap and warm water before leaving the laboratory.

______________________ ______________________

Student's Signature    Parent's or Guardian's Signature

______________________

Date

## ◆ Extended Laboratory Program

*Master Materials List*

This materials list was prepared by Ward's Natural Science Establishment, the official supplier for the laboratory program of the *BSCS Blue Version*, eighth edition. The list includes all materials required for the Extended Laboratory Program. Some of these materials you may already have purchased for the laboratory program in the textbook; check your supplies. Materials for the electrophoresis investigations are listed first. These materials are grouped according to the way in which they are used; expendables are listed first in each group. For other investigations, the list is divided into expendable, nonexpendable, and supplemental or optional materials. For some investigations, Ward's has developed custom kits that match the lab requirements. Items are listed alphabetically, followed by package size, Ward's catalog number, number of packages or items for one year per class of 30, and the investigations in which the material is used. "Local" means the item should be available locally. For your convenience, the Ward's catalog numbers for some locally available items are listed. To order or for questions concerning the use of Ward's materials, call toll free 1-800-962-2660.

### Materials for Electrophoresis Labs (Please see page 53 for further information.)

| ITEM | WARD'S CATALOG # | QUANTITY | ELP NO. |
|---|---|---|---|
| **Safety Equipment** | | | |
| Lab apron (disposable), pkg/100 | 15WI050 | 1 | 6, 8, 13 |
| Gloves (plastic, ambidextrous), pkg/100 | 15W1074 | 7 | 6, 8, 13 |
| Safety goggles (ANSI-approved) | 15W3046 | 30 | 6, 8, 13 |
| **Electrophoresis Equipment, Kits, Gels, Samples** | 36W3402 | 1 | 6, 8 |
| PAGE Gel electrophoresis lab set | | | |
| 3 dual gel electrophoresis tanks | | | |
| 1 variable voltage power supply | | | |
| Agarose Gel electrophoresis lab set | 36W5103 | 1 | 13 |
| 3 dual electrophoresis chambers | | | |
| 1 variable voltage power supply | | | |
| Bacteria ID Kit | 36W5337 | 1 | 6 |
| Cardiac Risk Assessment Kit | 36W5335 | 1 | 8 |
| DNA Restriction Mapping Kit | 88W8503 | 1 | 13 |
| 10 precast 10–20% SDS polyacrylamide gels | 36W5259* | 1 | 6 |
| 10 precast 3–8% borate polyacrylamide gels | 36W5268* | 1 | 8 |
| **Miscellaneous Labware** | | | |
| Beaker (600 mL) | 17W4060 | 6 | 6 |
| Graduated cylinder (250 mL) | 18W1740 | 6 | 6, 8 |
| Plastic bag (self-sealing), pkg/50 | local | 1 | 6, 8, 13 |
| Plastic rectangular dishpan | 18W9881 | 6 | 6, 8, 13 |
| Plastic stain/destain container w/lid | 18W3650 | 6 | 6, 8, 13 |
| Plastic waste receptacle (widemouth w/cap), gallon | 18W9706 | 6 | 6, 8, 13 |
| Scalpel with blade #22 | 14W0900 | 6 | 6, 8, 13 |
| **Pipets/Measuring** | | | |
| Plastic transfer pipets (disposable) | 18W2974 | 1 | 6, 8, 13 |
| Serological pipet (10 mL × 1 mL grad), pkg/100 | 17W4853 | 1 | 6 |
| Serological pipet bulb, pkg/10 | 15W0511 | 1 | 6 |
| Reagents | | | |
| Deionized/distilled water, 1 gal | 88W7005 | 10 – 14 | 6, 8, 13 |
| **Reagent Preparation Labware** | | | |
| Beaker (2 L), pkg/4 | 17W4122 | 1 | 6, 8, 13 |
| Graduated cylinder (500 mL) | 18W1750 | 1 | 6, 8, 13 |
| Plastic storage container (gal) | 17W0007 | 6 | 6, 8, 13 |
| **Sample Preparation** | | | |
| Biohazard bag (autoclavable), pkg/100 | 18W6905 | 1 | 6 |
| Filter paper (Whatman #1, 90 mm-diam), pkg/100 | 15W2835 | 1 | 6 |
| Index card (white), pkg/500 | 15W9820 | 1 | 6 |
| Marking pencil, glass | 15W1155 | 6 | 6, 8 |
| Microfuge tube w/cap (1.5 mL), pkg/500 | 18W1361 | 1 | 6 |

*Specify delivery date when ordering; gels have a shelf life of 4 months under refrigeration.

## Materials for Electrophoresis Labs, continued

| ITEM | WARD'S CATALOG # | QUANTITY | ELP NO. |
|---|---|---|---|
| **Sample Preparation, continued** | | | |
| Beaker (1 L), pkg/6 | 17W4080 | 1 | 6 |
| Bubble rack (20-hole) | 18W4216 | 3 | 6 |
| Funnel (plastic, to fit microfuge tube), pkg/36 | 18W1421 | 1 | 6 |
| Hot plate | 15W7999 | 3 | 6 |
| Incubator | 15W0060 | 1 | 6 |
| Lab burner (natural gas) | 15W0603 | 6 | 6 |
| Lab burner (LP) | 15W0797 | 6 | 6 |
| Test-tube rack (for 18 × 150 mm), 6-tube capacity | 18W4231 | 6 | 6 |
| **Supplemental or Optional Materials** | | | |
| Centrifuge | 15W2400 | 1 | 6 |
| Cuvette, box/12 | 14W5554 | 1 | 6 |
| Heat sealing equipment | | | |
| Polyethylene bags (6 x 10 in), pkg/100 | 18W6993 | 1 | 6 |
| Spectrophotometer | 14WA000 | 1 | 6 |
| Tube (centrifuge, 15 mL), pkg/12 | 18W2620 | 1 | 6 |
| Water bath | 14W5759 | 1 | 6 |

## Expendable Materials

| ITEM | WARD'S CATALOG # | QUANTITY | ELP NO. |
|---|---|---|---|
| Agar (plain), btl/100 g | 38W0015 | 1 | 11 |
| Apples | local | 20 | 18 |
| Carbonated water, 12-oz can | local | 1 | 1 |
| Corn syrup, pt | local | 1 | 9 |
| Cotton fibers, pkg/1 lb | 15W3830 | 1 | 7 |
| Coverslip, box/100 | 14W3555 | 1 | 5, 7, 18 |
| Cheesecloth, pkg/5 yd | 15W0015 | 1 | 11 |
| Cricket (live), pkg/10 | 87W6100 | 1 | 1 |
| Cricket (preserved), jar/10 | 68W4062 | 1 | 1 |
| Daphnia culture, class/30 | 87W5210 | 1 | 7 |
| Detain II™ (slowing agent), btl/1/2 oz | 37W7952 | 1 | 7 |
| Disposable sterile pipets, pkg 500 | 18W2972 | 1 | 18 |
| Distilled water, btl/1 gal | 88W7005 | 4 | 11, 20 |
| DNA Extraction kit | 88W8112 | 1 | 9 |
| Dropping pipet, box/12 | 17W0230 | 3 | 5, 7, 9, 11, 19 |
| Ethyl alcohol (99%), btl/500 mL | 39W0280 | 1 | 9 |
| Euglena culture, class/30 | 87W0100 | 1 | 19 |
| Fertilizer pkg, btl/28 oz | 20W7951 | 1 | 19 |
| Filter paper (15 cm diam), pkg/100 | 15W2821 | 1 | 20 |
| *Gloves (plastic ambidextrous), box/100 | I5W1074 | 2 | 1, 7, 9, 11, 12, 14, 18, 19 |
| Glucose-1-phosphate, btl/1 g | 39W1465 | 1 | 11 |
| Grape juice, 1/2 gal | local | 1 | 5 |
| Graph paper, pkg/100 | 15W3835 | 1 | 14, 21 |
| Grasshopper (preserved), pkg/100 | 68W4054 | 2 | 14 |
| Hydrochloric acid, 0.1 N, btl/1 L | 37W9561 | 1 | 3 |
| Limewater, btl/500 mL | 37W2630 | 2 | 1 |
| Lugol's iodine solution, btl/500 mL | 39W1685 | 1 | 11 |
| Marking pencil (glass) | 15W1155 | 8 | 4, 11, 19 |
| Mealworm larvae | 87W6250 | 1 | 17 |
| Microscope slide, box/72 | 14W3500 | 1 | 5, 7, 18 |
| Motor oil (SAE 20 or 30 nondetergent), qt | 37W0005 | 1 | 9 |
| Paper towels, roll | 15W9844 | 3 | 1, 9, 11 |
| Peas (wrinkled), pkg/100 | 86W8240 | 2 | 11 |
| Peas (smooth), pkg/100 | 86W8245 | 2 | 11 |

*Starred items also are listed with the electrophoresis materials; check required quantities before ordering.

 

## Expendable Materials, continued

| ITEM | WARD'S CATALOG # | QUANTITY | ELP NO. |
|---|---|---|---|
| Petri dish (sterile plastic), pkg/20 | 18W7101 | 1 | 11, 14 |
| Phenol red, btl/1 g | 3SM9432 | 2 | 1, 3 |
| Plastic wrap, roll | 15W9858 | 1 | 4 |
| Potato (white) | local | 10 | 4 |
| Rhizopus Stolonifer culture | 85W4901 | 1 | 18 |
| Sabouraud dextrose agar, btl/500 mL | 38W0471 | 1 | 18 |
| Safety matches, box | 15W9427 | 8 | 2 |
| *Scalpel with blade #22 | 14W0900 | 20 | 11 |
| Seeds (sunflower), pkt/30 | 86W8320 | 6 | 1 |
| Seeds (pea, smooth), pkt/1 lb | 86W8247 | 5 | 14 |
| Sodium fluoride (0.1 M), btl/1 L | 37W5516 | 1 | 11 |
| Straw (wrapped), pkg/500 | 15W9869 | 1 | 1 |
| String, ball/400 ft | 15W9863 | 1 | 14 |
| Sucrose, btl/500 g | 39W3182 | 1 | 4 |
| Tape (clear), roll | local | 1 | 18 |
| Tape (cloth-backed), roll | local | 1 | 2 |
| Transformation of *E. coli* kit | 88W8231 | 1 | 12 |
| Tubing (rubber), roll/50 ft | 15W1127 | 1 | 5 |
| Tubing (glass), 400 cm | 17W0931 | 1 | 5 |
| Vinegar (white), btl/500 mL | 39W0138 | 1 | 1 |
| Yeast, pkt/10g | 88W0929 | 3 | 1, 5, 21 |

## Nonexpendable Materials

| ITEM | WARD'S CATALOG # | QUANTITY | ELP NO. |
|---|---|---|---|
| Balance (triple beam) | 15W6057 | 10 | 4, 11 |
| Battery (6-volt, w/terminals) | 15W3263 | 10 | 2 |
| Beaker (100 mL), pkg/12 | 17W4020 | 2 | 20 |
| Beaker (250 mL), pkg/12 | 17W4040 | 2 | 11, 21 |
| Beaker (600 mL), pkg/6 | 17W4060 | 1 | 9 |
| *Beaker (2 L), pkg/4 | 17W4090 | 1 | 3 |
| Clock or watch with second hand | 15W1492 | 1–15 | 7, 17 |
| Cork borer (5–10-mm diameter) | 15W1665 | 10 | 4 |
| Dissecting needle | 14W0650 | 10 | 4 |
| Dropping bottle | 17W6011 | 30 | 7, 12 |
| Electrolysis apparatus | 15W8200 | 5 | 2 |
| Flask (250 mL), pkg/12 | 17W2982 | 1 | 3, 5 |
| Filtering flask w/ sidearm, 250 mL | 17W3305 | 20 | 3 |
| Forceps | 14W1001 | 10 | 18 |
| *Funnel (plastic), pkg/36 | 18W1421 | 1 | 20 |
| Grad cylinder (50 mL) | 17W0172 | 10 | 4, 5 |
| Grad cylinder (100 mL) | 17W0173 | 8 | 3 |
| *Hot plate | 15W7999 | 3 | 9 |
| Incubator | 15W0060 | 1 | 12 |
| Jar (screw top, 4 oz), pkt/10 | 17W2090 | 1 | 7, 9 |
| Lab apron (rubber) | 15W1015 | 30 | 1, 2, 3, 4, 5, 7, 9, 11, 12, 17, 18, 19 |
| Metric ruler (15-cm transparent) | 15W4655 | 30 | 3, 4, 14, 16 |
| Microscope (compound) | 24W2310 | 15 | 5, 7, 11, 18, |
| *Plastic rectangular dishpan | 18W9881 | 4 | 5 |
| Protractor | 15W4067 | 30 | 17 |
| *Safety goggles (ANSI-approved) | 15W3046 | 30 | 1, 2, 3, 4, 5, 7, 9, 11, 12, 14, 18, 19 |
| *Scalpel with blade | 14W0900 | 20 | 4, 11, 18 |

*Starred items also are listed with the electrophoresis materials; check required quantities before ordering.

## Nonexpendable Materials, continued

| ITEM | WARD'S CATALOG # | QUANTITY | ELP NO. |
|---|---|---|---|
| Screw (brass, for 18 x 150-mm test tubes) | local | 70 | 1 |
| *Stirring rod (glass), pkg/12 | 17W6005 | 2 | 5, 9 |
| Stopper (cork #5, 18-mm), pkg/100 | 15W8365 | 1 | 1 |
| Stopper (2-hole #6), lb/21 | 15W8516 | 1 | 5 |
| *Test tube (18 × 150 mm), pkg/72 | 17W1543 | 2 | 1, 2, 19 |
| Test tube (25 × 200 mm), pkg/48 | 17W0660 | 1 | 4 |
| Test-tube clamp, pkg/5 | 15W0841 | 8 | 9 |
| *Test-tube rack (for 18 × 150 mm, 6-tube capacity) | 18M4231 | 20 | 10, 12, 16 |
| Test-tube rack (for 25 × 200 mm, 40-tube capacity) | 18M4209 | 10 | 11 |
| Tube (screw top, 13 × 100 mL) | 17M1338 | 6 | 16 |
| *Tube (centrifuge), 15 mL | 17W1330 | 1 | 11 |
| Thermometer (nonmercury) | 15W1460 | 10 | 3, 5 |
| Vacuum bottle (8-oz) | local | 10 | 5 |
| Vinyl tubing | 18W5083 | 1 | 3 |
| Wooden board, (30 × 30 cm) | local | 10 | 17 |
| Wooden block | local | 30–40 | 17 |
| Wooden splint, pkg/500 | 14W0105 | 1 | 2 |

## Supplemental or Optional Materials

| ITEM | WARD'S CATALOG # | QUANTITY | ELP NO. |
|---|---|---|---|
| Density indicator strip, pkg/25 | 15W1996 | 10 | 19 |
| Power supply (6–12 volt DC) | 15W1995 | 5 | 2 |
| *Water bath | 14W5759 | 1 | 9, 12 |

*Starred items also are listed with the electrophoresis materials; check required quantities before ordering.

# ◆ Extended Laboratory Program

*Preparation of Solutions*

The table that follows provides directions for the preparation of all reagents and solutions used in the Extended Laboratory Program. Before preparing any of the materials listed in the table, familiarize yourself with the safety procedures and storage information in the Guidelines for Laboratory Safety of the *BSCS Blue Version Teacher's Annotated Edition.* Also refer to individual reagent labels and Material Safety Data Sheets for further information about hazards and precautions. The table consists of four columns:

**Substance**—All substances are listed alphabetically.* Each listing includes a description of the appearance and, in the case of a pure substance, the formula and the Chemical Abstract Service (CAS) reference number if it exists. (CAS abstracts and indexes chemical substances by a reference number used worldwide and published in *Chemical Abstracts.*) The table also identifies the investigations in which the substance is used. Personal protective equipment other than safety goggles and lab apron, which you should wear when working with any chemical, also is noted in this column.

**Preparation**—Directions are provided for the preparation of specific reagents and solutions.

**Hazards**—Chemicals are rated in four hazard categories, on a scale of 0 to 4, as described in the Guidelines for Laboratory Safety. The categories are H (health), F (flammability), R (reactivity), and C (contact). The scale is 0 = minimal, 1 = slight, 2 = moderate, 3 = serious, and 4 = extreme. Examples: Agar is rated H = 0, F = 1, R = 0, C = 0; concentrated hydrochloric acid is rated H = 3, F = 0, R = 2, C = 3. Specific hazards are described in the table. Storage color is specified.

**Remarks**—This column contains information that is added specific to the substance for spills, disposal, storage, and first aid. (Such information is provided only when it differs from the general information in the Guidelines for Laboratory Safety.)

**Electrophoresis Investigations**

**Materials (not included in kits)**

- 2-L beaker or flask
- 500-mL graduated cylinder
- 6 1-gallon plastic containers for storing reagents (milk containers are excellent)
- 5–7 gallons deionized or distilled water

**Color Codes for Chemical Storage**

Red: All reds in the table are flammables. Store only in a flammable-storage cabinet away from all other chemicals. Store all chemicals coded red in the table with each other unless otherwise indicated in the Remarks column. Do not store any other chemicals in the flammable-storage cabinet.

Yellow: All yellows in the table pose reactivity hazards and should be stored with other chemicals coded yellow in the table and away from all other chemicals. See the Remarks column of the table for special instructions.

White: White indicates a corrosive/contact hazard. Store away from all other chemicals. *Note that some white-coded chemicals in this program cannot be stored with other white-coded chemicals.* See the Remarks column of the table for special instructions.

Blue: All blues pose toxic/health hazards. Store in a secured poison area and away from all other chemicals.

Green: All other chemicals are coded green. Store these green-coded chemicals together and away from all other chemicals. See the Remarks column of the table for special instructions.

*For convenience, all substances used in the electrophoresis investigations (Extended Laboratory Programs 6, 8, and 13) are listed first, followed by substances used in the other investigations in the Extended Laboratory Program.

## Preparation of Solutions

| SUBSTANCE | PREPARATION | HAZARDS | REMARKS |
|---|---|---|---|
| **Buffer, SDS Gradisol™**<br>Clear, colorless (ELP 1) or dark-blue (ELP 3) liquid<br>ELP 6, 8 | (premixed, ready to use) | H = 1, F = 1,<br>R = 1, C = 1<br>Storage: green | |
| **Buffer, SDS Gradisol™ with 2-mercaptoethanol**<br>Dark-blue liquid<br>ELP 6 | (premixed, ready to use) | H = 1, F = 1,<br>R = 1, C = 1<br>Storage: green | |
| **2-Mercaptoethanol**<br>Clear, colorless liquid<br>ELP 1 | (premixed, used for treating bacterial cultures) | H = 2, F = 2,<br>R = 2, C = 2<br>Storage: red | |
| **Stain solution (Kwik-Code™)**<br>Brown solution<br>ELP 6, 8 | (premixed, ready to use) | H = 0, F = 0,<br>R = 1, C = 1<br>Storage: green | |

## Preparation of Solutions, continued

| SUBSTANCE | PREPARATION | HAZARDS | REMARKS |
|---|---|---|---|
| **Agar, plain**<br>Tan granular powder; becomes tan gelatinous solid<br>CAS # 9002-18-0<br>ELP 11 | (used to prepare glucose-1-phosphate agar) | H = 0, F = 1,<br>R = 0, C = 0<br>Storage: green | |
| **Agar, glucose-1 phosphate**<br>Beige gelatinous solid<br>ELP 11<br>See also agar and glucose-1-phosphate. | To 400 mL of distilled water, add 8 g of plain agar and 2 g of glucose-1-phosphate. Boil vigorously until the mixture foams. Pour into 40–50 small petri dishes, adding just enough agar mixture to cover the bottom. Cover and refrigerate until one hour before use. | H = 0, F = 0,<br>R = 0, C = 0<br>Storage: green | |
| **Agar, Sabouraud dextrose**<br>Beige gelatinous solid<br>ELP 18 | To 1 L of distilled water, add 10 g of neo-peptone, 40 g of dextrose, and 15 g of agar. Boil vigorously until the mixture foams. Pour into 40–50 regular petri dishes, adding just enough agar mixture to cover the bottom. Cover and refrigerate until one hour before use. | H = 0, F = 0,<br>R = 0, C = 0<br>Storage: green | |
| **Alcohol, denatured ethyl, 99%**<br>$C_2H_5OH$<br>Volatile, colorless liquid; alcoholic odor<br>ELP 7, 9 | (used to prepare dilute solution) | H = 3, F = 3,<br>R = I, C = I<br>Storage: red | First aid: Before using alcohol, read the label/MSDS and follow the recommended procedures. |
| **Alcohol solution, 0.5%**<br>Colorless liquid<br>ELP 7 | Add distilled water to 0.5 mL 99% denatured ethyl alcohol to make 100 mL solution; mix well. | H = 0, F = 0,<br>R = 0, C = 0<br>Storage: green | |
| **Alka-Seltzer™ tablets**<br>White tablets<br>ELP 7 | (used to prepare dilute solution) | H = 2, F = 0,<br>R = 1, C = 0<br>Storage: blue | |
| **Alka-Seltzer™ solution**<br>Clear, effervescent liquid<br>ELP 7 | Dissolve 1 tablet in 100 mL distilled water. Retain 25 mL, and discard the remainder down the drain. | H = 0, F = 0,<br>R = 0, C = 0<br>Storage: green | Store no longer than one week. |
| **Ammonium sulfate**<br>$(NH_4)_2SO_4$<br>White crystalline solid<br>CAS # 7783-20-2<br>ELP 11 | To purify enzyme extract and concentrate starch phosphorylase, add approximately 4 g/20 mL filtrate. This will precipitate out the proteins. Centrifuge and use the supernatant; keep as cold as possible. | H = 2, F = 0,<br>R = I, C = I<br>Storage: green | |
| **Benadryl Elixir™**<br>Clear, light-orange liquid<br>ELP 7 | (used to prepare dilute solution) | H = 1, F = 0,<br>R = 0, C = 1<br>Storage: green | |
| **Benadryl solution**<br>Clear, pale-orange liquid<br>ELP 7 | Dilute each mL of Benadryl Elixir™ 1:10 with distilled water. Limit quantities to 25 mL total in the laboratory. | H = 1, F = 0,<br>R = 0, C = 1,<br>Storage: green | Store no longer than one week. |
| **Calcium hydroxide**<br>$Ca(OH)_2$<br>White crystals<br>CAS # 1305-62-0<br>ELP 1, 5 | (used to prepare limewater) | H = 1, F = 0,<br>R= I, C = 2<br>Storage: white | Do not store with acids; store with sodium hydroxide. |
| **Glucose-1-phosphate**<br>$C_6H_{11}O_5PO_4 \cdot 2H_2O$<br>White, crystalline powder<br>CAS # 5996-14-5<br>ELP 11 | (used to prepare glucose-1-phosphate agar) | H = 0, F = 0,<br>R = 0, C = 0<br>Storage: green | |

## Preparation of Solutions, continued

| SUBSTANCE | PREPARATION | HAZARDS | REMARKS |
|---|---|---|---|
| **Corn syrup**<br>Clear, viscous, straw-colored liquid<br>ELP 9 | Half-fill small (4-oz) jars. | H = 0, F = 0,<br>R = 0, C = 0<br>Storage: green | |
| **Limewater**<br>Milky liquid<br>ELP 1, 5<br>See also calcium hydroxide. | To distilled water, add more calcium hydroxide ($Ca(OH)_2$) than will dissolve when the container is stoppered and shaken. Allow the solution, tightly stoppered, to stand overnight. Decant the clear liquid into smaller containers for distribution to the class. Prepare about 500 mL per class for ELP 12; 250 mL for ELP 15. | H = 0, F = 0,<br>R = 0, C = 1<br>Storage: green | |
| **Macromolecular solution**<br>Opaque, viscous liquid<br>ELP 9 | (prepared, ready to use) | H = 0, F = 0,<br>R = 0, C = 0<br>Storage: green | Spills are very slippery. |
| **Motor oil, SAE 20 or SAE 30, nondetergent type**<br>Yellowish, viscous liquid<br>ELP 9 | Half-fill small (4-oz) jars. | H = 1, F = 1,<br>R = 0, C = 1<br>Storage: green | Spills are slippery. |
| **Nicotine solution**<br>(cigarette/water extract)<br>Tea-colored liquid; mild, sweet odor<br>ELP 7 | Boil the tobacco of one cigarette in 200 mL water for 1 minute. Cool and filter; then dilute 0.5 mL of the filtrate to 25 mL, mixing well. Discard the remainder. | H = 2, F = 0,<br>R = 0, C = 0<br>Storage: blue | Do not store longer than 1 week. Discard down the drain with plenty of water. |
| **Phenol red**<br>Red-brown powder<br>CAS # 143-74-8<br>ELP 1, 3 | (used to prepare solution) | H = 0, F = 1,<br>R = 0, C = 0<br>Storage: green | |
| **Phenol red solution**<br>Clear, burgundy-colored liquid<br>ELP 1, 3 | Dissolve 0.1 g of phenol red (phenol-sulfonephthalein) indicator in 1 L distilled water. If the solution is yellow, add a very small amount of a weak base such as sodium bicarbonate (powder or solution) until the solution becomes red. If the solution is too basic, the reaction during the experiment will take place quite slowly. Blow through a straw to test the solution. Adjust the phenol red solution with weak acids or bases so that the color change from red to yellow will occur with the absorption of a small amount of carbon dioxide. | H = 0, F = 0,<br>R = 0, C = 0<br>Storage: green | |
| **Sodium fluoride**<br>NaF<br>White, crystalline solid<br>CAS # 7681-49-4<br>ELP 11 | (used to prepare molar solution) | H = 3, F = 0,<br>R = 1, C = 1<br>Storage: blue | |
| **Sodium fluoride, 0.1 *M* solution**<br>Clear solution<br>ELP 11 | Dissolve 4.2 g NaF in 500 mL distilled water; add another 500 mL distilled water and mix well (available from Ward's, already prepared). | H = 2, F = 0,<br>R = 1, C = 0<br>Storage: blue | |

## Preparation of Solutions, continued

| SUBSTANCE | PREPARATION | HAZARDS | REMARKS |
|---|---|---|---|
| **Sucrose**<br>$C_{12}H_{22}O_{11}$<br>White, crystalline powder<br>CAS # 57-50-1<br>ELP 4 | (used to prepare percent solutions) | H = 0, F = 1,<br>R = 0, C = 0<br>Storage: green | |
| **Sugar solutions, 10%, 20%**<br>Clear, colorless liquids<br>ELP 4 | 10%: To 10 g of sucrose, add enough distilled water to make 100 mL of solution.<br>20%: To 20 g of sucrose, add enough distilled water to make 100 mL of solution. | H = 0, F = 0,<br>R = 0, C = 0<br>Storage: green | |
| **Unisom™ tablets**<br>Light blue tablet<br>ELP 7 | (used to prepare dilute solution) | H = 3, F = 1,<br>R = 0, C = 1<br>Storage: blue | |
| **Unisom™solution**<br>Colorless liquid with suspended particles<br>ELP 7 | Crush 1 tablet using a mortar and pestle, and dissolve in 100 mL distilled water. Discard all but 25 mL down the drain. | H = 1, F = 0,<br>R = 0, C = 0<br>Storage: blue | Store no longer than 1 week. |
| **Yeast-sugar solution**<br>Milky, beige, foaming liquid; yeasty odor<br>ELP 1 | Dissolve 100 g of sucrose or 50 mL of corn syrup in enough water to make 1 L of solution. Add one-half cake of fresh yeast or one-half package of dried yeast. Shake or stir, and allow to stand for 1 day. Shake or stir again before using. | H = 0, F = 0,<br>R = 0, C = 0<br>Storage: green | |
| **Yeast-sugar solution, boiled**<br>Milky, beige liquid; yeasty odor<br>ELP 1 | Suspend a container of yeast-sugar solution in a boiling water bath, or heat the solution in a double boiler for 10 minutes. Make sure the solution boils so that the yeast cells are killed and the excess carbon dioxide is driven off. | H = 0, F = 0,<br>R = 0, C = 0<br>Storage: green | |
| **Yeast-water mixture**<br>Milky, beige liquid; yeasty odor<br>ELP 5 | Mix 5 packages dry yeast in 500 mL tap water; stir well. Use half the mixture as is. To the other half, add another 50-100 mL tap water and boil in a water bath or double boiler for 10–15 minutes to kill the yeast. | H = 0, F = 0,<br>R = 0, C = 0<br>Storage: green | |

## Care and Release of Life-Forms

| LIFE-FORMS | CARE | RELEASE |
|---|---|---|
| Bacteria (Eubacteria) | Slant or broth cultures may be stored for extended periods (up to 6 months) at refrigeration temperatures. | Discard only after autoclaving cultures or contaminated materials. |
| Butterflies/Moths | House in a perforated box covered with muslin netting. Keep caterpillars dry. Provide a continuous supply of fresh leaves. Use egg cartons as surface for population. Keep humidity at 60%. | May be freely released shortly after emergence. |
| *Chaeopterus* | Requires either an established 25-gallon marine aquarium or an equal amount of conditioned seawater. Acclimate the animals to the seawater. Aerate. Feed by introducing hatched brine shrimp. If care is exercised, the animals may be returned to their tubes and left in the aquarium at the conclusion of the investigation. Do not attempt to maintain these animals except in an established marine aquarium. | Release only in Florida Gulf area. Otherwise, euthanize using MS-222 (tricane methanesulfonate). |
| Crickets/Grasshoppers/ Small Insects | House in plastic containers that are sealable. Screen in air holes. Use egg cartons as "apartments." Keep temperature at 21–31°C (70–90°F); maintain humidity at around 60%. Feed dried food pellets. Provide a number of watering devices (pieces of wet sponge or cotton in plastic petri dish). | May be freely released. |
| Earthworms/Other Worms | Place commercial earthworm bedding or rich soil in a plastic washtub. Introduce worms. Allow approximately 150 $cm^3$ per worm. Cover with sphagnum moss or leaves. Cover container with muslin. Long-term storage is best under refrigeration. | May be freely released. |
| Elodea | Place in established freshwater aquarium. Provide a photo period of 16 light:8 dark using plant grow lights (fluorescent or incandescent). Light source should be placed approximately 16 – 18 inches over plants. | Do not release. |
| Frogs | Southern species (*Rana belandearei*): Place up to five animals in a 10-gallon aquarium or similar container that is slightly tilted. Provide about 1–2 inches of water in lower end. Feed crickets 2–3 times/week. Keep tank covered to prevent escape. Place tank in a suitable location away from direct sunlight and excessive temperature. Northern species (*R. pipiens*): These animals may be stored at refrigeration temperatures for no longer than 5 days. They may be housed for longer periods as described above. | May be released into appropriate habitats. |
| Geraniums/Coleus | Pot in potting soil. Maintain a 16 light:8 dark photoperiod. Use of plant grow lights is recommended. | May be transplanted. |
| Gerbils | For up to three gerbils, provide a commercial enclosure or use a plastic washtub with screened cage top. Use a commercial water bottle introduced through the cage top. House at room temperature of 21–23°C (70–74°F). Use sawdust bedding changed at least weekly. May be fed commercial pellet food (rat/mice) or a mixture of seeds and cereals. | Not recommended. Usually, pet stores will accept these animals. |
| Goldfish | 15–25 gallon aquarium; allow 60 $cm^3$ of water for each fish. Use aeration. Maintain water at 20–24°C (68–75°F). Use of aquatic plants is recommended. Goldfish are omnivorous; feed them commercially prepared food daily. Remove uneaten food. | Not recommended. Usually, pet stores will accept these animals. |

## Care and Release of Life–Forms, continued

| LIFE–FORMS | CARE | RELEASE |
|---|---|---|
| Guppies | 10–25 gallon aquarium; allow 10 $cm^3$ of water for each fish. Use aeration. Maintain water at 24–27°C (70–85°F). Use of aquatic plants is recommended. Guppies are omnivorous; feed them commercially prepared food daily. Remove uneaten food. | Not recommended. Usually, pet stores will accept these animals. |
| Hermit Crabs | Place in a plastic dish pan with 1–2 inches of dry aquarium gravel. Add two finger bowls, one containing dry pellet rat chow and the other filled with moderate-sized rocks submerged in water. Hermit crabs eat apples, will climb on a small branch, and need some cover under which to hide. | Release only in southern United States, subtropical climates. |
| *Hydra* | Store in jars at refrigeration temperature; change water (spring or pond) weekly. Feeding is not necessary unless animals are stored at room temperature; animals can be stored safely for up to 3 weeks at refrigeration temperatures. If stored at room temperature, feed weekly by introducing *Daphnia* into jar. | May be freely released. |
| Mice | For up to three mice, provide a commercial enclosure or use a plastic washtub with screened cage top. Use a commercial water bottle introduced through the cage top. House at room temperature of 21–23°C (70–74°F). Use sawdust bedding changed at least weekly. May be fed commercial pellet food (rat/mice) or a mixture of seeds and cereals. | Not recommended. Usually, pet stores will accept these animals. |
| Microinvertebrates | Short-term holding: Unscrew jar caps and place in a cool area of lab away from direct sunlight. For long-term culture, place in jar containing 1/2 inch of pond mud that also contains sprigs of elodea. Place in an area that will receive southern light exposure but not direct sunlight. Gentle aeration may be applied. | May be freely released. |
| Planaria | Store in jars at refrigeration temperature; change water (spring or pond) weekly. Feeding is not necessary unless animals are stored at room temperature; animals can be stored for up to 3 weeks at refrigeration temperatures. If stored at room temperature, feed weekly by introducing small strips of fresh liver into jar. | May be freely released. |
| Protists | Short-term holding: Unscrew jar caps and place in a cool area of lab away from direct sunlight. Refer to culture instructions for long-term culture. | May be freely released. |

## ◆ Extended Laboratory Program 1

# Chapter 1 The Chemistry of Life

*A Substance Given Off by Living Things*

**Teacher's Notes**

This investigation, along with ELP 4, gives students an opportunity to follow a sequence of activities in which they practice a scientific method of problem solving, rather than merely reading about it and memorizing the steps. In this activity, students learn to use indicators to identify unknown substances. The investigation is useful for introducing the idea of controls in experimentation. See whether students can determine which test tubes (1–7) serve as controls. It demonstrates the importance of scrupulous observations. It should improve your students' understanding of the difference between observation and inference, which even good students often confuse. This investigation should help students recognize what observations are and what conclusions can logically be drawn from those observations. Students should learn to be critical of the logic they use in making interpretations and drawing conclusions.

Teams of two to four work well, depending on the availability of supplies.

**Time**

Part A of the investigation, once the procedures are started, must be completed in one period. The materials cannot be held over until the next day once the experiment is started. However, Part B can be postponed until the second day if necessary. If the students have reviewed the investigation and understand what they are to do, they should be able to do Parts A and B in the same period.

**Materials (per class of 30—teams of 3)**

- 30 lab aprons
- 30 pairs safety goggles
- 30 pairs plastic gloves
- 130 18-mm × 150-mm test tubes
- 70 cork stoppers to fit test tubes
- 20 test-tube racks
- 20 wrapped straws
- 1 roll paper towels
- 70 small brass screws to fit into test tubes
- 10 dropping bottles phenol red solution (500 mL total)
- 500 mL limewater
- 10 dropping bottles carbonated water (250 mL total)
- 10 dropping bottles distilled vinegar (100 mL total)
- 500 mL yeast-sugar solution
- 500 mL boiled yeast-sugar solution
- 70 dry seeds (melon, radish, sunflower, etc.)
- 70 sprouted seeds of the same kind
- 10 small live insects
- 10 small dead insects of the same kind

**Safety**

Caution students to blow gently into tubes 10 and 13 so that the solutions do not splash onto their faces. Use corks rather than rubber stoppers. Rubber stoppers fit too tightly, and $CO_2$ pressure could build up enough to burst a tightly stoppered test tube. Caution students not to share drinking straws, and have students discard straws when finished.

**Preparations**

Directions for preparing the solutions are provided in the Preparation of Solutions section. Prepare sprouted seeds in advance. Melon or sunflower seeds will sprout in 48 hours if placed between pieces of moist paper and kept in a warm, dark place. Radish seeds will sprout in 24 hours in the same conditions. The larger the seeds, the fewer you need. Students can supply their own insects, or you can order insects from a biological supply company. Nonflying insects such as crickets and small grasshoppers are easiest to work with. Do not permit the use of insects that sting or bite.

Brass screws do not cause a color change in the phenol red solution, but screws of other metals are coated with an antirust chemical that might affect the phenol red. The top of the screw should be large enough that the seeds and other materials cannot fall into the phenol red solution. If brass screws are not available, students can insert cotton plugs or form aluminum foil into a shape that will support the materials above the level of the phenol red solution.

The paper towels prepared for tubes 2 and 3 should be moist only with the yeast-sugar solutions. If too wet, the paper towels will leak liquid into the phenol red solution and cause erroneous results.

Students should record the time required for the phenol red to change color in each tube. Exact times are not important. The reason for having students record the time is to encourage them to make more careful observations.

Different groups of students can use different kinds of acids in tube 11. This will show that any acid can cause phenol red to change to yellow.

Avoid too detailed a discussion of acids and bases at this time. A later investigation will cover that subject. Also, save the discussion of respiration for later.

**Analysis**

1. When carbon dioxide is dissolved in water, an acid is formed since the carbon dioxide produces the same color change in phenol red as do the different kinds of acids.
2. According to the test performed in this investigation, any acid will cause phenol red to turn yellow. Therefore, when an unknown substance is added to phenol red and it turns yellow, students can be relatively sure that this substance contains an acid.

   Because all acids turn the phenol red to yellow, students cannot be sure that an unknown substance gives an acid test because of carbon dioxide.

3. The phenol red turned yellow when the breath was bubbled through it. Therefore, the breath contains something that forms an acid.
4. Yes. The breath contains carbon dioxide because limewater turns milky only when substances containing carbon dioxide are added to it. Carbon dioxide could be the acid-forming substance that causes the phenol red to change color.
5. The materials that caused a change in tubes 2–7 showed active life by moving or growing, whereas the ones that caused no change were either dead or dormant.
6. Tube 1 is a control for all the other tubes. It determines whether phenol red will change color in the presence of only the brass screw and the enclosed air.
7. **a.** Tube 3 is a control for test tube 2 to determine whether inactive as well as active yeasts will cause a reaction.
   **b. and c.** Tube 4 is a control for tube 5, and tube 7 is a control for tube 6, in each case again testing whether unknown variables associated with inactive or dead organisms will cause a reaction.
8. Active living things give off carbon dioxide.
9. Limewater could be used as an indicator to test the hypothesis that living things give off carbon dioxide.
10. The materials that caused no change in the indicators were used as controls for comparison. They were used to rule out every variable but one activity by living organisms.

## ◆ Extended Laboratory Program 1

# Chapter 1 The Chemistry of Life

### *A Substance Given Off by Living Things*

Do living things give off a common substance when they respire? Qualitative observations from chemical tests with carefully planned controls can reveal that you change the composition of the air you breathe. In this investigation, you will use a pH indicator to make qualitative observations of liquids exposed to living and nonliving things. The indicator solution will display one color following exposure to an acid and a different color following exposure to a base. Do you expect the indicator solution to respond differently to living versus nonliving things?

**Materials (per team of 3)**

3 lab aprons
3 pairs safety goggles
3 pairs plastic gloves
13 18-mm × 150-mm test tubes
7 cork stoppers to fit test tubes
2 test-tube racks
wrapped straws
paper towels
7 small brass screws to fit into the test tubes
phenol red solution in dropping bottle
limewater
carbonated water in dropping bottle
vinegar in dropping bottle
yeast-sugar solution
boiled yeast-sugar solution
5–10 dry seeds (melon, radish, sunflower, etc.)
5–10 sprouted seeds of the same kind
1 small live insect
1 small dead insect

**SAFETY Put on your safety goggles, lab apron, and plastic gloves.**

### PROCEDURE

#### PART A Testing Materials with Phenol Red

1. Label seven test tubes 1, 2, 3, 4, 5, 6, and 7. Place them in a test-tube rack, and add 5 drops of phenol red solution to each.

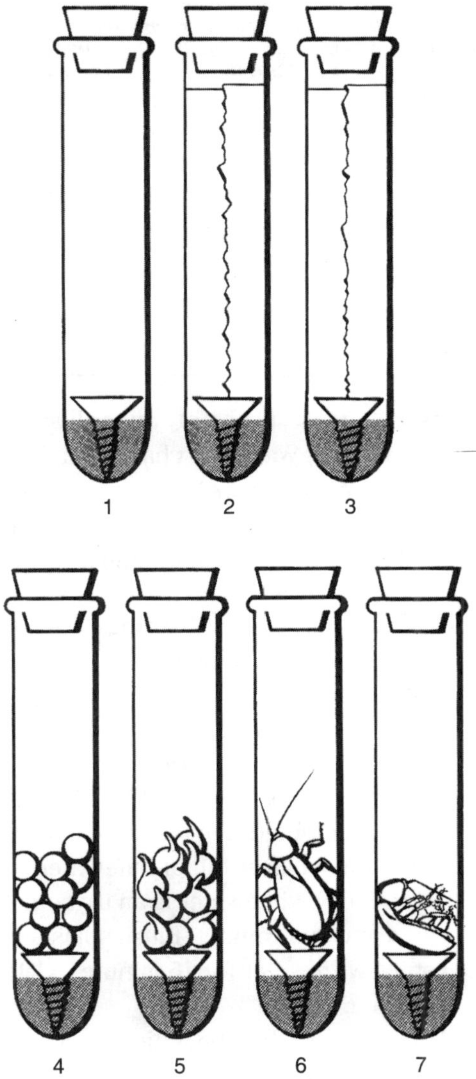


**FIGURE 1.1** The test-tube setups

2. Tilt each test tube slightly, and gently slide a screw, point first, to the bottom.
3. Add the following materials to the test tubes, as indicated in Figure 1.1.

| | |
|---|---|
| Test tube 1: | nothing |
| Test tube 2: | a small, rolled piece of paper towel moistened with yeast-sugar solution |
| Test tube 3: | a similar piece of paper towel moistened with boiled yeast-sugar solution |
| Test tube 4: | 5–10 small dry seeds of one kind |
| Test tube 5: | like number and kind of sprouted seeds |
| Test tube 6: | a live insect |
| Test tube 7: | a dead insect |

| TABLE 1.1 RECORD OF OBSERVATIONS | | | |
|---|---|---|---|
| TEST TUBE NUMBER | MATERIAL ADDED | INDICATOR CHANGE | TIME FOR CHANGE |
| 1 | | | |
| 2 | | | |
| 3 | | | |
| 4 | | | |
| 5 | | | |
| 6 | | | |
| 7 | | | |

4. Stopper the test tubes only after all seven have been prepared. Watch for changes in the phenol red solution, and record the approximate time required for each change to take place. Record your observations in a table similar to Table 1.1.

**PART B Determining the Meaning of Phenol Red Change**

5. The following series of tests will help you interpret the results of Part A. Set up six test tubes in a rack and label them 8, 9, 10, 11, 12, and 13. In test tubes 8, 9, and 10, place 10–12 drops of phenol red. Fill test tubes 11, 12, and 13 about one-quarter full of limewater.

   CAUTION: Limewater is an irritant. Avoid skin and eye contact. Flush splashes and spills with water for 15 minutes. Call your teacher.

6. Record in your table the indicator changes as the following substances are added:

   Test tube 8: 1–5 drops of vinegar (3–5% acetic acid)

   Test tube 9: 5–10 drops of carbonated water (water with carbon dioxide added)

   Test tube 10: your breath blown through a straw for 30 seconds into the phenol red solution

   CAUTION: Never ingest solution through a straw.

   Test tube 11: 15–20 drops of vinegar

   Test tube 12: 5–10 drops of carbonated water

   Test tube 13: your breath blown through a straw for 30 seconds into the limewater

7. Wash your hands thoroughly before leaving the laboratory.

**Analysis**

1. According to the results of the test in test tubes 8 and 9, what kind of substance does carbon dioxide form when it is dissolved in water?
2. If a certain substance causes a color change in phenol red like that in test tube 9, can you be sure the substance contains an acid? Can you be sure it contains carbon dioxide?
3. How can you know that your breath contains a substance that forms an acid when mixed with the water of the phenol red solution?
4. Judging from the results of the tests in test tubes 10 and 13, does your breath contain carbon dioxide? Explain.
5. How do the materials that caused a change in test tubes 2–7 differ from those that did not?
6. For which tubes is test tube 1 a control?
7. How does each of the following pairs of test tubes form a controlled pair?
   a. tubes 2 and 3
   b. tubes 4 and 5
   c. tubes 6 and 7
8. Develop a hypothesis that accounts for the color change in some of the first seven test tubes.
9. What indicator could be used to give support to your hypothesis?
10. Explain why some materials that caused no change in the indicator were included in this investigation.

## ◆ Extended Laboratory Program 2

# Chapter 1 The Chemistry of Life

*The Composition of Water*

**Teacher's Notes**

This investigation is more interesting and effective if it is performed before the students read about the composition of water in the text. Carried out before the reading, the investigation becomes one of discovery.

If enough electrolysis setups are available, groups of four to six students can work together. If equipment is limited, the lab can be performed as a demonstration.

**Time**

One class period should be sufficient.

**Materials (per class of 30—teams of 6)**

5 electrolysis setups
5 6-volt to 12-volt DC power supplies or 10 6-volt DC electrochemical cells with terminals
water
**For tests:**
30 lab aprons
30 pairs safety goggles
2 18-mm × 150-mm test tubes wrapped with cloth-backed tape
wooden splint
matches

**Safety**

Collect the gas samples, and perform the tests for the students. Be sure you and the students are wearing safety goggles and lab aprons when observing the tests.

DC power supplies must be equipped with three-wire grounded wiring. Lab electrical sockets must be protected with a GFI-type circuit breaker. See the safety remarks about electrical safety in the teacher's notes for Extended Laboratory Program 6.

**Preparations**

Use the 12-volt DC setting on a power supply or two 6-volt cells connected in series. Electrolysis proceeds much more slowly at 6 volts, and not much gas will evolve during one class period. You can obtain cells with terminals at hardware stores.

Electrolysis will proceed faster if you add 50 mL of 1 *M* $H_2SO_4$ per 500 mL water. If you do this, caution students about handling the solution, and have them wear plastic gloves as well as lab aprons and safety goggles.

Review the students' observations as they work to see whether they are confusing assumptions with observations; for example, a student who knows the formula of water ($H_2O$) may list as an observation that oxygen and hydrogen are bubbling out of the water. This is a good time to emphasize the difference between observations and inferences made from observations.

After students have considered Analysis questions 1 through 3, perform tests on the gases produced. To collect the oxygen (the one with the lesser volume), put a rubber tube on the stopcock outlet. Hold the collecting tube in an upright position; oxygen, being heavier than air, will remain in the upright tube. Collect the hydrogen by inverting a test tube over the stopcock outlet—let the gas flow into the tube. Hydrogen is lighter than air and will remain in the inverted tube.

Perform the glowing-splint test with the test tube containing the oxygen. As the test is performed, ask students why the splint is bursting into flame. To show the presence of hydrogen in the other tube, bring a burning match near the open end of the inverted tube. A mild explosion that makes a whistling sound or a pop demonstrates that hydrogen is being burned rapidly. Figure T2.1 illustrates the standard oxygen and hydrogen tests.

Student observations during Procedure 3 may be as follows:

1. The two tubes are filled initially to the same level.
2. Bubbles start to rise inside both tubes as soon as the current is turned on.
3. Bubbles rise faster in one tube than in the other. (This may be a point of argument.)
4. As bubbles collect and rise, the liquid level starts to drop in both tubes and continues to drop as long as the current flows.
5. The liquid level drops more rapidly in one tube than in the other.
6. The rate of gas production seems to be quite constant in both tubes.
7. As the experiment progresses, the water level in one tube drops lower than that in the other.
8. As the gases continue to collect, the ratio of one gas to the other approaches 2:1. (It will not be exactly 2:1. Some of the gas molecules dissolve in the liquid, and oxygen is about twice as soluble in water as hydrogen.)
9. The bubbles of gas rise from the electrodes only when the current is turned on.
10. More bubbles are produced at both terminals when the voltage is increased.

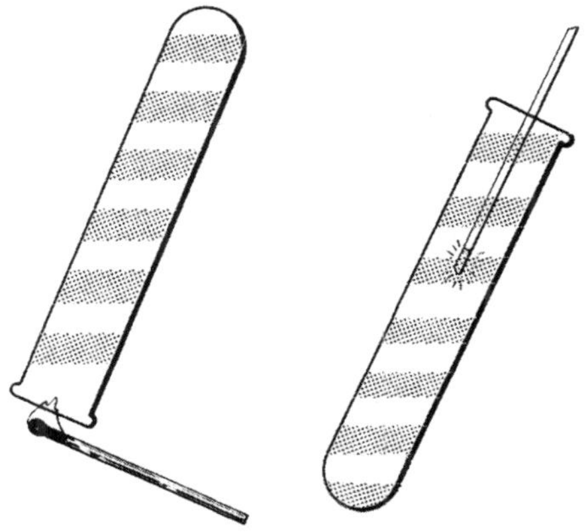

**FIGURE T2.1** Testing for hydrogen (left) and oxygen (right)

**Analysis**

1. Yes. In one tube, the bubbles are larger and rise faster.
2. The water level goes down as the bubbles evolve. If the amount of liquid is decreasing, then some of the liquid is decomposing to form gases.
3. The ratio of the volumes is about 2:1. Yes. (It may not be exactly 2:1 because oxygen dissolves in water more rapidly than hydrogen.)
4. The probable ratio of hydrogen particles to oxygen particles in a water molecule is two hydrogen to one oxygen.
5. The electricity provides the energy needed to decompose the water molecule.
6. A test tube was held upside down to collect gas from the electrolysis tube containing the greater volume of gas. This gas is lighter than air. The gas collected from the electrolysis tube with a higher water level is heavier than air.
7. The gas produced at the negative terminal is approximately twice the volume of the gas produced at the positive terminal. The fact that it is lighter than air and explodes with a loud pop when a burning splint is brought close to it indicates that the gas is hydrogen. The gas produced at the positive terminal is half the volume of that produced at the negative terminal. The fact that it is heavier than air and causes a glowing splint to burst into flame indicates that this gas is oxygen.
8. $2H_2O \xrightarrow{\text{electrolysis}} 2H_2 + O_2$ or

   $2H_2O \xrightarrow{\text{electrolysis}} 4H + 2O$

 

## ◆ Extended Laboratory Program 2

# Chapter 1 The Chemistry of Life

*The Composition of Water*

To understand many of the important biochemical processes of life, you need a basic knowledge of atoms, molecules, elements, and compounds. In this investigation, you will use electricity to separate water into its elements. The objective is to use results of some simple tests to identify the elements.

**Materials (per team of 6)**

6 lab aprons
6 pairs safety goggles
electrolysis apparatus
DC power supply (6–12 volts)
water

**SAFETY** Put on your safety goggles and lab apron.

**PROCEDURE**

1. With the stopcocks open, pour water into the reservoir of the electrolysis apparatus until the liquid level reaches the bottom of the stopcocks (see Figure 2.1). Now close the stopcocks.
2. Connect the wires from the electrodes to the terminals of the DC source. If you use a DC power supply, start with a low voltage and increase to a higher voltage. If you use an electrochemical cell, connect each pole to one of the electrodes. Watch for any reaction. Next, connect another cell in series with the first. Use only one wire to connect the two cells, the positive pole of one being attached to the negative pole of the other. Connect the remaining poles directly to the electrodes, as in Figure 2.1.
3. Observe the process closely for at least 15 minutes. In your logbook, list at least ten observations during this time.
4. Answer Analysis questions 1 through 3. If necessary, observe the electrolysis process again.
5. Your teacher will collect samples of the gases from each stopcock and perform tests on them. After observing the tests, answer Analysis questions 4 through 8.
6. Wash your hands thoroughly before leaving the laboratory.

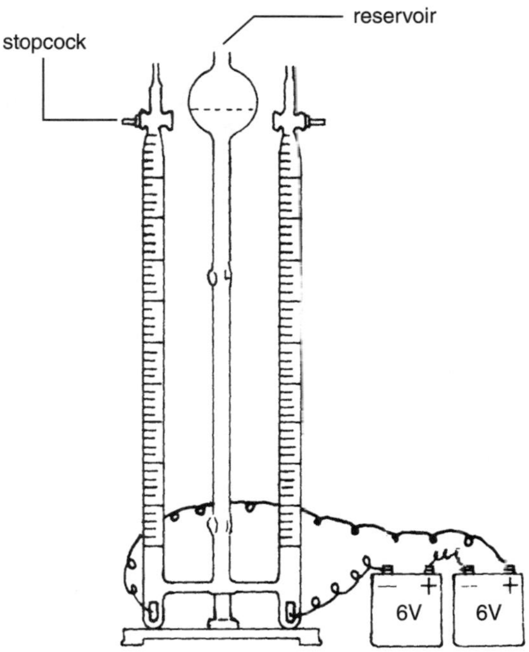


**FIGURE 2.1** Apparatus for the electrolysis of water

**Analysis**

1. Do bubbles rise from the electrodes? How do the bubbles differ in each tube? Do they rise to the surface at the same rate?
2. Explain what happened to the water in each tube. Develop a hypothesis to explain your observations.
3. What is the ratio of the volumes of gas in the two tubes? Did this ratio remain constant during the observation time?
4. What is the probable ratio of hydrogen particles to oxygen particles in a molecule of water?
5. What is the function of the electric current in the electrolysis of water?
6. Which test tube was held upside down as the stopcock was opened to collect the gases? What does this technique indicate about the gases?
7. Write a short paragraph summarizing the results of the tests.
8. Complete the equation to show how many atoms of each would be produced by splitting two molecules of water.

$$2H_2O \xrightarrow{\text{electrolysis}} ? + ?$$

## Extended Laboratory Program 3

# Chapter 3 Exchanging Materials with the Environment

*Rate of Diffusion*

**Teacher's Notes**

This is a simple but revealing experiment about diffusion. Encourage students to write their predictions about time and the effect of temperature before they actually carry out the experiment. This investigation stresses two critical process skills: making predictions and interpreting graphs. Encourage students to try their own temperature-related diffusion experiments.

Teams of four work well for this investigation.

**Time**

One class period.

**Materials (per class of 30—teams of 4)**

30 lab aprons
30 pairs safety goggles
8 100-mL graduated cylinders
8 250-mL beakers
16 125-mL side-arm flasks
8 10-cm pieces of clear plastic tubing that fit onto the side–arm flasks
8 tubing clamps
8 thermometers
8 rulers
8 stopwatches or watches with second hands
8 pot holders
0.1 *M* HCl solution
phenol red indicator solution
cold tap water
hot tap water

**Safety**

Caution students about dispensing and handling hot water. Temperatures may exceed 50°C, which can cause second-degree burns. They should handle the beakers with pot holders.

If a computer is used in the laboratory, the electrical sockets must be protected with a ground fault interrupter (GFI) circuit breaker. Every electrical outlet in the lab must be the three-hole type, and every set of three holes must be checked with a circuit tester before use to ensure that the wiring has been correctly connected to that set. (In other words, there must be no open ground, open neutral, open hot, hot/ground reverse, or hot/neutral reverse wiring.) A circuit tester indicates these improper wiring conditions by various configurations of red and green lights. Circuit testers are widely available at hardware and electronic-supply stores for about $5 each. Do not use extension cords; they create tripping hazards that could result in falls.

**Preparations**

Plan ahead and make sure that an adequate supply of side–arm flasks and plastic tubing is available.

Prepare 400 mL of the phenol red indicator solution and 2.5 L of the 0.1 *M* HCl solution as described in Preparation of Solutions. To save time, you can prepare the diluted phenol red solution in advance and store it in the refrigerator until you need it. You may wish to heat some of the phenol red dilution and have it available to the students as they reach step 16.

**Analysis**

1. Predictions may vary widely from the actual time.
2. Predictions will vary, but they should be much closer to the actual time after running the experiment once.
3. The second prediction should be more accurate because it is based on the results of a previous experiment and not just a guess concocted without data. You may wish to discuss the importance of this with the students.
4. Diffusion occurs more rapidly at higher temperature.
5. Diffusion is a relatively slow process, and it is ineffective in transporting nutrients more than a few cell lengths; therefore, it is used by microorganisms. Larger plants and animals have evolved transport systems that move nutrients and wastes over much greater distances.
6. Diffusion is required for microorganisms to grow and divide. Because diffusion occurs more slowly at lower temperature, the growth rate of the microorganisms on the food will slow down, leading to a longer shelf life.

## ◆ Extended Laboratory Program 3

# Chapter 3 Exchanging Materials with the Environment

*Rate of Diffusion*

Diffusion is the movement of molecules from an area of greater concentration to one of lesser concentration—from higher energy to lower energy —and is caused by the random movement of molecules. In this experiment, you will monitor the rate of diffusion of water molecules. You can make the extent of diffusion visible by allowing a weakly acidic solution to mix with water containing a pH indicator called phenol red. A color change from red to yellow indicates diffusion between the two solutions.

**Materials (per team of 4)**

- 4 lab aprons
- 4 pairs safety goggles
- 1 100-mL graduated cylinder
- 1 250-mL beaker
- 2 125-mL side-arm flasks
- 1 10-cm piece of clear plastic tubing that fits onto the side-arm flasks
- 1 tubing clamp
- 1 dropping pipet
- 1 thermometer
- 1 ruler
- stopwatch or watch with a second hand
- 0.1 *M* HCl solution
- phenol red indicator solution
- water
- pot holders

**SAFETY Put on your safety goggles and lab apron.**

**PROCEDURE**

1. First prepare the phenol red solution. Use the graduated cylinder to add 180 mL of water to the 250-mL beaker. Next, use the graduated cylinder to add 20 mL of phenol red indicator solution. Gently stir with the dropping pipet until the solution has a uniformly red color.
2. Assemble the diffusion apparatus, as shown in Figure 3.1. Moisten one end of the plastic tubing with water, and carefully attach it to one of the 125-mL side-arm flasks.

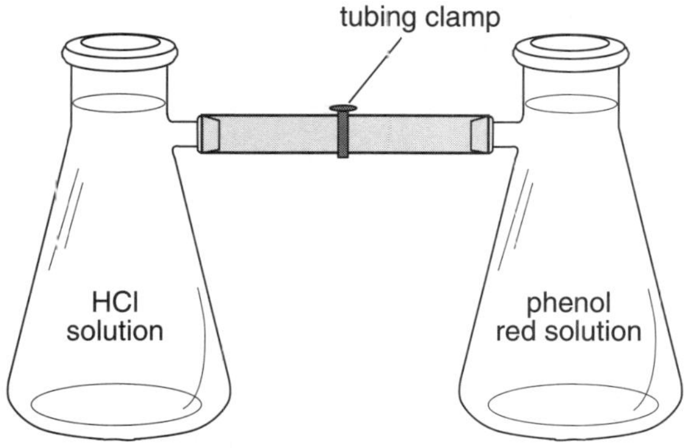


**FIGURE 3.1** Diffusion apparatus

3. Place the tubing clamp around the plastic tubing at its center point.
4. Fill the flask with the HCl solution. Tilt the flask to allow the solution to enter the plastic tubing until it is stopped by the clamp. If you have trouble getting the HCl solution into the tubing, use the dropping pipet to squirt the solution into the tubing.
5. Moisten the other end of the plastic tubing, and carefully attach it to a second 125-mL side-arm flask.
6. Fill the second flask with the phenol red solution up to the level of the side arm.
7. Use a dropping pipet to add phenol red solution to the empty side of the plastic tubing. Start adding the solution next to the clamp, and continue filling the tubing until it drips into the flask.
8. Continue filling the flask with phenol red until the level reaches the top of the flask.
9. If necessary, add more HCl solution to the other flask until the level reaches the top.
10. Use a ruler to measure the distance (in millimeters) between the tubing clamp and the side-arm entrance to the flask containing the HCl.
11. Before you remove the tubing clamp that separates the solutions in the two flasks, predict how long it will take for the colored solution to leave the tubing and enter the flask containing the clear HCl solution. Record your prediction in your logbook.
12. Remove the tubing clamp and begin timing the diffusion process. Observe the color change as the two solutions begin to mix.

13. Note the time it takes for the colored solution to move from the tubing until it enters the flask containing the clear HCl solution. Record the time in your logbook.
14. Carefully pour out the solutions from the two flasks, and disassemble the apparatus.
15. Rinse the flasks and tubing with tap water and reassemble as in Steps 1–4.
16. Prepare more phenol red solution as in step 1.
17. Place the beaker of phenol red solution on the hot plate and heat until the solution reaches 50°C.
    CAUTION: Use pot holders to hold the beaker. Do not let the beaker touch anyone's skin. If the beaker comes into contact with bare skin, immediately place the burned area under *cold* running water and call your teacher.
18. Add the heated phenol red solution to the apparatus as in steps 5–7.
19. Before you remove the tubing clamp, predict how long it will take the colored solution to leave the tubing and enter the flask containing the clear 0.1 *M* HCl solution. Record your prediction in your logbook.
20. Once again, remove the tubing clamp, and time how long it takes for diffusion to move the colored solution from the tubing to the entrance of the flask containing the 0.1 *M* HCl solution. Record the time in your logbook.
21. Disassemble the diffusion apparatus, and thoroughly rinse the flasks and tubing with tap water.
22. Wash your hands thoroughly before leaving the laboratory.

**Analysis**

1. Examine your data. How did your prediction about the time it would take for the room-temperature phenol red solution to diffuse into the flask containing the 0.1 *M* HCl solution compare with the actual time?
2. How did your prediction about the time for the hot phenol red solution to diffuse into the flask containing the 0.1 *M* HCl solution compare to the actual time?
3. Which of your predictions was more accurate? Explain.
4. What is the effect of temperature on the rate of diffusion?
5. Is diffusion an efficient mechanism for transporting nutrients throughout the body of an organism? Explain.
6. We are advised to keep food in the refrigerator to retard the growth of bacteria and other microorganisms that lead to spoiling. Relate this practice to what you have learned about diffusion and temperature.

## ◆ Extended Laboratory Program 4

# Chapter 3 Exchanging Materials with the Environment

*Effects of Three Solutions on Potato Cores*

**Teacher's Notes**

Researchers use two kinds of observations, qualitative and quantitative, to solve many biological problems. This investigation, which deals with measurement of materials, emphasizes the use of quantitative observations. By contrast, ELP 1 used indicators to identify unknown substances, thereby illustrating the use of qualitative observations.

Although students should realize that the changes in the measurements of the potato cores result from the movement of water into or out of the tissue of the potato, this activity provides an opportunity for you to hold to a complete discussion of this topic. Students should understand not only the direction of the water movement but also the reason for the movement.

Teams of three would be ideal. If materials are limited, teams of four could work satisfactorily.

**Time**

Allow two periods for student activity and another 20–30 minutes to discuss results and establish a format for written reports if required.

**Materials (per class of 30—teams of 3)**

30 lab aprons
30 pairs safety goggles
10 50-mL graduated cylinders
30 25-mm × 200-mm test tubes
10 metric rulers
10 balances
10 glass-marking pencils
10 cork borers (5-mm to 10-mm diameter)
10 scalpels
10 dissecting needles
10 test-tube racks
1 roll aluminum foil or plastic wrap
paper towels
500 mL 10% sugar solution (90% water)
500 mL 20% sugar solution (80% water)
500 mL distilled water (100% water)
10 white potatoes

**Safety**

Caution students to use care when pushing the cork borer through the potato so that they do not cut themselves.

**Preparations**

Students may bring their own potatoes, but they will obtain better comparisons if all the potatoes are the same type. Each potato should be large enough to provide three cores, each 30–40 mm long and at least 6 mm in diameter. The three cores for each group should come from the same potato.

The balances should measure the mass accurately to 0.1 g or, better yet, to 0.01 g. If the number of balances is limited, some groups can carry out the mass-measuring procedure while the other groups make the remaining measurements.

If distilled water is not available, use tap water. However, if the mineral content of the tap water is high, the change in the cores will be less than with distilled water.

To prevent evaporation, use small pieces of foil or plastic wrap to cover containers. If large test tubes are used, stoppers can be substituted for the foil.

If the room is warm, store the bottles or tubes with the potato cores in the refrigerator to prevent spoilage.

**Analysis**

1. Answers will vary, but students should notice a change in the hardness of the cores. Some cores may show a color change.
2. When the volume increases, the mass also increases.
3. In distilled water, cores gain mass. In 90% water, most cores will show little change. In 80% water, cores lose mass.
4. Graphs will vary depending on data collected by each group.
5. Answers will vary depending on graphs. Students should predict the water concentration that corresponds to the point on their plotted line where the mass change is zero.
6. Prepare sugar solutions in a range of concentrations near the predicted point. Experiment with the potato cores by using all the procedures described in this investigation.
7. Human error and/or differences in the accuracy of the measuring instruments could account for the variations in data.
8. Yes. If the water concentration is high (90–100%), the potato cores will increase in size; if the water concentration is only 80%, the potato cores will decrease in size. The line on the graph indicates that water concentration under 80% would cause an even further decrease in the size of potato cores.

## ◆ Extended Laboratory Program 4

# Chapter 3 Exchanging Materials with the Environment

*Effects of Three Solutions on Potato Cores*

What will happen to the size of potato cores placed in water or in sugar solutions for 24 hours? An experiment to answer this question requires two sets of data. First, you must determine the initial size of the potato cores. Then you must determine the size after 24 hours in water or in sugar solutions. The following materials and procedures will help you collect information.

**Materials (per team of 3)**

3 lab aprons
3 pairs safety goggles
50-mL graduated cylinder
3 25-mm × 200-mm test tubes
metric ruler balance
glass-marking pencil
cork borer (5-mm to 10-mm diameter)
scalpel
dissecting needle
test-tube rack
aluminum foil or plastic wrap
paper towels
10% sugar solution (90% water)
20% sugar solution (80% water)
distilled water (100% water)
white potato

**SAFETY Put on your safety goggles and lab apron.**

**PROCEDURE**

1. Use the cork borer to cut three cores from a potato. Use the scalpel to trim each core so that it is at least 30 mm long. Make all cores the same length, and keep them separated. Identify them as core A, core B, and core C.
   CAUTION: Cork borers and scalpels are sharp—handle with care.
2. Copy Table 4.1 in your logbook. Measure the length and diameter of each core to the nearest millimeter, and record them in the table.

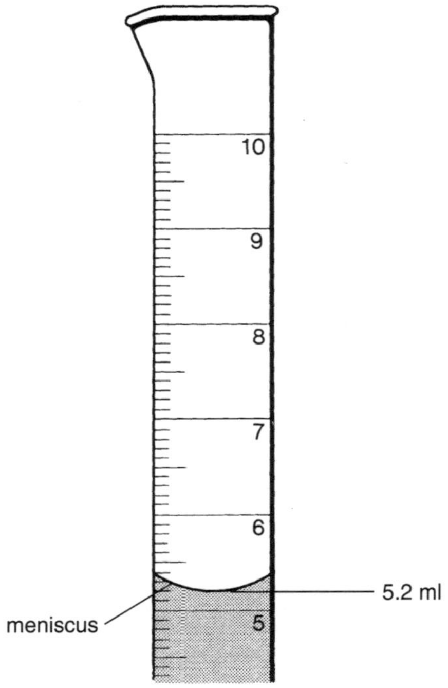


**FIGURE 4.1** Read water volume from the bottom of the meniscus.

3. Measure the volume of each core by the following method. Pour water into the graduated cylinder until it is about half full. Hold the cylinder at eye level, and read the line at the level with the lowest part of the curved surface of the water, as in Figure 4.1. This curved liquid surface is called the meniscus. Record the exact amount of water at this level. Holding the core with a dissecting needle, sink it just under the water's surface, and record the new water level. The difference between the two water levels represents the volume of the core in milliliters. Record the volume of each core in your table.
   CAUTION: Dissecting needles are sharp—handle with care.
4. Your teacher will show you how to use the balance. Blot dry the cores with paper towels. Determine the mass of each core to the nearest 0.1 or 0.01 g. Record your results in the table.
5. Place each core in a different test tube, and label each tube A, B, or C according to the core identification. Pour distilled water (100% water) into test tube A until core A is covered. Pour a 10% sugar solution in water (90% water) into

**TABLE 4.1 POTATO CORE DATA**

| MEASUREMENTS | CORE A (100% WATER) | | | CORE B (90% WATER) | | | CORE C (80% WATER) | | |
|---|---|---|---|---|---|---|---|---|---|
| | FIRST DAY | SECOND DAY | DIFFERENCE (+ OR -) | FIRST DAY | SECOND DAY | DIFFERENCE (+ OR -) | FIRST DAY | SECOND DAY | DIFFERENCE (+ OR -) |
| length (mm) | | | | | | | | | |
| diameter (mm) | | | | | | | | | |
| volume (mL) | | | | | | | | | |
| mass (g) | | | | | | | | | |

test tube B until core B is covered; then pour a 20% sugar solution in water (80% water) into tube C in the same way. Cover each test tube with foil or plastic wrap. Store the tubes in a test-tube rack until your next class period.

6. Wash your hands thoroughly before leaving the laboratory.
7. In the next class period, remove the cores from the test tubes and repeat steps 2–4. Record your results in the table.
8. Wash your hands thoroughly before leaving the laboratory.

**Analysis**

1. In addition to the changes shown by your data, what other changes occurred in each core?
2. How does the change in volume of core A compare with the change in its mass?
3. What is the relationship between the concentration of water and the change in mass of the potato cores?
4. On a graph like the one in Figure 4.2, plot the changes in mass of the three cores.
5. Predict the water concentration at which a potato core would not change in mass.
6. Describe an experiment that would test your prediction.
7. Several students measured the same potato core and got the following data for length: 30 mm, 31 mm, 29 mm, 28 mm. How do you account for the variation in data?
8. Has this investigation answered the question in the introduction? Explain.

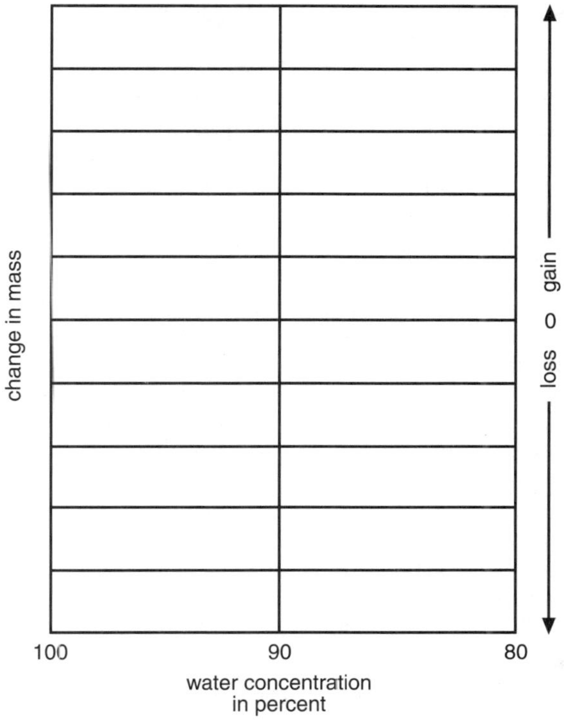


**FIGURE 4.2** Units of measure for mass changes can be entered on the left, and a graph can be plotted for changes in mass of potato cores at different water concentrations.

## ◆ Extended Laboratory Program 5

# Chapter 5 Cell Respiration: Releasing Chemical Energy

*Fermentation*

### Teacher's Notes

This investigation illustrates the relationship between an energy-releasing life process and an enzyme's activity. Relate this investigation to the study of the enzyme catalase. The enzymatic breakdown of the sugar fructose to carbon dioxide and alcohol by yeast is similar to the action of catalase on hydrogen peroxide. The release of energy in the form of heat and the production of identifiable by-products occur in both reactions. Discuss with your class the difference in the rate of the two reactions. See whether the class can bring out the fact that the rate difference occurs primarily because fermentation is an intracellular activity under the control of many enzymes that act in a specific sequence. On the other hand, the action of catalase is an extracellular activity, the rate of which depends mainly on temperature and on the abundance of one enzyme. Relate these rate differences to the need for cells to carry out chemical reactions in specific cell compartments.

This investigation can be carried out by teams of six or by student demonstrators, with the class sharing the data collected. All students can make temperature, odor, and microscopic observations.

### Time

Allow about 30 minutes to set up and another full class period to observe results, analyze data, and discuss results.

### Materials (per class of 30—teams of 6)

- 30 lab aprons
- 30 pairs safety goggles
- 10 stirring rods
- 30 microscope slides
- 30 coverslips
- 10 dropping pipets
- 10 50-mL graduated cylinders
- 10 250-mL flasks, each with 25 mL limewater
- 10 8-oz vacuum bottles stopper assemblies:
  - 10 two-hole stoppers for flasks
  - 10 two-hole stoppers for vacuum bottles
  - 10 30-cm pieces rubber tubing
  - 20 8-cm pieces glass tubing
  - 10 15-cm pieces glass tubing
  - 10 nonmercury thermometers
- 15 microscopes
- 1.25 L grape juice
- 250 mL yeast-water mixture
- 250 mL boiled yeast-water mixture

### Safety

As students prepare to set up their apparatuses, caution them to moisten the glass tubing with water before attaching rubber tubing. Do not permit students to taste any of the materials.

### Preparations

Plastic vacuum bottles can be purchased from local supermarkets and discount stores. Glass-walled vacuum bottles should be purchased only through a biological supplier. (Those found in stores are too fragile.) Glass-walled vacuum bottles should be completely covered around the outside surface with cloth-backed adhesive tape.

Prepare the stopper assemblies before class. Insert one 8-cm and one 15-cm piece of glass tubing into each flask stopper. Insert one 8-cm piece of glass tubing and one nonmercury thermometer into each vacuum-bottle stopper. Protect both hands with several layers of cloth padding, and lubricate the glass tubing and thermometer with glycerin or water before inserting them into the stopper. Caution students not to take apart the stopper assemblies.

Check to make sure students have inserted stoppers tightly into the vacuum bottles and flasks. Directions for preparing limewater and yeast-water mixture are provided in Preparation of Solutions.

If you have a Trekker data collection system, you can adapt this investigation to use a temperature probe. The unit can be set up to collect automatic hourly temperature readings.

### Analysis

1. Graphs should show an increase in the temperature of the grape juice with live yeast and little change in the other.
2. The temperature change and the bubbling of carbon dioxide are evidence of chemical change.
3. The change in the limewater indicated the production of carbon dioxide.
4. Alcohol can be smelled in the bottle with live yeast.
5. The system was sealed so that a fresh supply of air could not enter, but the reactions continued at an increasing rate.
6. The sugar in the grape juice provided the yeast with energy.
7. No. Part of the energy was used to grow and multiply, as evidenced by the increase in the number of cells observed. Part of the energy was given off as heat.
8. Sugar ($C_6H_{12}O_6$) was converted to ethyl alcohol ($C_2H_5OH$) and carbon dioxide ($CO_2$).
9. Dead yeast cells did not produce heat or carbon dioxide. This is evidence that the reaction in the bottle with live yeast was enzymatic. Enzymes are responsible for chemical activity in living cells.

## ◆ Extended Laboratory Program 5

# Chapter 5 Cell Respiration: Releasing Chemical Energy

*Fermentation*

All living things need energy. The source of energy for living cells is food. To get energy from food, cells must change the food to release the stored energy. Fermentation is an energy-releasing process. What kind of cells carry on fermentation? Into what does food change when the energy is released by fermentation? What forms of energy does fermentation produce? These questions may be answered by doing the following investigation.

**Materials (per team of 6)**

- 6 lab aprons
- 6 pairs safety goggles
- 2 stirring rods
- 6 microscope slides
- 6 coverslips
- 2 dropping pipets
- 100-mL graduated cylinder
- 2 250-mL flasks, each with 25 mL limewater
- 2 vacuum bottles
- 2 two-hole stopper assemblies for flasks
- 2 two-hole stopper assemblies for vacuum bottles
- rubber tubing
- 3 microscopes
- 250 mL grape juice
- 50 mL yeast-water mixture
- 50 mL boiled yeast-water mixture

**SAFETY** Put on your safety goggles and lab apron.

## PROCEDURE

1. Assemble two apparatuses as shown in Figure 5.1.
2. Pour 125 mL grape juice into each vacuum bottle. About 7 cm of space should remain in each vacuum bottle.
3. Add 50 mL of yeast-water mixture to the grape juice in one of the vacuum bottles. Add 50 mL of boiled yeast-water mixture to the other vacuum bottle.

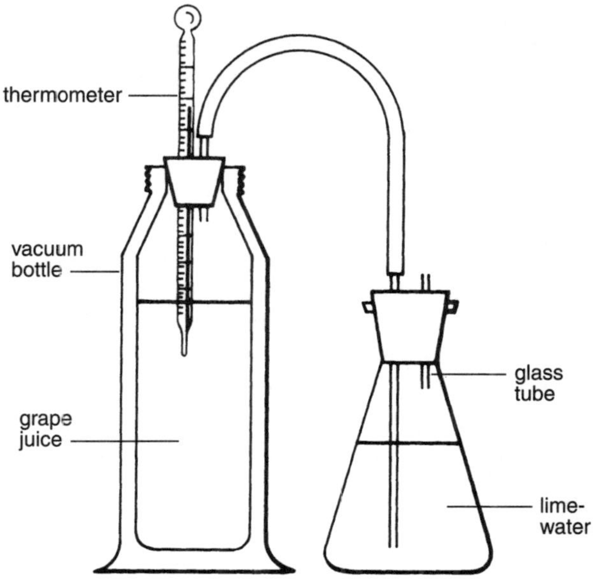


**FIGURE 5.1** Apparatus for fermentation

4. Stir the contents of both bottles well. With separate dropping pipets, remove a few drops from each bottle. Examine the drops on separate slides under the high-power objective of the microscope. (Use coverslips.) Note the concentration of cells on both slides. In your logbook, sketch the fields as they appear and keep the drawings for comparison 48 hours later.
5. Set the apparatus shown in Figure 5.1 in operation. Be sure the thermometer, but not the glass tube, is in the grape juice when the stopper is in place. You will have two such apparatuses. Store them in a safe place as directed by your teacher.
6. For each apparatus, record the temperature every hour for the remainder of the school day and again the next day. Note changes, if any, in the temperature and in the limewater during this observation period.
7. Wash your hands thoroughly before leaving the laboratory.
8. At the end of the 48-hour period, open both vacuum bottles and compare the odor of their contents. Stir the contents of each bottle with separate stirring rods. Using separate droppers, remove a drop from each bottle and examine these drops under the high-power objective of the microscope. (Use coverslips.) Sketch the fields again, showing the concentration of cells

on each slide. Also note any differences in the size and shape of the cells on the two slides.

9. Wash your hands thoroughly before leaving the laboratory.

**Analysis**

1. Plot a line graph of the temperature changes in both bottles. Put both lines on the same graph for comparison. Use the horizontal axis for time and the vertical axis for temperature. Compare your graph with those of other groups.
2. What evidence is there that a chemical change occurred in the bottle with the live yeast?
3. What did the change in limewater indicate?
4. What new substance do you smell?
5. What evidence indicates that yeast can live, multiply, and derive energy from food without using oxygen?
6. What was the source of energy of the live yeast cells?
7. Was all the energy released from the food by yeast used to grow and multiply? Explain.
8. $C_2H_5OH$ is the chemical symbol for ethyl alcohol, and $CO_2$ is the symbol for carbon dioxide. What substance in the grape juice could have been converted to these products?
9. Recall that boiling destroys the action of enzymes. What evidence do you have that fermentation depends on enzymes produced by living cells?

## ◆ Extended Laboratory Program

# Introduction to Electrophoresis

### *What Is Gel Electrophoresis?*

The technology of gel electrophoresis opens a window to the invisible, allowing you to see what was previously only theoretical. Using gel electrophoresis, one can visualize DNA, proteins, and other macromolecules that are essential for life.

In simple terms, gel electrophoresis is a method for separating macromolecules. The gel provides the substance in which the separation takes place. *Electro* refers to the energy of electricity. *Phoresis*, from the Greek verb *phores*, means *to carry across*. Gel electrophoresis, then, refers to the technique in which molecules are forced across a span of gel by an electrical current.

The separation of macromolecules depends on two physical properties: mass and electrical charge. When a biological sample is mixed in a buffer solution, applied to a gel, and electrophoresed, these two properties act in concert. DNA samples are negatively charged owing to their phosphate groups. The electrical current causes them to migrate toward the positive electrode (anode). Proteins can be either positively or negatively charged. Typically, they are mixed with a buffer containing the negatively charged detergent sodium dodecyl sulfate. The soap denatures the proteins and coats them with a negative charge so that, during electrophoresis, all proteins will migrate toward the anode.

The gel itself is usually made of agarose (derived from seaweed) or polyacrylamide (the same synthetic polymer used to make soft contact lenses). The gel contains microscopic pores that function as a molecular sieve, separating the molecules by their size, as shown in Figure T6.1. Varying the gel concentration changes the average size of the pores; therefore, the resolving range of the gel can be adjusted. Typically, when electrophoresing cell extracts that contain both large and small molecules, there is no single ideal gel concentration. In this case, gradient gels are used. Gradient gels have differing gel concentrations that form a gradient from top to bottom. As the percentage of gel increases, the average size of the pores decreases.

A tracking dye added to the sample moves ahead of the macromolecules and provides a visual indication of migration through the gel. When electrophoresis is finished—the process may take minutes to hours—the gel is removed from the apparatus and placed in a staining solution to visualize the separated macromolecules. To increase contrast between the stained samples and the gel, a destaining step is often included. Destaining removes stain from the gel background, helping the stained samples to stand out more. The separated macromolecules can be seen as patterns of bands resembling the bar codes on consumer products. Each band in a pattern consists of macromolecules of the same size that migrated at the same rate through the gel.

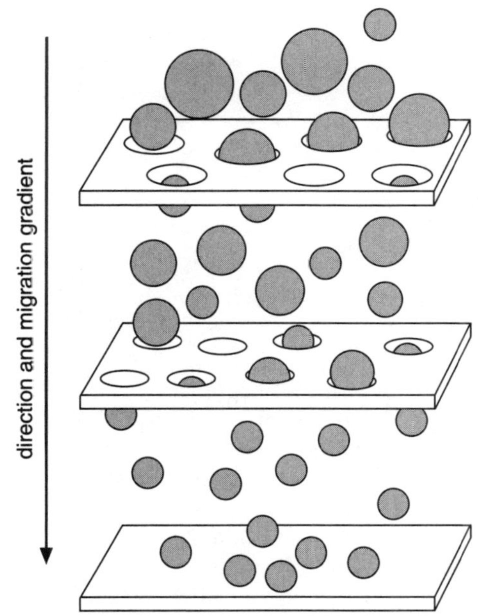


**FIGURE T6.1** As the percentage of gel increases, pore size in the gel decreases. Therefore, smaller macromolecules can migrate through the gel easily, whereas larger macromolecules get trapped in the pores and move more slowly.

The electrophoresis investigations in the Extended Laboratory Program are designed for equipment marketed by Ward's Natural Science Establishment. If other equipment is substituted, follow the manufacturer's directions. There is no guarantee that use of other reagents and equipment will be compatible with the procedures in the investigations.

Ideally, a class of 30 would require three dual electrophoresis tanks for polyacrylamide gels and six agarose gel electrophoresis chambers. If fewer setups are available, make every effort to involve as many students as possible in the activities. Each investigation requires its own experiment kit. In addition, Extended Laboratory Programs 6 and 8 require precast polyacrylamide gels. The Master Materials List includes the other materials needed for the electrophoresis investigations.

Timing is critical in all the investigations. Time schedules have been developed for each investigation based on lab periods of 55 minutes and 7 periods per day. There should be little difficulty in modifying these schedules to fit your particular situation.

| TABLE T6.1 TIME REQUIRED FOR EXTENDED LABORATORIES THAT USE GEL ELECTROPHORESIS | ELP 6 BACTERIA IDENTIFICATION | ELP 8 CARDIAC RISK ASSESSMENT | ELP 13 RESTRICTION OF LAMBDA DNA |
|---|---|---|---|
| Apparatus assembly | 30–40 minutes | 30–40 minutes | 10 minutes |
| Electrophoresis | 20–30 minutes | 30–40 minutes | 60 minutes |
| Staining | 15–30 minutes | | 15–30 minutes |
| Destaining | 1 hour to overnight | | 30 min. to overnight |
| Approximate time | 2–3 hours | 1–2 hours | 2–3 hours |

Table T6.1 indicates the major parts of the procedures and the time needed to complete each part. Staining and destaining times refer to elapsed time from the beginning to the end of the procedure. Actual hands-on time is only a few minutes.

A time schedule must be worked out before you begin an investigation. Sample schedules are provided in the teacher material for each investigation. Because the number of classes and the length of periods vary from school to school, you will need to determine a time schedule for your own situation.

The gradient gels used in Extended Laboratory Programs 6 and 8 are very fragile, especially the 3–8% gels. Handle them with great care. Use gloves and always handle them by the bottom half, the end with the higher concentration. When handling gels, support as much of the gel as possible with all four fingers of both hands. Transfers are easier if both gel and container or storage bag are immersed in water. One-gallon plastic milk containers are excellent for storing reagents; five-gallon plastic paint buckets make good waste containers.

## ◆ Extended Laboratory Program 6

# Chapter 6 Cell Structures and Their Functions

*Bacteria Identification*

**Teacher's Notes**

In this investigation, students use gel electrophoresis to separate protein molecules from three known species of eubacteria and two unknowns. By comparison of the known and unknown separation patterns, they can identify the unknowns. Figure T6.2 is a photograph of a gel showing banding patterns for the five samples used in this investigation. Gel electrophoresis technology clearly demonstrates the diversity of proteins present in different species of eubacteria. Since many of the proteins come from the cells walls, students can look for proteins that distinguish the gram-negative and gram-positive species. Because an SDS (denaturing) gel is used, students also can determine relative molecular masses of the various polypeptides seen in the gel lane. Finally, students gain experience culturing bacteria and determining densities of the cultures.

Two teams of five can share one dual gel electrophoresis tank, each team working with one gel.

**Time**

The approximate time for each step in the investigation is as follows:

| | | |
|---|---|---|
| Culture rehydration/slant inoculation (teacher) | | 30–45 minutes |
| Part A: | Culture inoculation | 30–45 minutes |
| | Culture incubation* | 24–48 hours |
| | Cell density measurements | 10–15 minutes |
| Part B: | Sample preparation | 30–60 minutes |
| Part C: | Apparatus assembly | 30–40 minutes |
| | Gel run | 20–30 minutes |
| Part D: | Processing the gel | 10–15 minutes |
| | Stain* | 15–30 minutes |
| | Destain* | 1 hour–overnight |

*Incubation and stain/destain times are total elapsed time. Hands-on time is only several minutes.

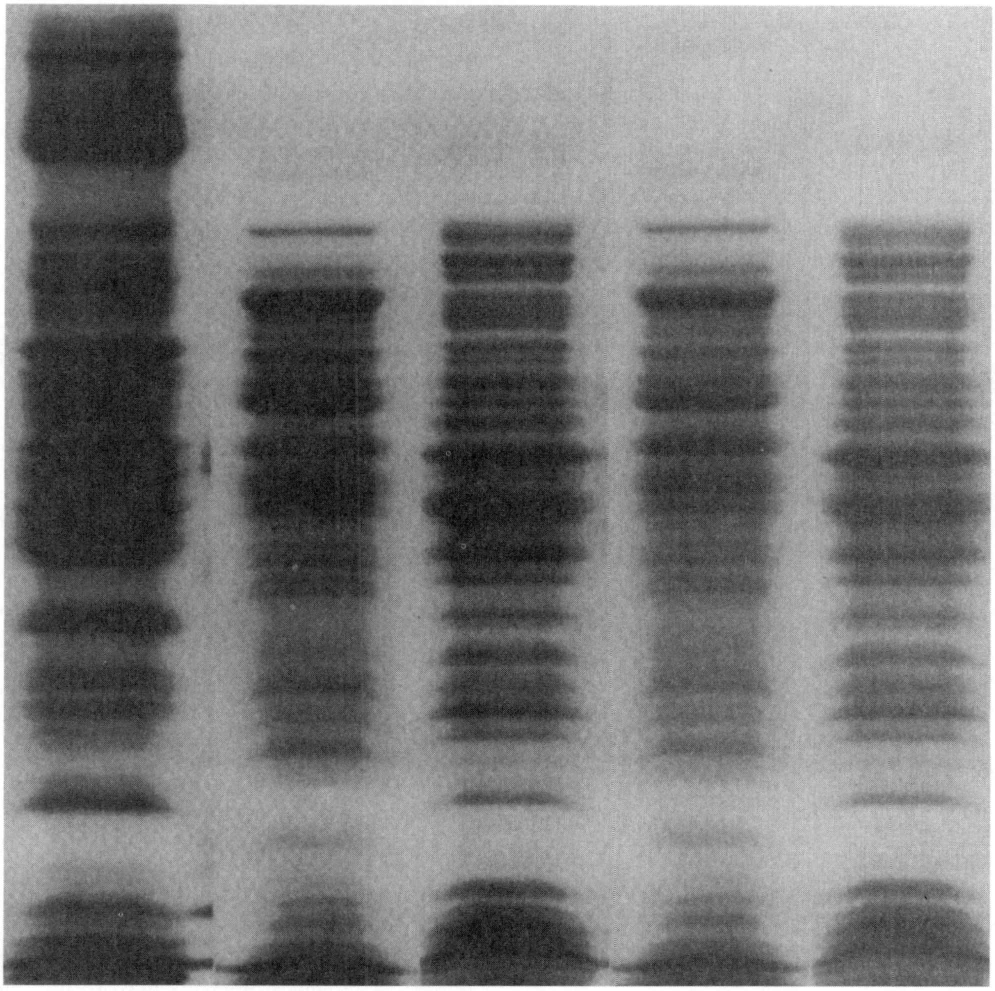

**FIGURE T6.2** Banding patterns produced by (left to right) *E coli, M. luteus, S. marcescens* D1, Unknown A (*M. luteus*), Unknown B (*S. marcescens* WCF)

| DAY FIVE PERIOD | PART C STEPS 16-30 | PART D STEPS 31-35 USING DESIGNATED GELS | COMPLETE ANY OF PART D STEPS 31-35 THAT PREVIOUS CLASS DID NOT FINISH | PART D STEPS 37-38 | PART D STEP 39 |
|---|---|---|---|---|---|
| 1 | gel 1 | | | | |
| 2 | gel 2 | gel 1 | | | |
| 3 | gel 3 | gel 2 | gel 1 | | |
| 4 | gel 4 | gel 3 | gel 2 | | |
| 5 | gel 5 | gel 4 | gel 3 | gel 1 | |
| 6 | gel 6 | gel 5 | gel 4 | gel 2 | |
| 7 | gel 7 | gel 6 | gel 5 | gel 3 | gel 1 |

TABLE T6.2 GEL PROCESSING SCHEDULE

To allow for completion of the entire investigation, you must work out a time schedule that will allow every student to participate in and observe the results of each step in the process. The following schedule shows how a school with 7 periods of 55 minutes each was able to complete the investigation in one school week.

| | | |
|---|---|---|
| Day 1 | Part A: | Inoculation of broth cultures |
| Day 2 | Part A: | Density measurements (*E. coli* fastest grower; hold in a refrigerator until other culture densities are ready for sample preparation.) |
| Day 3 | Part A: | Density measurements (Cultures may be held in a refrigerator 1–2 weeks.) |
| Day 4 | Part B: | Sample preparation (Samples may be held in a refrigerator up to 5 days.) |
| Day 5: | | Consists of a set of steps listed in Table T6.2 |

At the end of day 5, the gel from period 1 is completed. Gels from periods 2 and 3 are placed in the water destain overnight. Gels from periods 4 and 5 are placed in stain solution overnight. The gels from periods 6 and 7 must be removed from the cassettes at the end of their run and placed in stain solution overnight.

**Materials**
**(per class of 30—teams of 5, with 2 teams per tank)**

30 lab aprons
30 pairs safety goggles
150 pairs plastic gloves

**For sample preparation:**

60 microfuge tubes
30 plastic funnels to fit microfuge tubes
30 sterile plastic transfer pipets with bulbs
12 serological pipets
10 pipet bulbs
30 sterile cotton-swab applicators
6 glass-marking pencils
30 white index cards
1 box Whatman #1 filter paper (4-cm diameter)
30 Insta-Rate™ density indicator strips
6 test-tube racks
6 lab burners
3 1-L beakers (or 1 water bath)
3 20-hole bubble racks (to fit 1-L beakers)
3 hot plates
3 biohazard bags
incubator
*lysozyme solution
*SDS buffer with 2-mercaptoethanol
30 culture tubes containing tryptic soy broth
30 culture tubes containing tryptic soy agar slants

**For electrophoresis:**

3 dual gel electrophoresis tanks
1 three-cell power supply
6 10–20% precast polyacrylamide gels
6 sample pipets (5–50 µL delivery)
6 250-mL graduated cylinders
6 600-mL beakers labeled *Waste Stain*
30 plastic transfer pipets with bulbs
6 scalpels
6 self-sealing plastic bags
6 plastic stain/destain containers with lids
6 rectangular plastic dishpans
labeled plastic containers for reagents
5 gallons deionized or distilled water
*6 L running buffer solution (SDS)
*3 500-mL bottles staining solution
*3 packets neutralizer for stain
prepared bacteria samples

**Safety**

When electrical equipment is used in the lab, the electrical sockets must be protected with a ground fault interrupter (GFI) circuit breaker. Every electrical outlet in the lab must be the three-hole type, and every set of three holes must be checked with a circuit tester before use to ensure that the wiring has been correctly connected to that set. (In other words, there must be no open ground, open neutral, open hot, hot/ground reverse, or hot/neutral reverse wiring.) A circuit tester indicates these improper wiring conditions by various configurations of red and green lights. Circuit testers are widely available at hardware and electronic-supply stores for about $5 each.

*In BSCS Bacteria Identification Kit.

Do not use extension cords; they create tripping hazards that may result in falls.

The electrophoresis equipment was designed to be extremely safe. As with any equipment, however, misuse can result in serious injury. With your students, discuss the dangers inherent in using electrical equipment. The following list shows the effects that various levels of current can have on the human body. (These effects are only suggestive; individuals will vary.)

| | |
|---|---|
| 0.001 ampere: | Electricity can be felt |
| 0.005 ampere: | Pain |
| 0.010 ampere: | Paralysis (inability to release firm grip; with loose grip, victim may be thrown violently) |
| 0.050 ampere: | Respiratory paralysis (can be fatal) |
| 0.080 ampere: | Fibrillation of the heart (can be fatal) |
| 1.000 ampere: | Heart paralysis (often fatal) |

**Chemical disposal:** Have students discard the running buffer solution, used stain, and used destain solution down the drain with plenty of water. If you are using buffers and stains from other manufacturers, follow their disposal procedures.

Students are instructed to discard used cotton-swab applicators, transfer pipets, and serological pipets in a biohazard bag. Make certain the bags are available at convenient locations in the laboratory. Used slant cultures and broth tubes should be collected in a central location that you designate. The biohazard bags, slant cultures, and broth tubes must be sterilized before disposal. If an autoclave is not available, see the Guidelines for Laboratory Safety in the Teacher's Annotated Edition for alternative methods of sterilizing these waste materials.

Reagents and bacterial cultures should be stored in a special refrigerator devoted to biological materials only—no food or drinks should ever be stored in this refrigerator. The outside of the refrigerator should be labeled to indicate these rules.

Remind students to wear their safety equipment and to wash their hands before leaving the laboratory.

### Preparations

Before this activity, acquaint students with all parts of the electrophoresis apparatus and with how the various parts go together to form the working assembly. Before starting this investigation, read the Introduction to Gel Electrophoresis. Share with students the principles of electrophoresis so that they understand what is happening during the process.

Directions for preparing running buffer solution are provided in the BSCS Bacteria Identification Kit. Prepare in advance. To prepare the lysozyme solution, add all 10 mg of lysozyme powder to the 20-mL bottle of 10X SDS buffer. Mix by inversion, but do not shake the bottle. Allow the powder to dissolve completely. Mix again. The SDS buffer with 2-mercaptoethanol comes prepared and ready to use.

Lyophilized bacterial cultures require breakout or rehydration before students can prepare broth cultures. Follow the procedure in the next column to reestablish each lyophilized culture.

### Materials

- 5 lyophilized cultures: *Escherichia coli*, *Micrococcus luteus*, *Serratia marcescens* D1, Unknown A (*M. luteus*), and Unknown B (*S. marcescens* WCF)
- 1 tube tryptic soy broth
- 30 tryptic soy agar slants
- 5 sterile plastic transfer pipets with bulbs
- 5 sterile cotton-swab applicators

### Procedure

1. Lift the lid of the plastic outer vial, and remove the inner cryovial.
2. Using sterile technique, untwist the cap of the cryovial. Using a sterile pipet, add 0.5 mL of tryptic soy broth to the lyophilized pellet in the bottom of the cryovial.
3. Allow 10 seconds for the pellet to soften. Mix by drawing the suspension up and down through the pipet 8–10 times.
4. Clear the pipet of any rehydrated material. Discard the used pipet by inserting it back into its outer protective wrapping and placing it in a biohazard bag.
5. Dip a sterile swab applicator into the suspension, and wet the cotton swab thoroughly. Remove the swab, and streak 6 agar slants. Completely cover the entire slant area. This will create a carpet, or lawn, of luxurious growth. Label each slant with the species name or letter designating the respective unknown.
6. Repeat steps 1–5 for each of the remaining lyophilized cultures.
7. Loosen the caps of the slant tubes one-half turn. Incubate the *E. coli* slants at 37°C and all others at room temperature for 24–48 hours, until you observe a homogeneous lawn. (Cultures may be stored in a refrigerator for several weeks before student use.)

Information below about the steps in parentheses refers to the procedure in the student pages:

(*Step 1*) Before students begin the investigation, demonstrate the sterile techniques needed for handling bacterial cultures.

(*Steps 1–5*) Insta-Rate™* density indicator strips afford an easy method of determining the correct cell density required to ensure that enough cell lysate sample is obtained. (*Step 5*) Make sure that all students use the same general technique in determining the number of obscured bars when taking density readings. If possible, have them take readings using a lighted window as the backdrop. A spectrophotometer can also be used to determine cell densities; see Alternative Methods at the end of these notes, and modify steps 1–5 accordingly.

(*Step 4*) *E. coli* will grow out much faster than the other cultures (within 12–18 hours following inoculation). You may wish to monitor these cultures and arrest further growth by placing cultures (of correct cell densities) in a refrigerator until the others grow out. Once the correct cell density for all cultures is obtained (bar 4 obscured), the cultures may be

*Insta-Rate™ is a trademark of Kemtec, Inc.

held in a refrigerator up to 2 weeks, allowing ample time to coordinate the rest of the investigation.
(*Step 8*) This amount is somewhat critical. Students should be reminded that consistent results (band intensity) depend on having consistent cell densities. Too little material will produce faint banding.
(*Steps 9,10*) These amounts are critical. Remind students to use care when measuring.
(*Step 11*) Set up two or three hot plates with 1-L beakers containing about 2.5 cm of water. Bring the water to a slow boil. Turn off the heat just before immersion of the bubble rack containing the students' microfuge tubes. Overheating will coagulate the proteins. (If you have a water bath, students can set the bubble racks directly in it.)
(*Step 12*) You may wish to demonstrate the filtering process (using water). You may also wish to show how to fold the filter paper. Cut 9-cm rounds down to 4 cm before students form the cones. A small cone is used to minimize sample loss due to paper absorption. Alternative methods for sample preparation that require a tabletop centrifuge (4000 × g minimum) and a spectrophotometer are presented at the end of these notes. These methods provide crisper, more definable imaging results. If they are used, make modifications to steps 1–5 and 12 as necessary.
(*Step 13*) Samples may be refrigerated up to 5 days.
(*Step 19*) Filling the wells with running buffer solution before inserting the gel cassette into the electrophoresis tank helps to prevent the formation of bubbles in the wells. If bubbles are present, position the point of the pipet close to the air bubble and use a gentle discharge of running buffer solution to dislodge it.
(*Steps 21–24*) A microsyringe (Hamilton or other) with a 10-µL or 20-µL tip is the best tool for filling the wells. If the smaller tip is used, students should simply add two 10-µL measured amounts of each sample to the respective well. The sample pipet in the materials list is an alternative to this expensive precision device. Demonstrate the correct technique for withdrawing samples and filling wells with the pipet. If the pipet tip is inserted too far into the sample, some of the sample material will adhere to the outside of the tip and diffuse through the buffer into wells adjacent to the one being filled. One sample applicator can be used to apply all samples if it is flushed with running buffer between samples. After all wells have been filled, the applicator should be flushed with distilled water and stored for use by the next class. However, long storage of sample applicators is not advisable because the very fine bore tends to become clogged with minerals from water or other substances.
(*Step 35*) Notching indicates the orientation of the gel when the wells were filled and will enable students to identify the lanes and interpret the band patterns correctly.
(*Step 40*) Complete destaining is fairly rapid, often occurring within 1–2 hours. Students may replace blue-tinted water with fresh distilled or deionized water to decrease destaining time. If time is a problem, gels may be left overnight (or longer) in water, as only those areas of the gel not containing protein release excess stain.
(*Step 41*) Gels may be stored in the stain/destain containers. As an alternative, direct students to place gels in self-sealing plastic bags along with 5–10 mL deionized or distilled water. An optional storage method uses heat-sealed polyester bags, available separately. See the materials list.

**Alternative Methods of Sample Preparation**

If a spectrophotometer and centrifuge are available, the following methods may be used to achieve a more consistent preparation of samples.

**Determining Density with a Spectrophotometer**

**Materials**

Spectrophotometer and cuvettes

**Procedure**

A spectrophotometer can be used quantitatively to assess bacterial cell concentrations. From 0.2 optical density units (ODU) to 2.0 ODU, bacterial concentrations vary colinearly with optical density. Thus, a bacterial cell suspension of 2.0 ODU has about 10 times the cell concentration as a sample of 0.2 ODU. Set the spectrophotometer at a wavelength of 540 nanometers, and zero using a sterile tryptic soy broth blank. The ideal concentration of cells for preparing electrophoresis extracts is about 1.4 ODU.

**Concentrating Samples with a Centrifuge**

**Materials**

tabletop centrifuge (4000 × g minimum)
1.5-mL microfuge tubes with caps
15-mL standard centrifuge tubes (5 per gel run)

**Procedure**

1. Set up a boiling water bath (step 11).
2. Label five microfuge tubes *E. coli*, *M. luteus*, *S. marcescens* D1, Unknown A, Unknown B.
3. Transfer 0.18 mL (180 µL) of each bacterial suspension (density bars 1–4 obscured; Part A) to its appropriately labeled microfuge tube. Place the microfuge tube in a regular centrifuge tube.
4. Pellet the bacteria by centrifugation: 10 minutes at 4,000 × *g* (3 minutes at 12,000 × *g*). Discard the broth supernatant, and resuspend the bacteria in 0.18 mL (180 µL) of deionized or distilled water. (This procedure removes the tryptic soy broth, which may contain exuded bacterial digestive enzymes, waste products, or unwanted amino acids

and buffering agents that may affect the efficiency of lysozyme activity.)

5. Add 20 µL, of lysozyme solution to each of the bacterial suspensions. Gently swirl and incubate at room temperature for 10 minutes.
6. Following incubation, add 0.067 mL (67 µL) of 4X SDS solubilization buffer with 2-mercaptoethanol to each microfuge tube and cap the tube. Gently swirl each tube to mix.
7. Turn off the heat source of the hot plate. Place the microfuge tubes in the bubble rack, and place the bubble rack in the 1-L beaker water bath for 10 minutes.
8. Following incubation, remove the bubble rack from the water bath and allow the microfuge tubes to cool to room temperature (5 minutes).
9. Place the microfuge tubes in regular centrifuge tubes. Centrifuge for 10 minutes at 4,000 × *g* (3 minutes at 12,000 × *g*). Carefully decant supernatant into appropriately labeled microfuge tubes. Use a micropipet to apply 20-µL samples to the gel wells. (Samples may be applied directly from microfuge tubes if care is taken to avoid disturbing the pellets.)

**Analysis**

1. This enzyme serves an antimicrobial function and, with other antimicrobial proteins (antibodies), serves as the first line of defense against harmful microorganisms. Lysozyme breaks down the peptidoglycan layer of bacterial cell walls, resulting in cell lysis. Antibiotics, such as penicillin, inhibit the cross-linking of peptide chains crucial for peptidoglycan synthesis.
2. SDS solubilizes lipoprotein, thus attacking the plasma membrane of the bacterial protoplast. This action induces osmotic shock and results in rupture of the cell. The effect of SDS is minimal on bacterial cells whose cell walls have an intact peptidoglycan fabric.
3. 2 mercaptoethanol breaks disulfide bonds that hold together the polypeptide constituents of whole proteins, making possible direct comparison of polypeptides. Polypeptides are the direct products of gene expression.
4. Bromophenol blue is used as a tracking dye in each sample. Students should see a thin blue line slightly below the sample wells. Some students may say that gravity is the force responsible for the movement. Gravity contributes to the movement, but only to a very slight degree. The major driving force is the electrical current that is being generated through the gel from the top negative electrode (the cathode) to the bottom positive electrode (the anode).
5. When the bromophenol-blue dye reaches the bottom of the gel, the power supply should be turned off.
6. Using information given in preliminary discussion and in the introduction to this investigation, students should be able to include some, if not all, of the following points in their paragraphs.
   **a.** An electrical force is moving from the top tank electrode to the bottom tank electrode. Evidence for this is the downward movement of the tracking dye.
   **b.** The electrical force is pushing the proteins through the gel.
   **c.** Because of its many pores, the gel acts as a molecular sieve.
   **d.** The larger proteins will be nearer the top of the gel, and the smaller proteins will be nearer the bottom.
7. As the percentage of polyacrylamide increases, the pore size of the gel decreases. With this information, students should be able to reason that larger molecules will be trapped first and progressively smaller molecules will be trapped later. Thus, the bottom of the gel is the high-percentage end and is the region of smallest macromolecules.
8. The staining solution simultaneously fixes and stains separated proteins in the gel matrix. The gel matrix also will absorb the stain; therefore, at the end of the staining process, the entire gel will have a uniform light-blue color.
9. The gel must be destained so that the stained protein molecules will become visible. The proteins attract the stain more strongly than the gel matrix does.
10. These are the separated proteins from the bacterial samples that were placed in the sample wells of the gel.
11. (Note: Use of an overhead projector to examine the gels in the storage bags will help students assess the bands.) Lanes 1 and 2: *E. coli* (gram-negative); lanes 3 and 4: *M. Luteus* (gram-positive); lanes 5 and 6: *S. marcescens* D1 (gram-negative); lanes 7 and 8: Unknown A—*M. luteus* (gram-positive); and lanes 9 and 10: Unknown B—*S. marcescens* WCF (gram-negative). Students should observe that certain samples show more banding patterns overall than others. However, it is difficult to distinguish gram-negative from gram-positive bacteria solely on the basis of protein patterns. This is partly because we are observing not only the proteins that are found in the cell wall but also those proteins found within the interior of the cell.
12. Unknown A is *Micrococcus luteus.* Students should observe an exact correlation of protein patterns in lanes 3 and 4 (the known standard) with lanes 7 and 8 (the unknown). No other bacterium has this exact pattern. (Caution students to exercise care in their comparison; some variation in band definition will occur owing to differences in applied sample volumes and protein concentrations.) Thus, electrophoresed unknowns can be compared with known strains to identify microorganisms without the use of gram staining or morphological parameters. The process of using electrophoresis to separate bacterial-cell proteins and comparing these separations to separations of reference strains stored in a computer database has important clinical applications. Alert students may notice that one unknown was the same color as one of the reference strains. Point out that other bacteria produce yellow pigments as well.

13. The band pattern of Unknown B (lanes 9 and 10) bears a very close resemblance to that of *Serratia marcescens* D1 (lanes 5 and 6). Students may note that this relationship is not a perfect match. Unknown B is a colorless mutant strain of *S. marcescens* (strain WCF) and is missing some bands that *S. marcescens* D1 possesses. Strain WCF cannot produce a red cellular pigment (prodigiosin), which is present in strain D1. If this difference is detected—it is difficult to see—students should conclude that the genome of strain WCF is missing certain gene sequences that code for this pigment. Some students may infer that Unknown B is *E. coli* on the basis of a lack of pigmentation. Such inferences are inconsistent with the banding patterns for both samples.
14. The consistently dark-staining band in each sample, located at the bottom of the gel, is the protein lysozyme. Students should observe that this is the darkest band in every sample. Students should also recall that this protein was consistently added in relatively large amounts (100 µL) during sample preparation. Some students may suggest that its appearance at the bottom of the gel is consistent with a relatively low molecular mass—the supposition being that most enzymes have relatively low molecular masses.
15. The gradient gels separate macromolecules semilogarithmetically. After doing the analysis, students should arrive at an approximate molecular mass value of between 13,300 and 15,000 for the dark-staining band. The *Merck Index* lists the molecular mass of lysozyme as 14,400. If you instruct students to research molecular mass values for the reagents added during sample preparation, they will find that the values for 2-mercaptoethanol (75) and SDS (349) do not agree with the data. These molecular masses are too small to be imaged by gel electrophoresis. Furthermore, these molecules are not visualized with the protein stain used in this experiment.

    If the tracking dye has migrated off the gel (that is, the gel has been run too long), have students use the value 70 mm (the gel height) in lieu of the tracking-dye migration distance.

**References**

Kreig, N. R., and J. G. Holt, Eds. *Bergey's Manual of Systematic Bacteriology,* vol. 1, 9th ed. Baltimore: Williams and Wilkins, 1984.

Starr, M. P., *et al.*, Eds. *The Prokaryotes: A Handbook on Habitats: Isolation and Identification of Bacteria,* 2 vols. Heidelberg: Springer-Verlag, 1981.

Volk, W. A. and M. F. Wheeler. *Basic Microbiology,* 5th ed. New York: Harper & Row, 1984.

## ◆ Extended Laboratory Program 6

# Chapter 6 Cell Structures and Their Functions

### *Bacteria Identification*

Gel electrophoresis is a method for separating macromolecules. The process involves using electricity to force molecules across a span of gel. The gel is made from a substance called polyacrylamide, the same material that is used in soft contact lenses. Because the gel contains submicroscopic pores, molecules can be separated by size. By controlling the percentage of the gel (from 3% to 30%), precise pore sizes can be obtained. So-called gradient gels have large pores in the low-percentage end of the gel to very small pores in the high-percentage end of the gel.

The separation of macromolecules depends on two physical properties: mass and electrical charge. In electrophoresing protein samples, they are first mixed with a buffer that contains the negatively charged detergent sodium dodecyl sulfate (SDS). The amount of SDS bound to an individual protein is proportional to its mass. When electrophoresis begins, the SDS-proteins are repelled from the negative electrode (the cathode) and attracted toward the positive electrode (the anode). The gel acts as a molecular sieve and separates the proteins by mass. A visible tracking dye added to the sample moves ahead of the proteins and provides a visual indication of their progress through the gel. When the gel run is complete, the gel is stained and the separated proteins can be seen. They appear as an irregular column of blue bands spread from the top of the gel to the bottom.

By using gel electrophoresis, it is possible to identify unknown bacteria by comparing their protein compositions with those of known bacteria. This technique provides an alternative identification technique to biochemical bacterial identification methods. The primary advantage of using gel electrophoresis to identify unknown bacteria is speed. Gel analysis can be completed in just hours, compared to days for traditional culturing methods.

The metabolic and biosynthetic pathways of eubacteria are more diverse than those of eukaryotes. In addition to being different from eukaryotes, they also are vastly different from one another. Eubacteria generally are separated into two taxonomic groups, based on the gram reaction. This universally applied test depends on whether bacterial cells stain purple (gram-positive) or pink (gram-negative). The outcome of the gram reaction is, to some extent, conditioned by the physiological state of the cell; but it also correlates with some major differences in the chemical composition of eubacterial cell walls. (It is interesting to note that mammalian cells are uniformly gram-negative.)

All prokaryotic cells (except mycoplasmas and archaebacteria) have in common a structural cell-wall component called peptidoglycan. The peptidoglycan layer of a bacterial cell wall is a macromolecular fabric that completely encloses the bacteria and provides the support necessary to maintain the integrity of the cell. In cell walls of gram-negative bacteria, peptidoglycan occurs as a single layer and in a small quantity (1–10%); in gram-positive bacteria, there are many layers and a much greater quantity (60–100%). In addition to peptidoglycan, each bacterial-cell wall contains a unique assortment of other macromolecules (phospholipids, teichoic acids, and polysaccharides) that are specific to it alone. Figure 6.1 illustrates the differences of macromolecular complexity in cell walls of gram-negative and gram-positive bacteria.

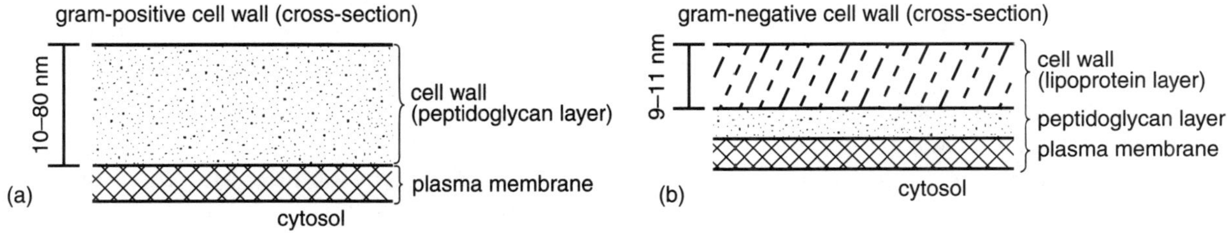


**FIGURE 6.1** Diagrammatic representation of cell wall cross-sections of a gram-positive (a) and a gram-negative (b) bacterium showing the differences in complexity

In this investigation, you will grow five bacterial cultures, prepare protein samples, and then use gel electrophoresis to analyze the samples. You will see examples of bacteria that are gram-positive and gram-negative. Finally, you will identify two unknown bacterial samples by comparing their protein patterns against known types.

**Materials (per team of 5)**

5 lab aprons
5 pairs safety goggles
25 pairs plastic gloves

**For sample preparation:**

10 microfuge tubes with caps
5 plastic funnels (to fit microfuge tubes)
2 serological pipets with bulbs
5 sterile plastic transfer pipets with bulbs
5 sterile cotton-swab applicators
glass-marking pencil
5 white index cards
10 pieces Whatman #1 filter paper (4-cm diameter)
5 Insta-Rate™ density indicator strips
test-tube rack
lab burner
water bath (1-L beaker with bubble rack)
hot plate
incubator
biohazard bag
lysozyme solution
SDS buffer with 2-mercaptoethanol
5 culture tubes containing tryptic soy broth
slant cultures of bacteria (*Escherichia coli*, *Micrococcus luteus*, *Serratia marcescens* D1, Unknown A, Unknown B)

**For electrophoresis:**

dual gel electrophoresis tank
three-cell power supply
10–20% SDS polyacrylamide precast gel
sample applicator (5–50 µL delivery)
5 plastic transfer pipets with bulbs
250-mL graduated cylinder
600-ml beaker labeled *Waste Stain*
scalpel
self-sealing plastic bag
plastic stain/destain container with lid
rectangular plastic dishpan
1,500 mL deionized or distilled water
prepared samples of bacteria
labeled plastic containers of reagents:
2 L working buffer solution
250 mL staining solution

**SAFETY** Put on your safety goggles, lab apron, and gloves; tie back long hair.

## PROCEDURE

### Part A Preparing Cultures of a Specific Cell Density

Part A includes the following steps:

- Transfer bacteria from agar slant to broth
- Incubate
- Check density
- Refrigerate cultures

1. Affix an Insta-Rate™ density indicator strip (with the darkest bar at the top) to each of five culture tubes containing tryptic soy broth. Label each bar on each density strip from 1 to 5, 5 being the darkest bar.
2. Using sterile technique as demonstrated by your teacher, carefully unwrap a sterile swab applicator and insert it into the slant culture tube containing a lawn (luxurious growth) of *E. coli*. Use the swab to collect as much of the lawn as possible. Do not rub so hard that you break the agar surface. Remove the swab applicator, flame the tube, and insert the applicator into a culture tube containing tryptic soy broth. Release collected bacteria from the swab into the broth by repeatedly pressing the swab against the tube wall and then using it to lightly stir the suspension. Flame the broth culture tube, and screw on the cap. Label this tube *E. coli*. Place the used swab applicator back in its original wrapping, and discard it in a biohazard bag as directed by your teacher.

   CAUTION: Lab burners have a pale blue flame that may be difficult to see. Keep hair and clothing away from the flame.
3. Repeat step 2 for each of the remaining slant cultures. Label each culture tube appropriately. Return used slant cultures to the location designated by your teacher.
4. Loosen the caps of the culture tubes one-half turn. Incubate the *E. coli* tube at 37°C and the other tubes at 25°C for approximately 24–48 hours, until all but the topmost bar (bar 5) is obscured. To read the density indicator strips,

tighten the cap on the tube, and resuspend the cells by gently inverting the culture tubes. Hold each tube at eye level (with the density strip away from you), and read through the tube against a white index card.

5. When each culture tube has reached the appropriate cell density (bars 1–4 obscured), store in a refrigerator until you are ready to do Part B.
6. Wash your hands thoroughly before leaving the laboratory.

**Part B Preparing Samples**

Part B includes the following steps:

- Transfer bacteria from broth to microfuge tube
- Add lysozyme
- Incubate at room temperature
- Add SDS buffer with 2-mercaptoethanol
- Incubate in water bath
- Filter
- Refrigerate filtrate

7. Use a glass-marking pencil to label ten microfuge tubes with your team symbol; then in sets of two, label them as follows: *E. coli*, *M. luteus*, *S. marcescens* D1, Unknown A, Unknown B.
8. Tighten the cap, and invert each broth culture tube, thoroughly mixing the cells. Use a new, sterile, plastic transfer pipet with bulb to transfer 1 mL (1000 µL) of each suspension (prepared in Part A) to one of the pair of appropriately labeled microfuge tubes. Discard used pipets in a biohazard bag as directed by your teacher. Return the used culture tubes to the location designated by your teacher.
9. Use a serological pipet to add 0.1 mL (100 µL) of lysozyme solution to each microfuge tube. Swirl gently, and incubate at room temperature for 10 minutes. Discard the serological pipet in the biohazard bag. Lysozyme is an enzyme that breaks down bacterial cell walls by attacking the peptidoglycan fabric. Bacterial cells whose cell walls have been removed by enzymatic action are called protoplasts. A bacterial protoplast consists of an intact plasma membrane, cytosol, and nucleoid. Protoplasts are highly sensitive to disruption by osmotic change.
10. Following incubation in the enzyme solution, use another serological pipet to add 0.33 mL (333 µL) of 4X SDS buffer with 2-mercaptoethanol to each microfuge tube, and swirl to mix. Discard the serological pipet in the biohazard bag. The reagent 2-mercaptoethanol breaks disulfide bonds of certain large proteins, yielding individual polypeptides. SDS (sodium dodecyl sulfate) is a soap. In addition to breaking up lipid constituents of the plasma membrane, it coats all proteins and polypeptides with an overwhelming negative charge so that these polypeptides can be separated in the gel on the basis of molecular mass. The mixture of disrupted cellular material is termed the lysate.
11. Place the five labeled microfuge tubes in the bubble rack and the bubble rack in the water bath (1-L beaker) for 10 minutes. Remove the bubble rack from the beaker, and allow the tubes to cool to room temperature (about 5 minutes).
12. Set up the other five empty microfuge tubes with filtering apparatus in a test-tube rack, as illustrated in Figure 6.2. The goal is to filter each bacterial-cell lysate to obtain a perfectly

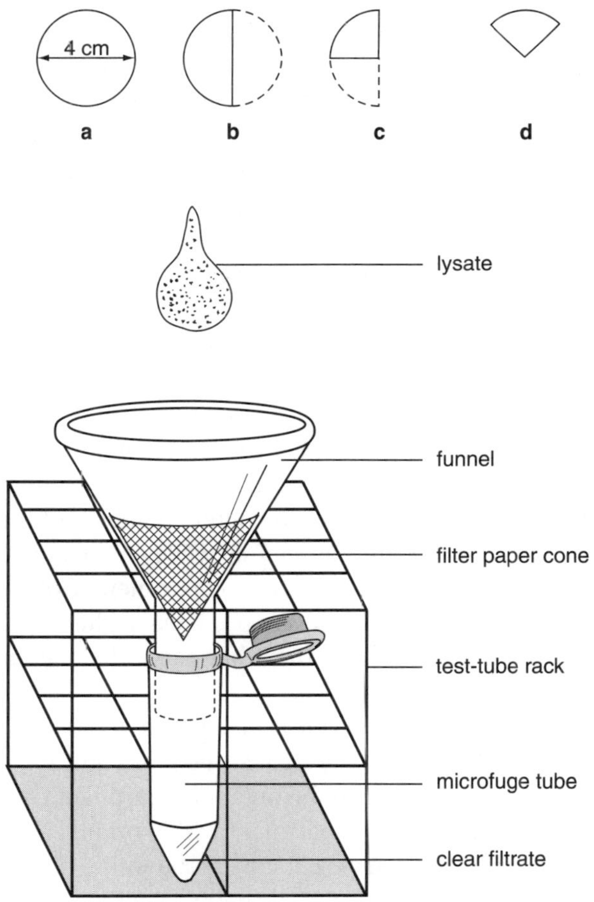


**FIGURE 6.2** Constructing a filter-paper cone (a)–(d); filtering apparatus

clear filtrate, free from large cellular debris that can interfere with electrophoresis. Use Figure 6.2(a)–(d) as a guide to make a cone from each 4-cm filter-paper disc. Place a cone into each of the five funnels, and insert a funnel into each microfuge tube. Each funnel stem should fit tightly in the microfuge tube so that the tube is suspended from the funnel when you place the filtering apparatus in the test-tube rack. Pour the incubated and cooled sample through the filter apparatus to drain into the lower microfuge tube. Sample yields will be small, about 0.6–0.8 mL. It may be necessary to refilter with a fresh piece of filter paper to obtain clear filtrates of the samples.

**13.** Cap the microfuge tubes and store in the freezer until your next lab period.

**14.** Answer Analysis questions 1–3.

**15.** Wash your hands thoroughly before leaving the laboratory.

**Part C Apparatus Assembly and Gel Run**

Before you start the procedure for running a gel, familiarize yourself with the electrophoresis apparatus. Depending on the type of equipment in your lab, your teacher may give you instructions for assembling and electrophoresing gels that are in some ways different from those presented here. Part C includes the following steps:

- Place precast gel into tank assembly
- Add working buffer
- Flush sample wells in gel
- Load samples into wells
- Electrophorese the samples

**16.** Remove gel from package, and carefully remove comb from the gel.

**17.** Place gel into one of the outer chambers of the tank. Make sure the sample wells are facing inward, toward the inner compartment. Note: If you are running a single gel, insert the Plexiglas™ plate in the other outer chamber in place of a second gel.

**18.** Insert the wedges into each corner, making sure the gel is snug against the gasket to prevent leaks into the outer compartment.

**19.** Fill the inner compartment with working buffer solution until it covers the wells. Using a pipet, flush out any air bubbles in the wells.

**20.** Slowly fill the outer compartment with working buffer concentrate until it reaches the same level as that in the inner compartment.

**21.** Use the sample applicator to add 20 µL of the *E. coli* sample to wells 1 and 2. *Take care to insert only the terminal centimeter of the applicator tip into the sample for withdrawal of the material.* Your teacher will demonstrate the correct technique for this step of the procedure.

**22.** Flush the sample applicator several times with running buffer solution, and then add 20 µL of the *M. luteus* sample to wells 3 and 4.

**23.** Use the procedure of step 22 to add 20 µL each of the *S. marcescens* D1 sample to wells 5 and 6, the Unknown A sample to wells 7 and 8, and the Unknown B sample to wells 9 and 10. Flush the sample applicator several times with distilled water.

**24.** Make sure the cover is dry, and attach it to the chamber. Align the cover properly. Wipe off any spills on the apparatus before proceeding.

**25.** Make sure that the patch cords attached to the cover, as well as the female jacks on the power supply, are completely dry. Connect the red patch cord to the red electrode on the power supply. Connect the black patch cord to the black electrode terminal on the power supply.

**26.** Plug in the power supply, and set it to 110 volts.

**27.** Turn on the power supply. The red power light will illuminate. WARNING: The power supply must be used only according to the exact instructions provided. Use of this equipment other than as instructed can lead to severe injury or even death by electric shock.

**28.** Five minutes after electrophoresis has begun, observe the area of gel just below the sample wells. Record your observations.

**29.** Electrophoresis should continue until the blue loading dye nears the bottom of the gel. Turn off the power supply when the process is complete. (The red power light will go out.) Unplug the power supply. Wait about 10 seconds and disconnect the patch cords, first from the power supply; then remove the cover from the chamber.

**30.** Answer Analysis questions 4–6.

**Part D Processing the Gel**

Part D includes the following steps:

- Discard running buffer
- Remove and open gel cassette
- Transfer gel to stain/destain container
- Rinse with water
- Stain

- Destain (water) overnight
- Store gel

31. Remove the cover assembly and pull out the wedges, then the gel.
32. Using the scalpel, cut the yellow tape along one side of the gel and open the cassette. Note: It may be necessary to insert the edge of the scalpel between the glass plates and gently pry them apart.

    CAUTION: Scalpels are sharp; handle with care.
33. With the scalpel, make a small notch in the lower left corner of the gel to serve as a marker of the gel's orientation when you filled the sample wells.
34. Lift the gel tray with the gel out of the chamber, and gently place the gel in the staining tray.
35. Pour enough stain into the tray so that the stain barely covers the gel. Let the gel stain for about 15–30 minutes.
36. After staining, carefully pour off the stain solution into a 600-mL beaker labeled *Waste Stain.*
37. Gently pour 250 mL of deionized or distilled water over the gel in the container. Swirl periodically for 1 minute. Pour the distilled water down the drain.
38. Add 250 mL deionized or distilled water to the container. As the destain solution becomes blue (20–30 minutes), replace it with fresh distilled water. Pour used destain solution down the drain. Repeat this procedure until the gel is clear, leaving light-blue bands only in the sample lanes. Complete destaining will take place within 1–2 hours. (Or let the gel stand overnight in distilled water without water changes.)
39. Following destaining, the gel can be stored indefinitely in deionized or distilled water either in the container or in a self-sealing plastic bag. Remove the gel from the destain solution by carefully placing all four fingers of both gloved hands under the bottom of the gel. Immerse the gel in a sink or rectangular plastic dishpan half-filled with water, and slip the bag over the gel. Drain excess water from the bag, leaving from 5 to 10 mL. The gel should be stored in clean water. Rinse carefully if necessary. Seal the bag.
40. Answer Analysis questions 7–15.
41. Wash your hands thoroughly before leaving the laboratory.

**Analysis**

1. Lysozyme is present in most body fluids (that is, tears, blood, and saliva). What might be its function in these fluids?
2. Why is SDS buffer added after lysozyme?
3. Why is it critical to add 2-mercaptoethanol?
4. What do you observe during electrophoresis? This observation is evidence that the proteins in the various samples are moving. What causes this movement?
5. How can you determine when an electrophoresis run has been completed and the power-supply module should be turned off?
6. While electrophoresis is in progress, write a short paragraph describing what is happening to the proteins in the gel.
7. How can you distinguish the high-percentage end of the gel from the low-percentage end?
8. What is happening during the staining process?
9. What must be done after the staining process? Why?
10. What are the light-blue lines that remain in the sample lanes after destaining?
11. Examine the electrophoresis lanes. On the basis of the band patterns in these lanes, the information in the introduction, and Figure 6.1, can you determine which samples are gram-positive and which are gram-negative? Explain.
12. In wells 7 and 8, you placed the sample Unknown A. Compare its separation pattern to the patterns of the three reference types (*E. coli*, wells 1 and 2; *M. luteus*, wells 3 and 4; *S. marcescens* D1, wells 5 and 6); identify the unknown. Check with your teacher to see whether your determination is correct.
13. In wells 9 and 10, you placed the sample Unknown B. Compare its separation pattern to the patterns in the other lanes. Can you identify this unknown?
14. In every sample lane, a consistently dark-staining band appears. Develop a hypothesis that explains what this band represents.
15. Test this hypothesis by determining the molecular mass of this protein. Measure the distance (in millimeters) from the top of the gel to the position of the tracking dye (bromophenol blue). Tracking dye is usually visible as the hazy, wide, light-blue pattern at the bottom of the gel. Also measure the distance that the dark-staining band has migrated from

the top of the gel to its relative position in each sample. Express this relationship as percent of distance migrated (distance migrated by dark-staining band divided by distance migrated by tracking dye). Use the graph in Figure 6.3 to determine the molecular mass of this protein. If possible, use the *Merck Index* to verify this measurement and your hypothesis. Report the data to your teacher for confirmation.

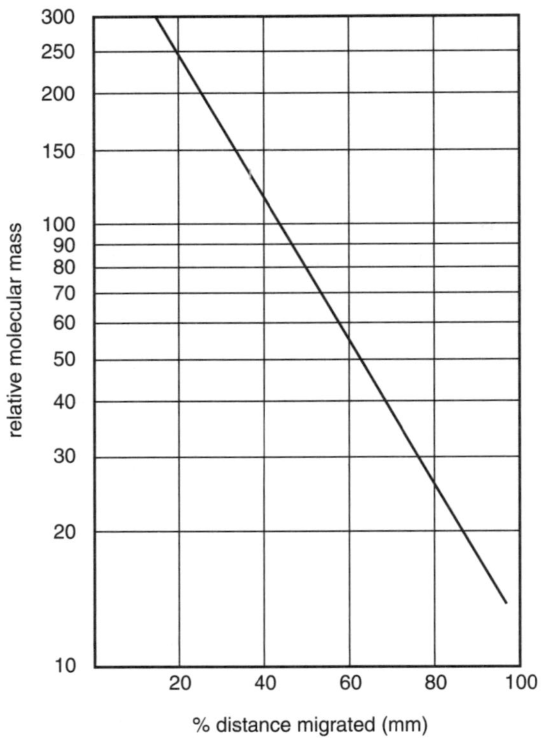


**FIGURE 6.3** Graph showing the relationship between molecular mass and mobility in a gel. Note that the greater the mass of a macromolecule is, the less distance it migrates.

## ◆ Extended Laboratory Program 7

# Chapter 7 Transport Systems

*Heartbeat and Daphnia*

**Teacher's Notes**

It is usually difficult to observe working systems in a complex animal. Daphnia is an exception because it is transparent. In this investigation, students will use microscopes to observe living daphnia; they will also experiment with the effects of stimulants and depressants on the rate of the animal's heartbeat. Direct correlations between the effects produced in daphnia and the effects the drugs would have on humans cannot be made, although the results with the daphnia raise useful questions.

Students should work in pairs. One can be the timer and the other the heartbeat counter. Be sure to have them alternate roles.

**Time**

One period should be sufficient to complete this lab.

**Materials (per class of 30—teams of 2)**

30 lab aprons
30 pairs safety goggles
30 pairs plastic gloves
15 microscope slides
30 coverslips
15 dropping pipets
5 jars for used daphnia
15 microscopes
clock or watch with second hand
cotton fibers
5 jars Detain II™ (slowing agent)
5 each dropping bottles of test solutions labeled A, B, C, and D
5 jars daphnia culture

**Safety**

Remind students not to taste or ingest any of the test materials.

**Preparations**

Daphnia can be collected from ponds and lakes during most months of the year, or cultures can be ordered from a biological supply house. Often they also can be obtained from aquarium supply stores or fish hatcheries. *Daphnia magna* is the largest and choicest species for observation. *Daphnia pulex* and *Daphnia longispina*, although about half the size of *Daphnia magna*, are also satisfactory for this investigation. This exercise can also be performed with mosquito larvae or *Tubifex* worms.

You can maintain the daphnia culture for 1–2 weeks in pond water at 20–25°C. The culture should be kept out of direct sunlight. To maintain the culture for a longer period, use a culture medium. One of the simplest techniques of culturing is the following. Place daphnia in tap water that has stood for several hours. Feed the culture a suspension of yeast in water; add enough suspension to make the water milky. When the water becomes clear, add more yeast suspension. Change the water once a month. Use an aerator, or stir the culture once a day.

The heartbeat may be too fast to count accurately. In that case, the observer should obtain a record of the rates by tapping a pencil on a piece of paper in rhythm with the heartbeat. The number of dots made by the pencil in one minute will provide a reasonably close count of the heartbeat rate.

Prepare a variety of solutions—some stimulants and some depressants—and label them A, B, C, D, and so forth. Provide one set of dropping bottles for every three lab teams. Limit the total amount of any of these solutions to 25 mL. This restriction limits the total amount of any stimulant or depressant available to students to a quantity that, if ingested by one student, would cause no more than temporary and minor harm. You may wish to identify the solutions after students have performed the investigations. Substances that have a stimulant effect include Alka Seltzer™, caffeine (strong tea—5 tea bags per cup of water), and nicotine (cigarette extract); for depressants, try Benadryl Elixir™, Unisom™, and 0.5% alcohol solution. Directions for preparing appropriate solutions of these substances are provided in the Preparation of Solutions section. Benadryl Elixer™, Unisom™, and Alka Seltzer™ can be purchased in any drugstore. Be sure each team uses at least one stimulant and one depressant.

**Analysis**

1. Determining the normal heartbeat rate before each application of a drug solution serves as a control.
2. Depressants result in a slower heartbeat rate; stimulants increase the rate. Specific identifications will depend on the solutions you prepare.

3.–4. Individual variation from the average heartbeat may be quite great. The graph to be constructed in answer to question 4 will show just how great.

5. Differences in heartbeat rate may have resulted from chance variations rather than from chemical treatment. This can be concluded if the bars for normal heartbeat rate and those for the rate after treatment with one of the chemicals overlap considerably and if the class average of the heartbeat rate for treated daphnia is similar to the class average for the normal rate. However, if the bars do not overlap and the averages differ markedly, the differences most likely result from the action of the chemical.

## ◆ Extended Laboratory Program 7

# Chapter 7 Transport Systems

*Heartbeat and Daphnia*

Daphnia is a small aquatic animal. Although it is commonly called a water flea, it is actually a crustacean related to crayfish and shrimp. Daphnia is a very useful laboratory subject because it is transparent. Using a microscope, you can see the internal organs at work.

In this exercise, you will observe the heart of daphnia to determine whether there is a correlation between the rate of its heartbeat and variations that you will introduce into its environment.

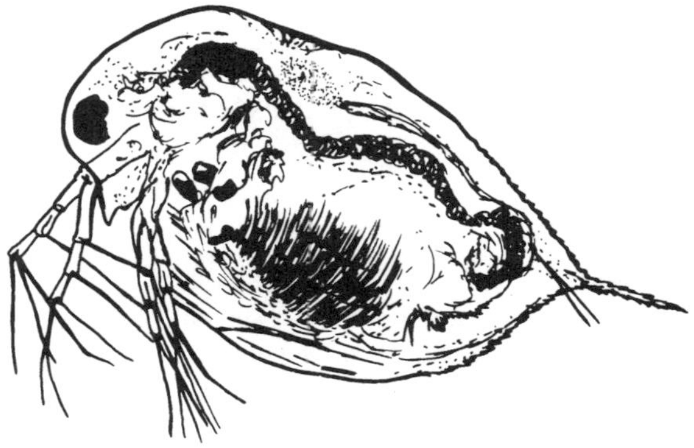

**FIGURE 7.1** Sketch of a daphnia as it appears under the compound microscope.

**Materials (per team of 2)**

- 2 lab aprons
- 2 pairs safety goggles
- 2 pairs plastic gloves
- microscope slide
- coverslip
- dropping pipet
- jar for used daphnia
- microscope
- stopwatch or watch with second hand
- cotton fibers
- Detain II™ (slowing agent)
- dropping bottles of test solutions labeled A, B, C, D, etc.
- daphnia culture

**SAFETY Put on your safety goggles and lab apron.**

**PROCEDURE**

1. Use a dropping pipet to transfer a daphnia specimen from the culture to a slide. Withdraw as much water as possible from the slide, leaving just enough to keep the organism alive. Add a drop of Detain II™ or a few cotton fibers under the coverslip to slow down the movements of the animal. Place the slide under the low-power objective of the microscope. Refer to Figure 7.1 to locate body parts.
2. You should be able to observe the heart beating very rapidly in the transparent animal. (Reducing the amount of light may help you see better. It also will prevent a temperature rise that can be fatal to the animal.) With either you or your partner as timekeeper, determine the rate of heartbeat under normal conditions by counting the number of beats per minute. Do this for three successive minutes. (Later you will average these counts, but merely record all three rates now.)
3. Your teacher will determine which solutions you should test. Add a drop of the first test solution (A) to the slide. Note the reaction of the daphnia. Determine three 1-minute counts of the heartbeat as before. Record these rates. Transfer the daphnia to the used daphnia jar.
4. Place another daphnia on a clean slide, and withdraw as much water as possible. Determine the normal heartbeat rate. Put a drop of the second test solution (B) on the daphnia, and observe the results. Determine the heartbeat rates again.
5. With a third specimen of daphnia, determine the normal rate of heartbeat and the rate after treatment with the third test solution (C). If time permits, test the effect of other solutions provided by your teacher. Record all the data in your logbook.
6. Determine the average of each set of three figures you obtained. Also determine the class average for the controls and the class average for each experimental treatment.
7. Wash your hands thoroughly before leaving the laboratory.

## Analysis

1. Explain what serves as a control for these experiments.
2. Describe the effect of each of the added chemicals. Which were stimulants? Which were depressants?
3. How great is the variation in the normal heartbeat rate of daphnia? In other words, do all the animals have about the same heartbeat rates or do some individuals vary a great deal from the average?
4. Using the class averages of the three (or more) treatments, prepare a bar graph. Plot heartbeat rate (in suitable intervals) on the horizontal axis and number of individuals on the vertical axis. Show the normal heartbeat rate in one color and the rates under the effect of each of the solutions in other colors.
5. What is the likelihood that differences noted in the heartbeat rate actually resulted from chance variations rather than from chemical treatment? Explain your answer on the basis of the graph you prepared.

## ◆ Extended Laboratory Program 8

# Chapter 7 Transport Systems

*Cardiac Risk Assessment*

**Teacher's Notes**

In this investigation, students use gel electrophoresis to determine the LDL/HDL ratios of three samples. To eliminate the risk of health hazard, bovine serum is substituted for human serum. The density of some of the stained bands produced during electrophoresis indicates the amount of LDL-bound or HDL-bound cholesterol. If you do not have a scanning densitometer, students can still assess the cardiac risk for each sample's source by visually comparing a sample's band density with those illustrated in Figure T8.1.

Two teams of five can share one gel electrophoresis tank, each team working on a single gel. If this is your first investigation using electrophoresis, refer to the Introduction to Electrophoresis on page 53 and to the notes in Extended Laboratory Program 6.

**Time**

The approximate time for each step in the investigation is as follows:

| | | |
|---|---|---|
| Part A: | Apparatus assembly | 30–40 minutes |
| Part B: | Loading and running the gel | 25–30 minutes |
| | Removing gel from apparatus | 5–10 minutes |

Neither staining nor destaining is necessary. After electrophoresis, the gel may stay in the glass cassette. The top and bottom of the cassette should be taped to prevent the gel from slipping out and drying. The gel fades rapidly, but it can still be read the next day. For a permanent record, photograph the gel.

**Materials**
**(per class of 30—teams of 5, with 2 teams per tank)**

30 lab aprons
30 pairs safety goggles
150 pairs plastic gloves
3 dual gel electrophoresis tanks
1 three-cell power supply
1 BSCS Cardiac Risk Assessment Kit
*1 set of three bovine serum samples
*3 vials lipoprotein prestain
*6 3–8% borate precast polyacrylamide gels
*2 borate 4X concentrate running buffer packs
*24 microcentrifuge tubes
*5 plastic transfer pipets
5 gallons deionized or distilled water
6 capillary micropipets (20 µL)
5 glass-marking pencils
1 roll clear tape
*1 "Student Study Sheet" copymaster
*1 "Student Analysis Sheet" copymaster

*Included in Cardiac Risk Assessment Kit.

**Safety**

See the teacher's notes in Extended Laboratory Program 6 regarding electrical safety in the laboratory, general cautions about the use of electrophoresis equipment, and procedures for discarding biological waste materials. Reagents should be stored in a special refrigerator devoted to biological materials only; no food or drinks should ever be stored in this refrigerator. The outside of the refrigerator should be labeled to indicate these rules.

Make sure students wear clean gloves when handling the gels and reagents. Remind students daily to wear their safety equipment and to wash their hands before leaving the laboratory.

**Preparations**

To make a stock solution of borate 4X concentrate running buffer, pour the contents of one pack into a 2-L beaker containing 1 L of deionized or distilled water. Stir until the powder has dissolved completely. Fill the beaker to the 1500-mL mark with distilled or deionized water.

To make each liter of 1X running buffer solution, mix 750 mL of deionized or distilled water with 250 mL of 4X concentrate buffer solution just before use. One liter of 1X running buffer solution is enough for two gels using a dual tank. Note: Although it is not usually necessary, filtration of running buffer helps ensure crisp band-imaging results. Store both running buffer solutions in a refrigerator.

Mix the low-risk bovine serum with the low-risk lipoprotein prestain, the normal-risk serum with the normal-risk prestain, and the high-risk serum with the high-risk prestain. Before electrophoresis, incubate the samples at 37°C for 30 minutes. Samples may be refrigerated for approximately 1 month.

Provide each student team with a set of labeled microcentrifuge tubes containing the samples that you prepared. Use a 20-µL micropipet to transfer 40 µL of each sample into the corresponding tubes provided to the students. Cap the tubes. Refrigerate unused samples.

Label student microcentrifuge tubes as follows:
Sample A: *Mr. Smith*
Sample B: *Ms. Jones*
Sample C: *Mr. Green*

Because students should be able to determine the cardiac risk associated with each sample from their electrophoresed gel, do not indicate on the label that Mr. Smith has a low cardiac risk, that Ms. Jones has a normal cardiac risk, and that Mr. Green has a high cardiac risk. Give students any combination of the three samples, but keep track of which samples you assign to each student team.

If you do not have a scanning densitometer, students must use the density chart in Figure 8.1 to assess each sample's source for cardiac risk. Tell students to place their gel next to the density chart and to match each lane's LDL and HDL bands with an appropriate density number. They can use the density numbers to determine the ratio of LDL-bound cholesterol to HDL-bound cholesterol; they can use the LDL/HDL ratio to assess the cardiac risk of the sample's source. For example, if a lane's LDL band

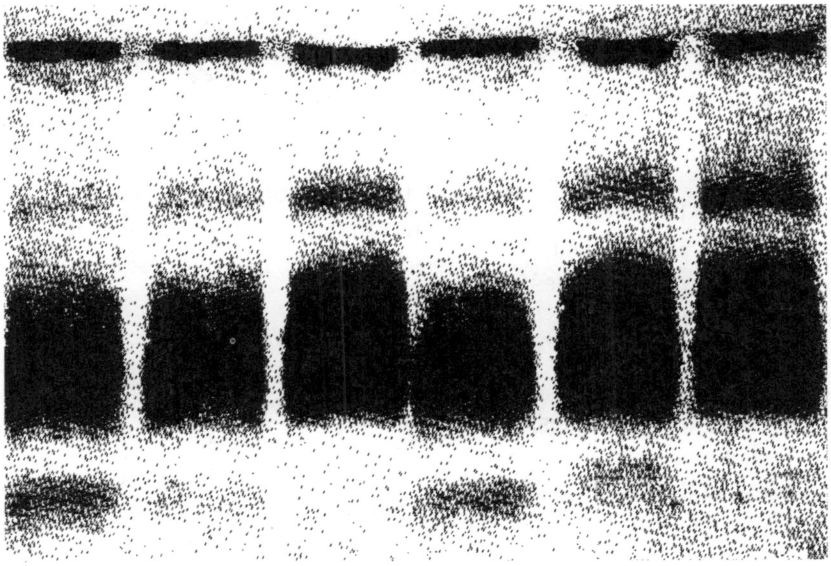


**FIGURE T8.1** Interpretation of experimental results

density is 4 and the HDL band density is also 4, then the LDL/HDL ratio is 4/4, or 1. According to the information provided in Analysis question 6, the sample's source would be assessed as having low cardiac risk. Note: Using an overhead projector to examine the gels will help students determine band densities.

**Analysis**

1. The electrical force pushed the lipoproteins through the gel toward the positive electrode. Because of its many microscopic pores, the gel acted as a molecular sieve. The larger lipoproteins separated at the top of the gel, and those of progressively lower mass separated below.
2. There are four stained lipoprotein bands (chylomicron, VLDL, LDL, and HDL) in each lane.
3. Students should observe the chylomicra, which are large lipoproteins that cannot migrate far down the gel, as a thick and densely stained band close to the top of the gel. Using a dark-field microscope, students can also observe chylomicra in the blood of people who have just eaten a meal with a high fat content.
4. The darkest band, the chylomicra, is located near the top of the gel. Other lipoprotein bands look fuzzy but are still distinct enough for students to observe.

The next two bands, VLDL and LDL, are located about one-quarter of the way down the gel. These first three bands constitute the low-density lipoproteins. The fourth HDL band is about two-thirds of the way down the gel. (Figure T8.1 shows an electronically enhanced image of an electrophoresed gel with typical banding patterns. Low-density lipoproteins consistently appear more darkly stained than HDLs.)

5.

| LANE 1 | LANE 2 | LANE 3 |
|---|---|---|
| LDL 5<br>HDL 5<br>LDL/HDL<br>Ratio: 5/5 = 1 | LDL 5<br>HDL 3<br>LDL/HDL<br>Ratio: 5/3 = 1.7 | LDL 5<br>HDL 1<br>LDL/HDL<br>Ratio: 5/1 = 5 |

6. Mr. Smith would have a low cardiac risk, Ms. Jones would have a normal cardiac risk, and Mr. Green would have a high cardiac risk.
7. An overall change in lifestyle, with emphasis on a low-cholesterol and low-fat diet combined with regular exercise, can be a preventive measure.

## ◆ Extended Laboratory Program 8

## Chapter 7 Transport Systems

*Cardiac Risk Assessment*

Heart disease affects an estimated 6 million Americans yearly and is the primary cause of death in the United States. One of the most common types of heart disease is atherosclerosis, characterized by the accumulation of cholesterol, fatty material, calcium, and cellular-waste products on the inner walls of arteries. The accumulations, called plaques, form in damaged areas of arterial walls where they gradually narrow arterial openings and reduce blood flow. A narrowed artery is vulnerable to total blockage by a blood clot, an event that causes either a heart attack or a stroke, depending on the blockage's location. Blockage also can occur when a plaque initiates the contraction of smooth muscle cells that line an artery, causing the artery to pinch off. Factors linked to arterial damage and plaque formation include cigarette smoking, high blood pressure, insufficient exercise, obesity, and improper diet (high in cholesterol and saturated fat). By altering these factors, individuals can decrease their risk of developing heart disease.

Plasma lipids, including both triglycerides and cholesterol, circulate in the bloodstream as lipoproteins, which are protein-lipid complexes manufactured in the liver. The major lipoproteins are chylomicra (fat-protein microdroplets), very low-density lipoproteins (VLDLs), low-density lipoproteins (LDLs), and high-density lipoproteins (HDLs).

The LDL groups (VLDLs and LDLs) act like delivery trucks, unloading cholesterol and triglycerides at body cells for energy and food storage. However, if the blood has high LDL levels, the LDLs will deposit their cholesterol in damaged arterial walls. HDLs are scavengers of cholesterol, collecting the excess and carrying it to the liver, which breaks it down into other products. When the blood level of HDL is high and the blood level of LDL is low, the amount of cholesterol that can accumulate as plaque decreases. On the other hand, if the blood level of LDL is high and the blood level of HDL is low, the accumulation of cholesterol as plaque increases.

In this investigation, you will use gel electrophoresis to determine the relative amounts of LDLs and HDLs in three samples. The lipoproteins in your samples have been stained to show as bands of varying densities in the polyacrylamide gel. The darker a band, the more lipoprotein-bound cholesterol is present. Conversely, the lighter a band, the less lipoprotein bound cholesterol is present. On the basis of the LDL/HDL ratios, you will assess the cardiac risk of the source of each sample.

**Materials (per team of 5)**

5 lab aprons
5 pairs safety goggles
5 pairs plastic gloves
1 set of three labeled microcentrifuge tubes containing serum samples
1 3–8% borate precast polyacrylamide gel
1 L 1X running buffer solution
1 capillary micropipet (20 µL)
1 plastic transfer pipet
1 glass-marking pencil
1 dual gel electrophoresis tank
1 three-cell power supply
clear tape

**SAFETY** Put on your safety goggles, lab apron, and gloves; tie back long hair.

**PROCEDURE**

Before you start the procedure for running a gel, familiarize yourself with the parts of the electrophoresis apparatus. Depending on the gel equipment being used, your teacher may give you instructions that differ somewhat from those presented here.

**Part A Apparatus Assembly**

Part A includes the following steps:

- Place precast gel into tank assembly
- Add working buffer
- Flush sample wells in gel

1. Remove gel from package, and carefully remove comb from the gel.
2. Place gel into one of the outer chambers of the tank. Make sure the sample wells are facing inward, toward the inner compartment. Note: If you are running a single gel, insert the

Plexiglas™ plate in the other outer chamber in place of a second gel.

3. Insert the wedges into each corner, making sure the gel is snug against the gasket to prevent leaks into the outer compartment.
4. Fill the inner compartment with working buffer solution until it covers the wells. Using a pipet, flush out any air bubbles in the wells.
5. Slowly fill the outer compartment with working buffer concentrate until it reaches the same level as that in the inner compartment.

**Part B Loading and Running the Gel**

Part B includes the following steps:

- Load samples into wells
- Electrophorese samples
- Discard running buffer
- Remove and open gel cassette
- Transfer gel to stain/destain container

6. Use the capillary pipet to load 20 µL of each sample into a separate well. If you are sharing a gel with another team, keep track of the well number into which each sample is placed. Your teacher will demonstrate the correct technique for this step of the procedure.
7. Make sure the cover is dry, and attach it to the chamber. Align the cover properly. Wipe off any spills on the apparatus before proceeding.
8. Make sure that the patch cords attached to the cover, as well as the female jacks on the power supply, are completely dry. Connect the red patch cord to the red electrode on the power supply. Connect the black patch cord to the black electrode terminal on the power supply.
9. Plug in the power supply, and set it to 110 volts.
10. Turn on the power supply. The red power light will illuminate. WARNING: The power supply must be used only according to the exact instructions provided. Use of this equipment other than as instructed can lead to severe injury or even death by electric shock.
11. Five minutes after electrophoresis has begun, observe the area of gel just below the sample wells. Record your observations.
12. Electrophoresis should continue until the blue loading dye has traveled halfway to the bottom of the gel. Turn off the power supply when the process is complete. (The red power light will go out.)
13. Unplug the power supply. Wait about 10 seconds; then disconnect the patch cords from the power supply, and remove the cover from the chamber.
14. Remove the cover assembly and pull out the wedges, then the gel. Take care not to change the orientation of the gel (left to right and top to bottom) when handling the cassette. Use a glass-marking pencil to mark the lower-left corner of the cassette.
15. Because the lipoproteins have been prestained, neither staining nor destaining is necessary. Leave the gel in the glass cassette. Tape the top and bottom of the cassette to prevent the gel from slipping out and drying.
16. Wash your hands thoroughly before leaving the laboratory.

**Analysis**

1. Describe what happened to the lipoproteins during electrophoresis.
2. How many bands can you observe in each lane?
3. Chylomicra are very large macromolecules, each composed of about 98% fat and 2% protein. Where would you expect to observe chylomicra in each lane sample? Explain your answer.
4. The bands should reflect the progressive decrease in molecular mass of the VLDLs, LDLs, and HDLs. Sketch and label all the bands. Describe the appearance of each band. Confirm your illustration with your teacher.

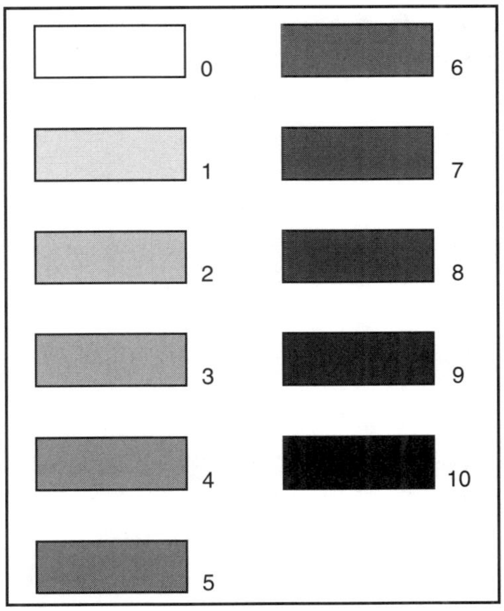


**FIGURE 8.1** Density chart

5. The density of each stained band indicates the amount of cholesterol bound to either an LDL or HDL band. Use the density chart in Figure 8.1, on the previous page, to assign a number to each LDL and HDL band. To do this, place your gel next to the density chart and match the LDL and HDL bands from each lane with the appropriate density number. Using the assigned density numbers, determine the ratio of LDL-bound cholesterol to HDL-bound cholesterol. Confirm your LDL/HDL ratios with your teacher.
6. Individuals with LDL/HDL ratios less than or equal to 1 are assigned low cardiac-risk status; those with ratios greater than 1 but less than 5 are assigned normal-risk status; and those with ratios equal to or greater than 5 are assigned high-risk status. Using the LDL/HDL ratio for each electrophoresed sample, assess the cardiac risk of Mr. Smith, Ms. Jones, and Mr. Green.
7. What, if anything, could a person who is assessed as having a high cardiac risk do to reduce the chance of developing heart disease?

## ◆ Extended Laboratory Program 9

# Chapter 8 The Cell Cycle

*DNA Extraction*

**Teacher's Notes**

This investigation provides a simple procedure for extraction and examination of DNA. Comparison of DNA with a macromolecular solution and with several other solutions enables students to observe some of the characteristics that DNA shares with other long-chain macromolecules. Ward's Natural Science Establishment has developed a kit (catalog #88 W 8112) for this investigation. The kit is designed for a team of five students, and it includes the special materials needed for this investigation.

**Time**

Allow two class periods for this investigation. During the first period, students complete Parts A and B; during the second period, students complete Part C and discuss their results.

**Materials (per class of 30—teams of 5)**

- 30 lab aprons
- 30 pairs safety goggles
- 30 pairs plastic gloves
- 12 18-mm × 150-mm test tubes
- 6 dropping pipets
- 3 600-mL beakers
- 3 nonmercury thermometers
- 12 glass stirring rods
- 1 roll paper towels
- 3 hot plates
- 6 test-tube clamps
- 6 test-tube racks
- 6 screw-top tubes, each containing 5 mL denatured ethyl alcohol
- 3 4-oz jars corn syrup
- 3 4-oz jars motor oil (SAE 20 or 30, nondetergent type)
- 3 4-oz jars water
- 1 DNA extraction kit:
  - 6 inoculating needles
  - 6 cotton applicator sticks
  - 6 tubes *Halobacterium salinarum* culture
  - 3 4-oz jars macromolecular solution

**Safety**

Denatured ethyl alcohol is flammable and toxic. Extinguish all flames in the laboratory; avoid sparks and excessive heat. The refrigerator used to store the DNA extract must be reserved for biological work and so labeled; no food or drink for humans should be stored in this refrigerator.

**Preparations**

When students have completed the investigation, you can destroy the *Halobacterium* cultures by adding distilled or tap water to the culture tubes. The bacterial cells rupture in the presence of water. Because these bacteria cannot survive in nonsalty environments, transfer of *Halobacterium* to other cultures is unlikely.

Baby-food jars work well for the solutions used in Part B. Extracted DNA (test tube B) may be stored in a refrigerator overnight. The DNA will cling to the surface of the inoculating wire at the interface of the distilled water and denatured ethyl alcohol. The extract from this procedure is not pure DNA; it consists of cellular debris along with proteins and nucleic acids. To purify this crude extract and remove extraneous proteins, it would be necessary to treat the material with hazardous chemicals such as phenol, chloroform, and isoamyl alcohol.

Long-chain macromolecules, such as DNA, have their own characteristic "signature" or "imprint." Following lysis of *Halobacterium* cells, the extracted DNA should appear viscous when placed in distilled water. Viscosity is the resistance to flow of a fluid due to internal friction resulting from the cohesion of molecules. Macromolecules have high viscosity values.

Another property of macromolecules is elasticity, the ability of a material to resume its original length, shape, or size after being deformed or stretched. Long-chain macromolecules display the property of elastic recoil. Students will observe that the extracted DNA easily forms thin, hairlike threads when placed on the lip of the test tube; however, when removal is attempted, the thread will recoil, or "snap," if pulled too far. Students will further investigate these two characteristics when they compare the viscoelastic properties of the macromolecular solution with that of the extracted DNA as well as with other common "viscous" substances, such as corn syrup and motor oil, and a nonviscous substance, such as water.

The macromolecular solution, like DNA, is a long-chain polymer. The molecular mass of this polymer exceeds 5 million amu. It is formulated expressly for student use in investigating the viscoelastic properties of long-chain macromolecules. This material possesses the repeating structural unit $[\text{-}CH_2\text{-}CH_2]_n$, where $n$ varies, with the degree of viscosity. (Chains may be broken by UV light, causing an observable loss of viscosity. As a side demonstration, you may wish to place a small sample of the material under a UV lamp and ask students to note changes in viscosity.)

When the extracted DNA material is heated, students should be able to observe a decrease in the viscosity of the material. This observed decrease signifies the unzipping of DNA. As temperature increases (around 45°C with this crude DNA extract from *Halobacterium*), double-stranded DNA becomes progressively converted to single-stranded DNA as a result of the breakdown of interstrand hydrogen bonds.

This "melting point" of DNA, the temperature at which it unzips or becomes denatured, is directly related to the guanine-cytosine nucleotide content. The higher the temperature at which unzipping occurs, the higher the G-C content of that DNA. (Hydrogen bonding between G-C pairs is stronger than hydrogen bonding between A-T pairs. This is a consequence of the fact that G-C pairs form three hydrogen bonds, whereas A-T pairs form only two.) The physical method of analysis described here provides an indication of the molecular heterogeneity of DNA. Such analysis (Mandel, 1969) has shown that there is a relatively narrow C-G content range (35–45%) among plants and animals, a wider variation among protists, and the widest variation among prokaryotes (20–80%). Thus, DNA base-composition analysis provides evidence for genetic divergence and evolutionary separation between and among major biological groups.

**Analysis**

1. Students should observe a cloudy, viscous, stringy mass of material.
2. DNA contains all of their genetic material and the instructions for carrying on all of their activities.
3. No. Animal cells will not rupture because they are not adapted to the salty conditions of the *Halobacterium* environment.
4. The macromolecular solution is more viscous than either engine oil or corn syrup; water is not apparently viscous by the qualitative measure used here. The evidence is the comparative length of time required for each material to drain completely to the opposite side of its jar. Another way to confirm this property is to note relative differences in resistance that each material offers when you gently stir the material with a glass stirring rod.
5. Yes. The stirring rod immersed in the macromolecular solution moves backward when released. A similar reversal motion is observed when the circular motion of the test tube containing extracted DNA material is stopped. This reversal motion is the property of elastic recoil.
6. Yes. In both instances, a thin strand of material is formed, indicating that both substances are viscoelastic. With the macromolecular solution, a tear-shaped drop will form. As it descends, a thin strand of attached and continuous material will appear. With the DNA material, a similar thin strand will form when the material is withdrawn from the lip of the test tube. The material will also "snap" away from the test tube if pulled too far or too quickly. The conclusion should be that the material is elastic.
7. Yes. Students should observe that both materials tend to wrap around themselves and climb up either the stirring rod or the connecting needle wire. This effect is due to the high viscoelasticity of both materials. Students should conclude that both materials (extracted DNA and macromolecular solution) exhibit similar properties of elasticity and elastic recoil. Both materials are highly viscous. Students should be able to conclude that both properties (viscosity and elasticity) are working in concert. Thus, both materials are long-chain macromolecules.
8. No. Although some of the test materials (corn syrup and motor oil) are viscous, they do not exhibit elastic recoil or viscoelasticity. Water does not exhibit either viscous or elastic properties. These other materials are not long-chain macromolecules.
9. Long-chain macromolecules share the physical characteristics of viscosity and elasticity.
10. Answers will vary. Macromolecules are giant molecules, often with molecular weights in the millions. Two types of macromolecules exist: functional proteins and polymers. A functional protein is an individual biochemical entity that cannot be subdivided without losing its functional (chemical) identity. A polymer is a combination of repeating chemical units linked together. Nylon is a synthetic polymer; cellulose is the most common example of a naturally occurring polymer.
11. Heating the extracted DNA causes it to become less viscous; chemical bonds that hold the two strands together are broken. Double-stranded DNA is being converted to single-stranded DNA. (The temperature at which the change in viscosity occurs will vary in this crude extract but is generally between 40°C and 50°C.)
12. Answers will vary. The significance is that DNA can be denatured at high temperatures. Some students may suggest that excessive heating can provide a clue to the makeup of the DNA molecule itself. Others may interpret temperature data as reflecting the upper thermal limits of life. That is not the case, for some thermophilic eubacteria live at temperatures exceeding 85°C, and at least one species of archaebacterium lives at 110°C.

**References**

Mandel, M. "New Approaches to Bacterial Taxonomy." *Ann. Rev. Microbiology*, Vol. 23, No. 239, 1969.

Schleifer, K. H. and E. Stackebrandt. "Molecular Systematics of Prokaryotes." *Ann. Rev. Microbiology*, Vol. 37, pp. 143–187, 1983.

 

## ◆ Extended Laboratory Program 9

# Chapter 8 The Cell Cycle

### *DNA Extraction*

The extraction of DNA is the first step in a set of procedures collectively termed recombinant-DNA technology. In this investigation, you will extract the DNA from *Halobacterium* and examine it.

*Halobacterium salinarum* (Figure 9.1) is an archaebacterium that is among the oldest organisms on the earth. These archaebacteria evolved in extremely salt-rich environments, such as the Dead Sea, in whose waters no other organisms can survive. Yet *Halobacterium* continues to thrive there. *Halobacterium* possesses the pink pigments that are visible on the medium supplied for this investigation. When *Halobacterium* is placed in distilled water, water rushes into each cell by osmosis and ruptures the cell wall. The DNA and other proteins from the cytosol are released into the distilled water.

**Materials (per team of 5)**

5 lab aprons
5 pairs safety goggles
5 pairs plastic gloves
2 18-mm × 125-mm test tubes
dropping pipet
inoculating needle
cotton applicator stick
test-tube clamp
test-tube rack
paper towels
1.5 mL distilled water
5 mL denatured ethyl alcohol, 99%, in screw-cap tube
1 tube *Halobacterium salinarum* culture

**(per two teams)**

600-mL beaker (water bath)
nonmercury thermometer
hot plate
4 glass stirring rods
4-oz jar macromolecular solution
4-oz jar corn syrup
4-oz jar motor oil
4-oz jar water

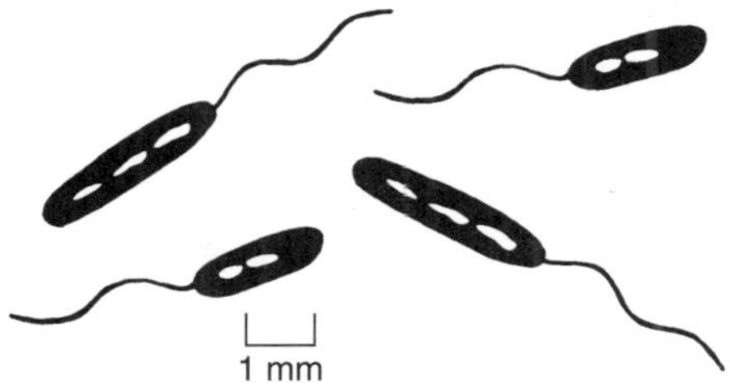


**FIGURE 9.1** Cells of *H. salinarum.* Note polar flagellum and gas vacuoles.

**SAFETY Put on your safety goggles and lab apron.**

**PROCEDURE**

**Part A Extraction of DNA**

1. Bend approximately 5 mm of the wire end of the inoculating needle upward to form a hook, as shown in Figure 9.2.
   CAUTION: The inoculating needle is very sharp; handle with care.
2. Using the dropping pipet, place approximately 1.5 mL of distilled water in each of two test tubes. Label the test tubes A and B.
3. Remove the cap from the culture tube of *Halobacterium salinarum.* Insert the cotton applicator stick into the culture tube, and gently swab the entire surface of the medium. As you work, rotate the cotton applicator stick so that all of its surface contacts the pink colonies of

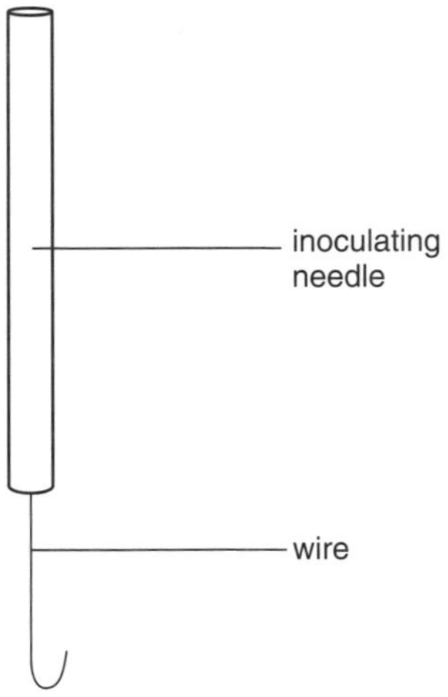


**FIGURE 9.2** Bending the wire of the inoculating needle

*Halobacterium*. Remove the cotton applicator stick from the culture tube, and replace the cap.

4. Transfer the cotton applicator stick to test tube A, and swirl it vigorously in the distilled water, occasionally pressing the cotton against the inner wall of the tube. When you remove the cotton applicator stick from the test tube, press the cotton against the tube wall to squeeze out as much fluid as possible. Discard the swab.
5. Observe the fluid in test tube A. Answer Analysis question 1.
6. Hold test tube A (containing the collected *Halobacterium* material) in one hand at about a 45° angle. Use the dropping pipet to add about 3 mL of denatured ethyl alcohol to the tube, one drop at a time, so that the drops run down the side of the tube, as shown in Figure 9.3. Slowly rotate the tube as you add the alcohol so that the alcohol can wash any *Halobacterium* material from the sides of the tube. *The alcohol should be in a distinct layer on top of the water. Do not shake the tube or attempt to mix the alcohol and distilled water.*

   CAUTION: Denatured ethyl alcohol is flammable and toxic. Extinguish all flames in the laboratory, and avoid heat and sparks. Avoid skin and eye contact; do not ingest. Flush spills and splashes with water for 15 minutes. If necessary, rinse mouth with water. Call your teacher.
7. Clean the inoculating needle with a paper towel moistened in denatured ethyl alcohol. Carefully insert the needle into the test tube so that the hook is at the boundary between the alcohol and the water. Gently rotate the needle in a circular motion to mix some of the water and alcohol at the boundary layer, as shown in Figure 9.4. You should notice some strands of material adhering to the wire and trailing off into the solution. These are strands of DNA from *Halobacterium*. Answer Analysis questions 2–3.

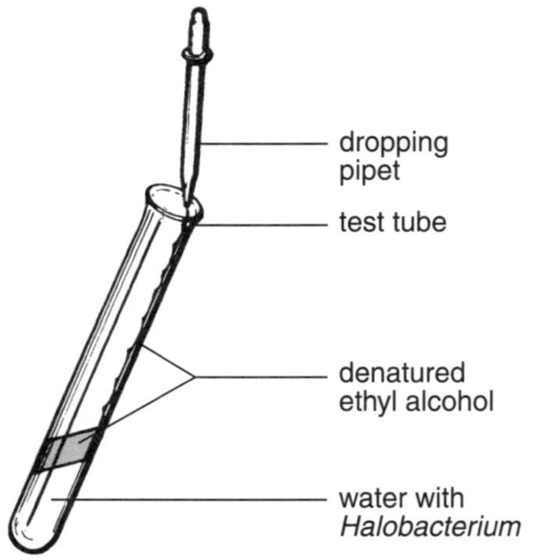


**FIGURE 9.3** Adding denatured ethyl alcohol to the *Halobacterium*-water mixture

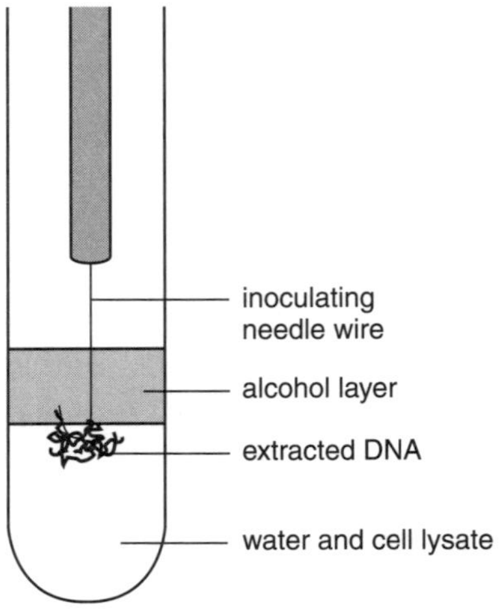


**FIGURE 9.4** Mixing the water and denatured alcohol at the boundary layer

8. Remove the needle from the test tube, and closely observe the material sticking to it. In your logbook, describe the material.
9. Insert the needle into test tube B (containing only distilled water), and deposit the extracted DNA. Extract as much additional DNA from test tube A as possible, depositing it in the distilled water of test tube B.

**Part B Physical Characteristics of Macromolecules**

Long-chain macromolecules, such as DNA, exhibit certain characteristics. Among these are viscosity (the resistance to flow of a fluid) and elasticity (the ability of a material to resume its original size after being stretched). You have already observed some of these characteristics during the extraction process.

10. Working together with another team, place sealed jars of macromolecular solution, motor oil, corn syrup, and water right side up on a flat surface. When the solutions have settled to the bottom of the jars, quickly invert each one. Note the time for each solution to drain completely after being inverted. Answer Analysis question 4.

11. Open the jar containing the macromolecular solution, and insert a stirring rod into it. Slowly move the stirring rod in a clockwise direction, about one-half revolution in the jar, and then let go. Note what immediately happens.
12. Swirl the extracted DNA in tube B by holding the top and gently rotating the bottom of the test tube. Then stop suddenly. Observe what happens when the circular motion of the tube is abruptly stopped. Answer Analysis question 5.
13. Lift the stirring rod out of the macromolecular solution, and observe what happens.
14. Insert the DNA collecting needle into test tube B, and remove a small amount of the extracted DNA material. Place the material (still attached to the needle) on the lip of the test tube and observe what happens when you try to remove the material from the lip. Answer Analysis question 6.
15. Reinsert the stirring rod into the macromolecular solution, and stir rapidly. Then release the stirring rod. Observe what occurs when the stirring rod is released.
16. Reinsert the collecting needle in test tube B, and slowly stir the collected DNA material. Answer Analysis question 7.
17. Repeat steps 11, 13, and 15 for each of the other three solutions. Answer Analysis questions 8–10.
18. Label test tube B with your team symbol, and store it in the refrigerator overnight. Wash your hands thoroughly before leaving the laboratory.

**Part C Effect of Heat on DNA**

19. Remove test tube B (containing the extracted DNA) from the refrigerator, and place it in a water bath at room temperature. Place a thermometer in the water bath, and record the temperature in your logbook.
20. Slowly heat the water bath on the hot plate. At regular intervals (about every 2°C to 5°C increase in water temperature), use a test-tube clamp to remove your test tube and closely examine the extracted DNA. After each observation, return the test tube to the water bath. In your logbook, record the temperature at which you observe any change in the DNA and describe the change. Stop your observations when the temperature of the water bath reaches 80°C. Answer Analysis questions 11 and 12.
21. Wash your hands thoroughly before leaving the laboratory.

**Analysis**

1. Describe the appearance of the fluid in test tube A.
2. What is the importance of DNA to *Halobacterium?*
3. Could the same extraction technique be used to remove the DNA from human or other animal cells? Why or why not?
4. Which of the solutions exhibits the greatest viscosity? What is the evidence for your conclusion?
5. Do the stirring rods in the macromolecular solution and in the extracted DNA material in test tube B exhibit the same general "reversal" response? Explain.
6. Describe your observations when comparing the macromolecular solution and the extracted DNA material. Are they similar?
7. Does the collected DNA material tend to behave in a way similar to that of the macromolecular solution? On the basis of your observations in steps 11–16, what can you conclude about DNA and the macromolecular solution?
8. Do any of the other test solutions behave the way the extracted DNA material or the macromolecular solution did? Explain.
9. On the basis of your observations, describe some of the physical properties of long-chain macromolecules.
10. On the basis of your definition of what a macromolecule is, give some examples of other macromolecules.
11. What change did you observe in the DNA as it was heated? What might have caused that change?
12. In what way is the temperature at which that change occurred significant?

## Extended Laboratory Program 10

# Chapter 9 Expressing Genetic Information

*Control of Molecule Synthesis*

**Teacher's Notes**

This investigation is a dry lab that describes investigation procedures and presents data. The Analysis questions about the data should lead students to the same conclusions that Beadle and Tatum reached: For each enzyme in *Neurospora* that is present and functioning in the synthesis of amino acids or vitamins, there is a particular gene or group of genes responsible for its formation. Doing this lab before consideration of the related material in the textbook would be helpful in developing students' understanding of the gene-enzyme relationship.

Students may work alone, but the investigation is most appropriate for small-group, cooperative learning.

**Time**

One to one-and-one-half periods will probably be sufficient for most students.

**Materials (per class of 30–teams of 3)**

none

**Analysis**

1. The best answer at this point is d because there is no way of knowing what is needed by the spores that would grow on complete medium but not on simple medium.
2. Prepare growth medium with amino acids but no vitamins and medium with vitamins but no amino acids. Add spores and see if they grow. If they grow on neither medium, then you know that the spores need both vitamins and amino acids.
3. The best answer would now be c because the irradiated spores did not grow with only vitamins. It may be that they need only amino acids.
4. Place the irradiated spores on a simple medium to which amino acids are added. Growth will show that the spores can make their own vitamins but not the amino acids.
5. No. Because all amino acids were added and the spores grew, it could be that all 20 are needed; however, it is possible that fewer than 20, and perhaps only one, would be necessary.
6. Prepare different cultures, each with a different amino acid added to the simple medium. Place the irradiated spores on each culture. Watch to see which ones produce growth.
7. The *Neurospora* spores have been affected by the X rays; therefore, they can no longer produce the amino acid arginine.
8. The mold would not grow.
9. The mold of mutant type 2 would not grow, whereas type 1 would grow.
10. Mutant type 2 would need citrulline.
11. Mutant type 1 would need ornithine.
12. Arginine is the most complex; ornithine is the simplest.
13. Ornithine could be converted to citrulline by adding carbon, oxygen, nitrogen, and hydrogen to the molecule at specific places. Citrulline could be converted to arginine by replacing the oxygen atom on the first carbon with an H atom and an N atom.
14. enzymes
15. b
16. Mutant type 1: The DNA that directs the synthesis of an enzyme necessary for the formation of ornithine is damaged by X rays in some way, with the result that it no longer functions. Mutant type 2: The DNA responsible for the enzyme that converts ornithine to citrulline is damaged by X rays. Mutant type 3: The DNA that directs the formation of the enzyme needed to convert citrulline to arginine is damaged by X rays.

## ◆ Extended Laboratory Program 10

# Chapter 9 Expressing Genetic Information

*Control of Molecule Synthesis*

In the 1940s, George W. Beadle and Edward L. Tatum of Stanford University directed their scientific research to this question: "What controls the synthesis of large molecules in an organism?" They used a common pink mold called *Neurospora* to determine whether the mold spores that start a new generation carry the instructions for building the large specialized molecules that characterize the organisms. It was known that *Neurospora* spores could be cultured on a medium containing only small and simple types of molecules, even though the mold made large, complex molecules during its growth.

This investigation describes several experiments similar to those conducted by Beadle and Tatum. For the sake of simplicity, the problems will be restricted to *Neurospora's* ability to synthesize amino acids and vitamins.

**Materials (per team of 3)**

none

**PROCEDURE**

Read the following descriptions of experiments with *Neurospora*, and analyze the results as directed in the Analysis questions.

**Experiment 1**

*Neurospora* spores were placed on agar containing only salts, sugar, and biotin (a vitamin). From now on, this will be referred to as the *simple medium*. At the spore stage of the life cycle, the spores were exposed to X-ray irradiation. Single irradiated spores were then placed on agar containing many vitamins and all 20 amino acids, in addition to the substances in the simple medium. This medium will be referred to as the *complete medium*. These spores developed and grew into new mold that produced spores. Some of these spores were then placed on simple medium and some on complete medium, as shown in the middle of Figure 10.1.

All the spores grew on the complete medium, but some did not grow on the simple medium. Answer Analysis questions 1 and 2.

**Experiment 2**

Spores from *Neurospora* that did not grow on the simple medium but did grow on the complete medium were then placed on agar to which all the vitamins but no amino acids had been added. This is represented at the bottom part of Figure 10.1. The spores did not grow. Answer Analysis questions 3 and 4.

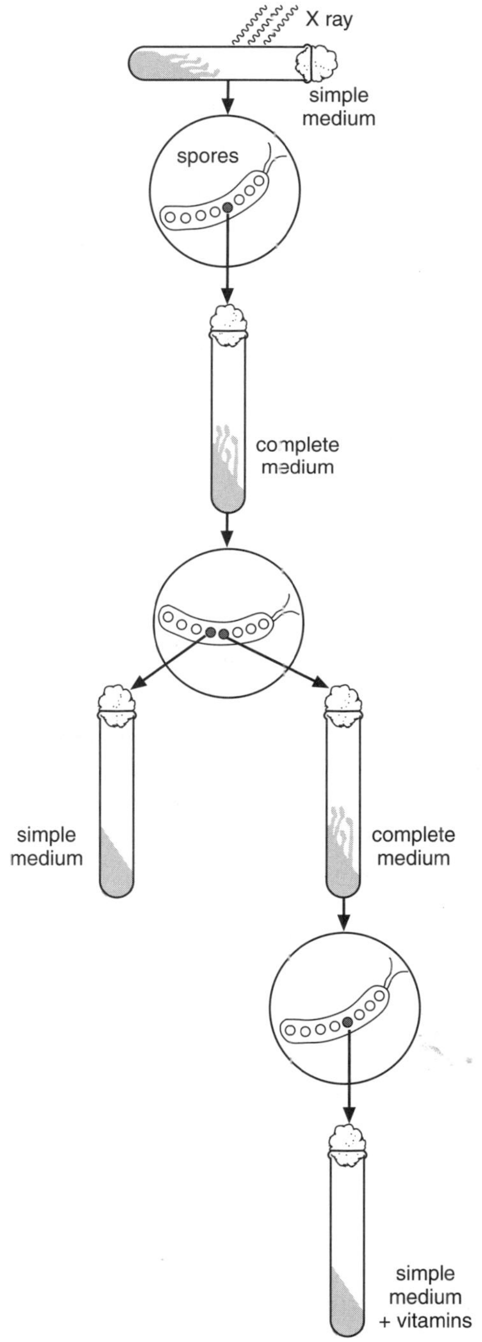


**FIGURE 10.1** Survival requirements for *Neurospora* following irradiation of spores.

**Experiment 3**

Beadle and Tatum decided that the X rays had changed, or mutated, the *Neurospora*, with the result that the spores no longer grew without an outside supply of amino acids. Further experiments showed that the irradiated spores would grow on a simple medium to which all 20 amino acids had been added. Answer Analysis questions 5 and 6.

**Experiment 4**

Twenty separate tubes were prepared, each with a simple medium and a different amino acid. Nineteen tubes showed no growth. In one of the tubes, the amino acid called arginine had been added to the simple medium. Growth occurred in this tube. Answer Analysis question 7.

**Experiment 5**

Beadle and Tatum discovered that some of the irradiated spores requiring arginine would grow if substances similar to arginine were added. In fact, they discovered three ways to get some of the mutated spores to grow (add ornithine, citrulline, or arginine) and two ways to get others to grow (add citrulline or arginine). But some would grow only if arginine was added to the medium. The first of these types, which would grow with any of the three substances, they called mutant type 1. The second, needing either citrulline or arginine, they called mutant type 2. The mutants that would grow only with arginine, they called mutant type 3. Figure 10.2 illustrates the three mutant types. The molecular structure of these three substances is quite similar, as you can see by examining Figure 10.3. Answer Analysis questions 8–16.

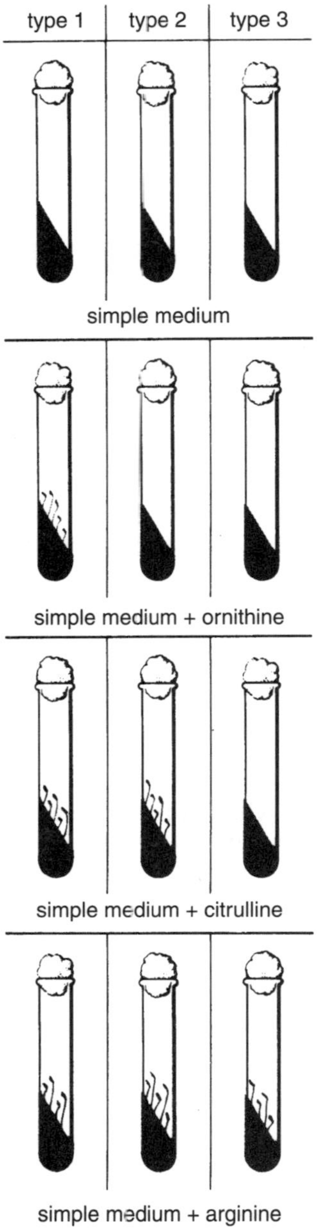


**FIGURE 10.2** Isolation of three types of mutant *Neurospora* spores by growth on simple media supplemented with different amino acids.

**Analysis**

1. Which statement best describes the results of this first experiment with *Neurospora?*
   a. X rays killed *Neurospora* spores.
   b. X rays changed something in the spores that interfered with their synthesis of vitamins.
   c. X rays changed something in the spores that interfered with their synthesis of amino acids.
   d. X rays changed something in the spores that interfered with their synthesis of vitamins or amino acids or both.
2. What experiment would now be necessary to determine the change that the X rays produced?
3. Which of the choices in Analysis question 1 now most completely describes the effect that X rays had on *Neurospora?*
4. Now that you know that the irradiated spores will grow on simple medium plus amino acids and vitamins but will not grow on simple medium plus vitamins alone, propose the next step in the experiment.
5. Is there any way of knowing from these experiments whether all amino acids are required? Explain.

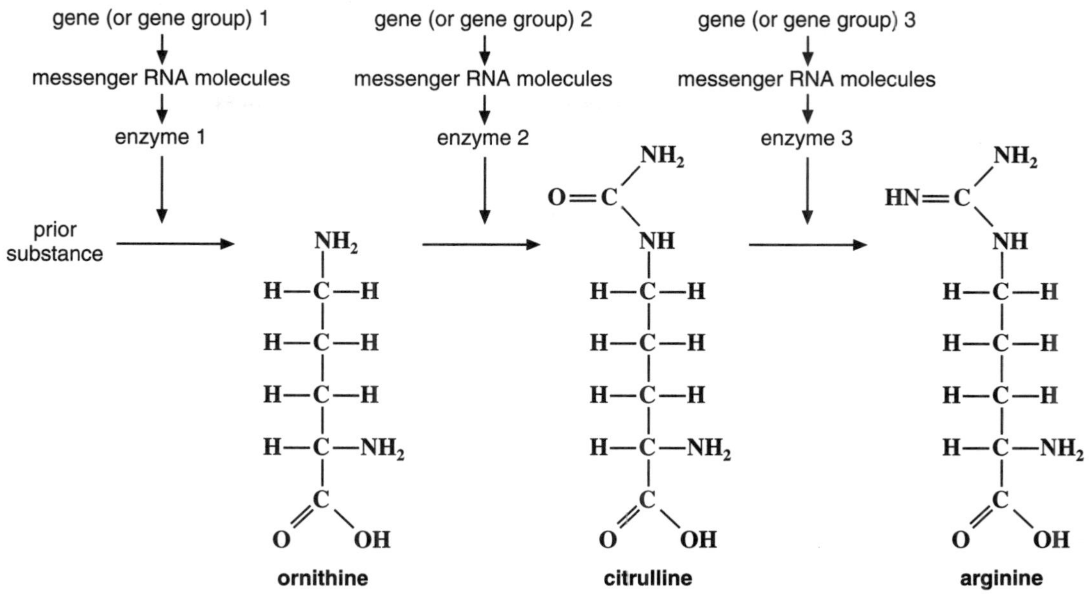


**FIGURE 10.3** One gene or a group of genes controls the production of an enzyme. In the diagram, each enzyme helps change an amino acid into a more complex amino acid. The result of each message from a gene or group of genes is the addition of more atoms to the amino acid structure. Many steps may be required to build a complex amino acid.

6. Describe the kind of experiment that would determine whether a certain amino acid is sufficient for the irradiated spores to grow.
7. Now describe the effect the X rays had on the *Neurospora* spores. Be more specific than you were in your answers to Analysis questions 1–3.
8. If ornithine were added to the simple medium and spores from mutant type 3 were placed on the medium, what results would you expect?
9. What would you expect if you used mutant type 2? mutant type 1?
10. What substance would mutant type 2 need to synthesize its own arginine?
11. What substance would mutant type 1 need to synthesize its own citrulline and arginine?
12. In Figure 10.3, which molecule is the most complex? Which is the simplest?
13. What modifications would be necessary to change ornithine into citrulline? citrulline into arginine?
14. The chemical changes discussed in question 13 could occur naturally but would take a long time. What substances in living cells bring about these changes at a more rapid rate?
15. Which of the following hypotheses would be more reasonable?
    **a.** DNA coding is responsible for the formation of arginine.
    **b.** DNA coding directs the formation of enzymes, which in turn direct the formation of arginine.
16. It seems that a substance in the medium is converted to ornithine by a particular enzyme. Next, other enzymes convert ornithine into citrulline, then citrulline into arginine. These different enzymes are present because of the particular coding of the DNA. Form some hypotheses concerning the effects of the X rays that produced the different mutant strains of *Neurospora*. Hint: Ask yourself where, in the sequence of molecular buildup, the mutation has occurred. What has been affected by the X rays?

## ◆ Extended Laboratory Program 11

# Chapter 13 Patterns of Inheritance

*Smooth and Wrinkled Pea Seeds*

**Teacher's Notes**

This investigation is like a biochemical treasure hunt. The hunt starts with the observation of the genetic difference between smooth and wrinkled seeds of peas. It ends with the discovery that this difference is due to a variation in the activity of a starch-branching enzyme.

In Part A, students learn that the difference between the two kinds of seeds is related to the amount of water each loses in maturing. After absorbing water, both types of seeds have smooth coats. A microscopic examination of seeds in Part B suggests that water absorption and loss may be related to the size and shape of starch grains contained in the seeds. Because the starch grains of wrinkled seeds have more surface area than those of smooth seeds, students should reason that wrinkled seeds can absorb more water than smooth seeds. Part C demonstrates that the seeds have considerably different capacities for the production of branched starch molecules from glucose-1-phosphate.

It is important that all of these findings be correlated with the fact that a single pair of alleles is responsible for the differences discovered in this investigation.

Teams of two or three students work well in this investigation.

**Time**

Allow a total of about two and one-half periods: about 20 minutes for steps 1 and 2 of Part A, one period for step 4 of Part A and all steps of Part B, and one period for Part C.

**Materials (per class of 30—teams of 3)**

30 lab aprons
30 pairs safety goggles
30 pairs plastic gloves
1 roll paper towels
10 glass-marking pencils

**Part A**

20 250-mL beakers
10 balances
150 dry wrinkled peas
150 dry smooth peas

**Part B**

10 dropping pipets
10 microscope slides
20 scalpels
10 microscopes
10 soaked wrinkled peas
10 soaked smooth peas

**Part C**

10 petri dishes with glucose-1-phosphate agar
30 dropping pipets
10 dropping bottles Lugol's iodine solution
40 mL wrinkled-pea extract (in centrifuge tubes)
40 mL smooth-pea extract (in centrifuge tubes)

**Extract Preparation**

centrifuge with tubes
hammer
mortar and pestle or grinder
cheesecloth
30 g dry wrinkled peas
30 g dry smooth peas
80 mL 0.1 *M* NaF

**Safety**

Caution students to wash their hands after handling the peas. Seed producers often treat seeds with a fungicide. No food should be handled or eaten during any investigation involving chemicals or pesticides. Scalpels should be stored in trays when not in use.

If the investigation extends to a second lab period, remind students at the start of each period to wear safety goggles and lab aprons and to wash their hands before they leave the laboratory.

**Preparations**

Purchase pea seeds from a seed company, biological supply house, or grocery store. Set out the petri dishes 1 hour before class. Directions for preparing glucose-1-phosphate agar and 0.1 *M* NaF are provided in the Preparation of Solutions section.

To prepare the extracts, measure 30 g dry wrinkled peas and 30 g dry smooth peas. Crush the peas with pliers or a hammer, and grind each kind separately in a mortar and pestle or grinder until no big pieces remain. You can do this before class.

At the beginning of class, add 40 mL of *cold* 0.1 *M* NaF to each mass of ground peas. Blend in a blender for about 30 seconds *but no more than 1 minute* to avoid breaking down the enzyme.

Squeeze the mash through cheesecloth. Centrifuge for 2 minutes to obtain a clear solution free of sediment. *Keep as cold as possible.* Have the students spot their plates with clean dropping pipets directly from the centrifuge tubes, or have student helpers circulate among the teams and place the drops of extract on the agar. Use the extract immediately because the enzyme deteriorates rapidly at room temperature.

If you wish to purify the filtrate further, add solid $(NH_4)_2SO_4$ (approximately 4 g/20 mL) filtrate. This will precipitate out the proteins. Centrifuge and use the supernatant. Again keep everything as cold as possible.

Caution students to blot each drop of each extract at 10-minute intervals.

Save some of the best results, and project the image of the petri dish with the starch spots through an overhead projector for the class to observe. The extract from the wrinkled seeds will produce a more intense color reaction in the 40 minutes than the extract from the smooth seeds. The color from the wrinkled extract will be blue-brown, and the color from the smooth extract will be brown. These results are consistent with the fact that the wrinkled-seed extract produces fewer branched starch molecules than does the smooth-seed extract. In fact, the wrinkled seeds have a mutation in one of two starch-branching enzymes and cannot produce the highly branched starch that is characteristic of the smooth seeds.

Soaked peas cannot be used for the extract preparation. Soaking the peas activates starch-digesting enzymes; therefore, starch-forming activity would not be evident.

**Analysis**

1. Wrinkled.
2. Yes. Plant wrinkled and smooth seeds in the same soil, and grow the plants in identical environments. If wrinkled seeds produce plants that make wrinkled seeds and if smooth seeds produce plants that make smooth seeds, then the differences must be genetic.
3. Yes. The wrinkled seeds absorbed more water than the smooth seeds; therefore, the wrinkled seeds can be expected to lose more water when they dry. The greater loss of water is the most likely cause of the wrinkling of the seeds.
4. Wrinkled: compound, divided, irregular; smooth: simple, whole, oval.
5. Yes.
6. The difference in pea seeds is more likely due to genes. To investigate, a researcher could grow plants from the different kinds of seeds in the same environment. Using a microscope, the researcher could examine the starch grains of the new seeds produced by those plants.
7. Yes. The iodine test showed less highly branched starch molecules in the extract from the wrinkled peas. This extract gave a stronger and bluer color reaction than the extract from the smooth seeds did.
8. The action of the starch enzyme is closest to being under the direct influence of the gene.

## ◆ Extended Laboratory Program 11

# Chapter 13 Patterns of Inheritance

*Smooth and Wrinkled Pea Seeds*

Mendel discovered that the shape of pea seeds is under genetic control. Purebred plants with factors (genes) for wrinkled seeds always produced offspring with wrinkled seeds, and purebred plants with factors for smooth seeds always produced offspring with smooth seeds.

This investigation deals with the question of how the gene affects the seed to cause it to be either wrinkled or smooth. What can a coded message in the nuclei of the seed cells do that would cause the seed either to shrink and wrinkle upon drying or *not* to shrink and wrinkle upon drying?

Read through the procedures, and develop a hypothesis that accounts for the two types of pea seeds.

**Materials (per team of 3)**

3 lab aprons
3 pairs safety goggles
3 pairs plastic gloves
paper towels
glass-marking pencil

**Part A**

2 250-mL beakers or glass jars
balance
15 dry wrinkled peas
15 dry smooth peas

**Part B**

dropping pipet
microscope slide
2 coverslips
2 scalpels
microscope
soaked peas: 1 wrinkled and 1 smooth

**Part C**

petri dish with glucose-1-phosphate agar
2 dropping pipets
dropping bottle Lugol's iodine solution
wrinkled-pea extract
smooth-pea extract

**SAFETY** Put on your safety goggles and lab apron.

**PROCEDURE**

**Part A Effects of Water on Smooth and Wrinkled Pea Seeds**

1. Examine the two kinds of peas. Describe the obvious difference between them.
   CAUTION: The peas may have been treated with a fungicide to prevent mold growth. Do not eat or taste the peas; wash your hands after handling. If you get any fungicide in your mouth, spit it out and rinse your mouth with water. Call your teacher.
2. Determine the dry mass of the 15 smooth peas and then the dry mass of the 15 wrinkled peas. Record these masses in a table in your logbook. Place the smooth and wrinkled peas in separate beakers or jars, and add about 4 times their volume of water. Label your jars for the two kinds of peas, and put the jars aside for 24 hours.
3. Wash your hands thoroughly before leaving the laboratory.
4. After soaking the peas overnight, pour off the excess water from each jar and empty the peas into separate piles on paper towels. Blot off all the surface water from the peas, and again determine their masses. Record the masses of the soaked peas in your table. Use the equation that follows to determine the percentage of increase in mass for each kind of pea.

   $$\frac{\text{difference in mass}}{\text{dry mass}} \times 100\% = \%\ \text{increase}$$

   Record the percentage increase in your table. Answer Analysis questions 1–3.

**Part B Starch Grains in Smooth and Wrinkled Pea Seeds**

5. Label one end of a microscope slide *W* for wrinkled. Label the other end *S* for smooth. Place a drop of water at each end.
6. Cut through a soaked wrinkled seed with a scalpel. Gently scrape a small amount of the material from the cut surface into the appropriate drop of water. Mix the scrapings into the drop completely. Add a coverslip. Wipe the blade of the instrument clean, and repeat the process with the smooth seed.
   CAUTION: Scalpels are sharp; handle with care.
7. Carefully observe both drops of water under low power. In your logbook, make sketches

showing the shapes of the starch grains in each drop. Examine a few slides prepared by other class members, and compare them with yours. Answer Analysis questions 4–6.

8. Wash your hands thoroughly before leaving the laboratory.

**Part C Rate of Starch Production in Smooth and Wrinkled Pea Seeds**

The large starch granules observed in Part B are composed of many starch molecules formed when smaller, simpler sugar molecules combined. In this part of the investigation, an iodine reaction is used to assess the extent to which extracts from wrinkled and smooth seeds can form branched starch molecules. Lugol's solution forms a brownish color with highly branched starch molecules and a blue-brown color with less-branched starch molecules. Also, the intensity of the color changes as a function of branching. With Lugol's solution, less-branched starch reacts more strongly than highly branched starch does.

9. Obtain a petri dish containing glucose-1-phosphate agar. Label the outside bottom of the petri dish *W* on one half and *S* on the other, as shown in Figure 11.1. On the *W* side of the petri dish, place one drop of the wrinkled-pea extract directly on the agar-glucose surface. Do the same with the smooth-pea extract on the *S* side of the dish. Record the time that the extract meets the agar. After 10 minutes, use a paper towel to blot up the drop of extract on the *W* side and the drop on the *S* side.

   CAUTION: The extract is prepared in sodium fluoride, a poison. Do not ingest. If you get any in your mouth, spit it out and rinse your mouth with water. Call your teacher.

10. At 10-minute intervals, repeat step 9 until you have deposited four drops on each side of the dish, as shown in Figure 11.1.

11. Use a dropping pipet to flood the surface of the plate with Lugol's iodine solution. Record your observations.

    CAUTION: Lugol's iodine solution is an irritant and may stain clothing. Avoid skin and eye contact; do not ingest. Flush spills and splashes with water for 15 minutes. If necessary, rinse mouth with water. Call your teacher.

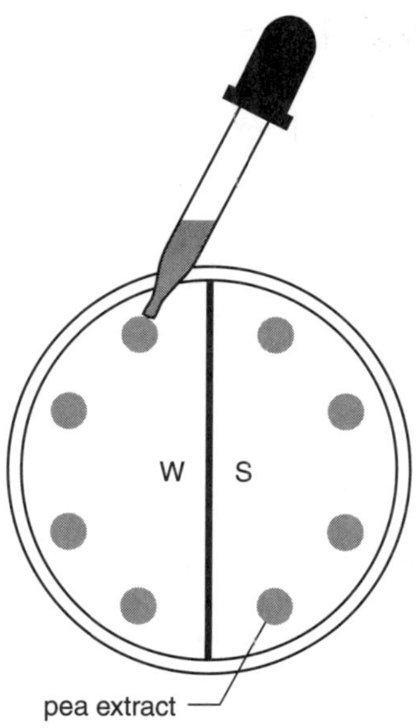


**FIGURE 11.1** Spacing drops of pea extracts on a petri dish

12. Wash your hands thoroughly before leaving the laboratory.

**Analysis**

1. Which seed type, wrinkled or smooth, had the greater ability to absorb water?
2. Is there any conclusive way to find out whether this difference is genetic or environmental? Explain.
3. If both kinds of soaked peas were placed in a drying oven and left for 24 hours, do you think they would again show the differences in seed coat? Explain your answer.
4. The following list contains descriptive words: compound, simple, whole, divided, oval, round, irregular. Which combination of three words from the list best describes the appearance of the starch grains of the wrinkled seeds? of the smooth seeds?
5. Do starch grains of pea types differ? Explain.
6. Is this difference more likely due to genes or to environment? Describe an experiment you could perform to make sure.
7. Was there any difference in the type of starch formed by the wrinkled and smooth peas? How could you tell?
8. One gene pair controls all the variations seen in this investigation. Is the texture of the seed coat, the shape of the starch grain, or the action of the starch enzyme closest to being under the gene pair's direct influence?

## Extended Laboratory Program 12

# Chapter 14 Other Forms of Inheritance

*Transformation of* Escherichia coli

### Teacher's Notes

This investigation gives students the opportunity to genetically engineer bacterial cells by having them take up plasmid DNA and express a gene for resistance to ampicillin. In the context of recombinant-DNA technology, plasmids are used as vehicles for transporting genes of interest into bacterial cells where they can be cloned or, in some cases, expressed. When a foreign gene is cloned into bacteria, it is often difficult to detect which bacteria have taken up the foreign gene. For this reason, plasmids used in cloning have one or more selectable markers that permit scientists to know which cells have taken up plasmid molecules. Usually, a selectable marker is a gene that encodes resistance to an antibiotic. By plating the transformed bacteria in the presence of the appropriate antibiotic, only those cells that have taken up the plasmid and expressed the resistance gene will be able to grow on the plate. Because the foreign gene resides on the same plasmid, it too must be present.

Most transformation protocols consist of four steps:

1. **Preincubation:** Cells are resuspended in a salt solution and placed on ice. Cations from the salt associate with negatively charged phosphates of lipids in the *E. coli* cell membrane.
2. **Incubation:** Following addition of plasmid DNA, the suspension is incubated in ice. The cations are thought to neutralize negatively charged phosphates in the DNA and the bacterial-cell membrane.
3. **Heat shock:** The bacteria/DNA mixture is briefly exposed to 42°C and then returned to ice. The temperature change creates a thermal gradient from one side of the membrane to the other. The mechanism by which this thermal imbalance sweeps the plasmids into the cell is not well understood.
4. **Recovery:** Luria broth is added to the bacteria/DNA mixture and incubated at 37°C (ideally with shaking). Transformed cells recover from their treatment and have time to amplify the transformed plasmid and express the resistance gene.

Transformation efficiencies using this rapid protocol range from $5 \times 10^3$ to $5 \times 10^4$ colonies per microgram plasmid DNA. More elaborate procedures can attain transformation efficiencies up to 200 times greater. This rapid protocol is not suitable for use with ligated DNA. The reason is that ligated DNA is composed of relaxed circular and linear plasmids. These plasmid conformations transform much less efficiently than the supercoiled plasmids used in this experiment.

Ampicillin is the most common antibiotic resistance gene used in transformation protocols. Ampicillin interferes with the ability of the bacteria to make new cell walls; therefore, it kills only cells that are replicating and assembling new cell-wall envelopes. Nonreplicating cells are not affected by the drug. As a result, cells can recover in the presence of the antibiotic because most cells will express the resistance protein before replicating.

Teams of two or three are ideal for this investigation.

### Time

One period for the activity and 20–30 minutes to discuss results.

### Materials (per class of 30—teams of 3)

30 lab aprons
30 pairs safety goggles
30 pairs plastic gloves
20 Luria broth (LB) agar plates
20 Luria broth agar plates with ampicillin (LB/amp)
10 *E. coli* starter plates
10 inoculating loops
10 bacterial spreaders
40 1-mL sterile graduated pipets
10 rubber pipet bulbs
20 microfuge tubes
10 mL calcium chloride solution
0.1 mL plasmid pUC8 solution
10 mL Luria broth
10 125-mL beakers containing about 50 mL of 70% isopropanol
1 water bath
1 incubator
1 hot plate
10 beakers of ice

### Safety

The refrigerator containing the agar plates and plasmid solution must be reserved for biological work and be so labeled; no food or drink should ever be stored in that refrigerator.

### Preparations

Make sure that the students are ready to work as soon as class begins.

The starter *E. coli* cultures should be inoculated the day before the experiment and incubated overnight at 37°C so they will be fresh for students to use. The ampicillin solution can be made well in advance and stored in the freezer until needed. To prepare plates using prepared media, loosen the bottle cap and place the bottle in a large Pyrex™ beaker or saucepan. Add water until the level rises above the level of medium in the bottle. Place a thermometer in the water, and heat the water on a hot plate until it boils. Boil gently for several minutes. Swirl the bottle occasionally to ensure that the agar is completely liquified.

CAUTION: Always wear heat-protective gloves when handling heated bottles. Turn off the heat, and remove bottle from the water bath.

When adding antibiotics to a hot agar solution, allow the media to cool to about 50°C before adding the ampicillin. If the agar solution is too hot, the high temperature will inactivate the ampicillin.

To pour the plates, first line them up in your work area. Remove the cap from the medium bottle, and flame the mouth of the bottle. Lift the lid of the petri dish just enough to admit the neck of the bottle. Pour just enough agar solution to cover the bottom of the dish (about 20 mL). Replace the lid. Swirl the plate gently to distribute the agar solution. The agar will solidify in about 20–30 minutes. Store the plates in the refrigerator until they are needed. For refrigerated plates, the ampicillin will remain active for at least 1 month.

**Analysis**

1. If growth is observed where it is not expected (nontransformed cells on LB/amp plate) or if growth is not present where expected (transformed cells on LB/amp plate), the most likely reason is that the wrong cell type was spread. Inactive ampicillin is another possible cause for the growth of nontransformed cells on an LB/amp plate. If a lawn of growth appears on all plates, an already transformed strain may have been used by mistake to prepare the starter plates.
2. **a.** These are both positive controls that confirm that the cells used in the experiment are viable.
   **b.** The negative control (−LB/amp) shows that the bacteria fail to grow in the presence of the ampicillin. The positive control shows that they are viable and will grow in the absence of the antibiotic.
   **c.** The negative control and the experimental plate (+LB/amp) show that only transformed cells grow in the presence of ampicillin and confirm that the plasmid pUC8 is responsible for the resistance.
   **d.** The experimental plate and the positive control demonstrate that transformation is a rare event. The lawn on the positive control indicates that the total number of cells spread, while the experimental plate indicates the small fraction of cells that were transformed.
3. Example calculation: Assume that the plasmid solution has a concentration of 0.005 µg/µL and that 100 colonies were counted on the +LB/amp plate.
   Total mass of plasmid (concentration × volume) is
   $0.005\ \mu g/\mu L \times 10\ \mu L = 0.05\ \mu g$
   Total volume of suspension is
   $250\ \mu L$ calcium chloride + $250\ \mu L$ Luria broth + $10\ \mu L$ plasmid pUC8 = $510\ \mu L$
   Fraction of suspension spread on plate is
   $100\ \mu L$ out of $510\ \mu L$, or 0.196
   Mass of plasmid in suspension spread is
   $0.05\ \mu g \times 0.196 = 0.0098\ \mu g$
   Transformation efficiency is
   $100$ colonies $\div\ 0.0098\ \mu g = 1.02 \times 10^4$ transformants per µg plasmid
4. Students may design experiments testing a number of potential variables. Some typical ones include the effect of the heat shock (both temperature and duration), the type of salt used, the strain of bacteria used, and the size and composition of the plasmid.

## ◆ Extended Laboratory Program 12

# Chapter 14 Other Forms of Inheritance

*Transformation of* Escherichia coli

In addition to their single chromosome, many bacteria possess small, circular DNA molecules called plasmids. Plasmid DNA encodes one or more genes that give the host bacteria some type of selective advantage, such as using a new food source or surviving in an environment that would normally prove fatal. Genes that confer resistance to antibiotics are often found on plasmids.

DNA can be exchanged between bacteria in three ways. One method, conjugation, has been described as bacterial sex. It involves a donor cell that transfers DNA to a recipient cell across a temporary bridge that forms between the two cells. A second method is called transduction. In this process, bacterial DNA is transferred as a consequence of infection by a bacteriophage. Sometimes bacterial DNA is inadvertently packaged inside a phage particle and, upon infection, is transferred to another cell. The third method is transformation: DNA that is released from one bacterial cell, often as a consequence of the cell's death and breakdown, may be picked up by another living bacterial cell.

The phenomenon of bacterial transformation was first described in the 1920s, and the concept played an important part in establishing the role of DNA as the molecule of heredity. More recently, bacterial transformation has become an important tool of the genetic engineer. It provides a simple method for inserting foreign genes into bacteria for cloning. An understanding of transformation also is critical to addressing the increasing problem of bacteria that are resistant to multiple antibiotics.

Bacterial transformation occurs in nature though only at a low rate. In the laboratory, we can use physical and chemical manipulations that greatly increase the probability that transformation will occur. In this experiment, you will manipulate *E. coli* cells to make them "competent" to accept plasmid DNA. The objective is to insert the plasmid pUC8, which bears a gene for resistance to the antibiotic ampicillin, into some *E. coli* cells.

**Materials (per team of 3)**

3 lab aprons
3 pairs safety goggles
3 pairs plastic gloves
2 Luria broth (LB) agar plates
2 Luria broth agar plates containing ampicillin (LB/amp)
1 starter culture of *E. coli*
1 inoculating loop
1 bacteria spreader
4 sterile graduated pipets
1 rubber pipet bulb
2 microfuge tubes
1 mL calcium chloride solution
1 mL Luria broth
10 µL plasmid pUC8
1 125-mL beaker with about 50 mL of 70% isopropanol
ice

**(per class)**

1 water bath
1 incubator

**SAFETY** Put on your safety goggles, lab apron, and gloves; tie back long hair.

**PROCEDURE**

1. Obtain two microfuge tubes, and label one "+" and the other "–." Later the "+" tube will have plasmid added to it.
2. Use a sterile pipet to add 0.25 mL (250 µL) cold calcium chloride solution to each tube.
3. Place both tubes on ice.
4. Obtain a starter plate containing the *E. coli* culture.
5. Use a sterile inoculating loop to transfer a large colony of bacteria from the starter plate to the "+" tube. Place the loop directly into the calcium chloride solution, and twirl the loop to dislodge the bacteria.
6. Use a sterile pipet to resuspend the bacteria in the calcium chloride solution.
7. Return the "+" tube to ice.

8. Repeat steps 5–7 to resuspend bacteria in the "−" tube.
9. Use a capillary pipet to add 10 µL of the plasmid pUC8 solution to the "+" tube.
10. Gently tap the "+" tube to mix the plasmid into the solution.
11. Incubate both tubes on ice for 15 minutes.
12. While the cells incubate on ice, obtain 2 LB plates and 2 LB/amp plates. Label one LB plate "+" and the other "−." Do the same for the LB/amp plates.
13. Following the 15-minute ice incubation, heat shock the cells by placing them in a 42°C water bath for 90 seconds. Immediately return the tubes to ice for at least 1 minute.
14. Use a sterile pipet to add 0.25 mL (250 µL) Luria broth to each tube. Tap with your finger to mix. Place the tubes at room temperature.
15. Use a sterile pipet to add 0.1 mL (100 µL) of the cell suspension from the "−" tube to the "−" LB plate and another 0.1 mL (100 µL) to the "−" LB/amp plate.
16. Sterilize the bacterial spreader by dipping it into a beaker of 70% isopropanol. Let it air-dry (about 10 seconds) before spreading cells.
17. Spread the cells over the entire surface of both "−" plates in succession.
18. Use a sterile pipet to add 0.1 mL (100 µL) of the "+" cell suspension to the "+" LB plate and another 0.1 mL (100 µL) to the "+" LB/amp plate.
19. Spread the cell suspensions as in steps 16 and 17.
20. Let the plates sit for a few minutes to allow the cell suspensions to be absorbed into the agar.
21. Label the plates with your team identification, and incubate them in an inverted position at 37°C overnight.
22. Wash your hands thoroughly before leaving the laboratory.

**Analysis**

1. Record the number of colonies growing on each plate in the following table. If individual colonies can't be counted, record "lawn." Were results as expected? Explain any variations from the expected results.

| | + PLASMID | − PLASMID |
|---|---|---|
| LB/amp plate | | |
| LB plate | | |

2. Compare the growth on the following pairs of plates, and explain what you can conclude from each pair.
   a. −LB and +LB
   b. −LB and −LB/amp
   c. −LB/amp and +LB/amp
   d. +LB/amp and +LB
3. Only a small fraction of the bacteria become competent, take up plasmid, and become resistant to ampicillin. The efficiency of transformation is defined as the number of transformants (resistant colonies) per 1 µg of plasmid DNA. Using the directions below, calculate the transformation efficiency for your experiment.

   Total mass of plasmid used (mass = concentration × volume) __________

   Total volume of cell suspension __________

   Fraction of cell suspension spread on plate (volume spread ÷ total volume) __________

   Mass of plasmid spread on plate (mass of plasmid × fraction spread) __________

   Number of transformants per µg plasmid (no. colonies on +LB/amp ÷ mass of plasmid spread) __________

4. Design an experiment to investigate which factors may influence transformation.

## ◆ Extended Laboratory Program 13

# Chapter 15 Advances in Molecular Genetics

*Restriction Digestion of Lambda DNA*

**Teacher's Notes**

Gel electrophoresis is an essential tool in recombinant-DNA technology. Depending on the type of gel used and its concentration, DNA molecules ranging in size from a few base pairs to millions of base pairs can be resolved. When electrophoresing DNA samples, two types of gels can be used: agarose and polyacrylamide. As long as agarose provides the required resolution, it is the gel matrix of choice, owing to its ease of use. Agarose gels can effectively resolve DNA fragments from 100 to 40,000 base pairs, though not on a single gel. The concentration of the agarose is varied to select for resolution over a particular range. Gels made with high concentrations of agarose (for example, 2%) have a smaller average pore size and are suited to separating smaller DNA fragments (several hundred base pairs). Gels made with less agarose (for example, 0.8%) are better suited to resolving DNA fragments of thousands of base pairs. Once the fragment sizes get larger than about 40,000 base pairs, conventional gel electrophoresis cannot be used. Instead, a variation called pulsed-gel electrophoresis is employed. This technique involves switching the direction of the electric field back and forth many times per second. The large DNA fragments (up to about 10 million base pairs) reorient themselves in a size-dependent fashion that permits their resolution in agarose gels.

The important technique of DNA sequencing depends on the ability of electrophoresis to resolve DNA fragments that differ from one another by a single base pair. Agarose gels cannot meet this ultimate requirement for resolution. For this application, polyacrylamide gels must be used. Polyacrylamide is the synthetic polymer used to make soft contact lenses. It can be used to make high-percentage gels that resolve the products of the sequencing reactions and withstand the heat generated by high-voltage electrophoresis.

In this investigation, students will gain experience working with agarose gels and interpreting the results of DNA restriction analysis. The objective is to have students prepare a graph relating the distance migrated by restriction fragments to their base-pair size. This graph is used as a standard curve to estimate the sizes of other unknown DNA fragments. Although the inverse relationship between distance migrated and base-pair size is linear, students will see a curve on their graphs. This is because the 0.8% gel cannot resolve DNA fragments of different sizes equally well. Under the conditions of this experiment, the large DNA fragments (10,000–20,000 base pairs) will not fall on the line. Instruct students to connect the data points to produce a slightly curved line.

**Time**

The approximate time for each step in the investigation is as follows:

| | | |
|---|---|---|
| Part A: | Casting an agarose gel | 30 minutes |
| Part B: | Loading the gel | 10 minutes |
| | Running the gel | 60 minutes |
| Part C: | Staining the gel | 30 minutes |
| | Destaining the gel | 30 minutes or overnight |

**Materials (per class of 30—teams of 5)**

30 lab aprons
30 pairs safety goggles
30 pairs plastic gloves
3 horizontal dual gel electrophoresis chambers
1 DC power supply that can run three chambers
6 capillary pipets or micropipets (20 µL)
30 metric rulers
30 calculators
1 Restriction Mapping of Lambda DNA Kit
  restriction digests of Lambda DNA
  bottle of 0.8% agarose
  bottle of 10X TBE running buffer
  bottle of Ward's DNA stain concentrate
  1 gel staining tray
  1 gel handling spatula

**Safety**

See the teacher's notes in Extended Laboratory Program 6 regarding electrical safety in the laboratory, general cautions in the use of electrophoresis equipment, and procedures for discarding biological waste produced during investigations.

Reagents should be stored in a refrigerator for biological materials only; no food or drink should ever be stored in that refrigerator. The outside of the refrigerator should be labeled with this information.

Make sure that students wear clean gloves when handling gels and reagents. Remind students daily to wear their safety equipment and to wash their hands before leaving the laboratory.

**Preparations**

Prepare 1X TBE buffer by adding 315 mL distilled water to 35 mL 10X TBE concentrate to obtain 350 mL of 1X TBE buffer. The bottle of prepared agarose provides enough material to pour eight 25-mL gels. Use a hot water bath or a microwave oven to melt the agarose. Always handle agarose bottle with heat-protective gloves. Be sure to loosen the cap of the bottle before placing it in the water bath or the microwave oven. Note: When using a microwave oven, if the agarose starts to boil before melting completely, stop the microwave, gently swirl the agarose, and continue.

Prepare dilute stain by adding 5 mL Ward's stain concentrate to 95 mL warm distilled water to obtain 100 mL of dilute stain. Optional: It is helpful to heat the DNA samples at 65°C for 5 minutes before students load them on the gel. This ensures that most of the Lambda DNA molecules will be in a linear conformation (rather than circular) and helps with band imaging.

You have the option of casting the gels in advance or having the students cast them during class.

**Analysis**

1. The loading dye serves two purposes. First, it contains glycerol or sometimes sugar that, when mixed with the DNA sample, makes it dense; this enables the sample to sink to the bottom of the well. Second, it contains a colored dye that electrophoreses ahead of the DNA fragments and serves as a tracking dye. When it reaches the far end of the gel, separation is complete and the power may be turned off.
2. Figure T13.1 is a template of semilog graph paper that you may copy for students to use to construct their standard curves.
3. The actual sizes of the DNA fragments in the *Eco*RI and *Hind*III digests of Lambda DNA appear in the table at the right.
4. The accuracy of the size estimations depends on a number of factors. First, if the gel is not run long enough, the restriction fragments will be close together and difficult to measure. Also, errors are possible in trying to find the leading edge of each band, especially when the gel is overloaded or when the bands are smeared. When two fragments are sufficiently similar in size, they will appear as a single band in the gel. As students construct their standard curves, make sure they are accurately assigning the sizes to the bands in their pattern. Students will find that large fragments are sized least accurately and that fragments in the middle of the banding pattern are sized most accurately. This reflects the optimal resolution for a 0.8% agarose gel.

| ***Eco*RI** **TUBE 1** | ***Hind* III** **TUBE 2** | ***Eco*RI / *Hind* III** **TUBE 3** |
|---|---|---|
| 21,226 | 23,130 | 21,226 |
| 7,421 | 9,416 | 5,148 |
| 5,804 | 6,557 | 4,973 |
| 5,643 | 4,361 | 4,268 |
| 4,878 | 2,322 | 2,037 |
| 3,530 | 2,037 | 1,894 |
| | (564) | 1,584 |
| | (125) | 1,375 |
| | | 947 |
| | | 831 |
| | | (564) |
| | | (125) |

Figures in parentheses indicate fragment sizes too small to be imaged on the gel.

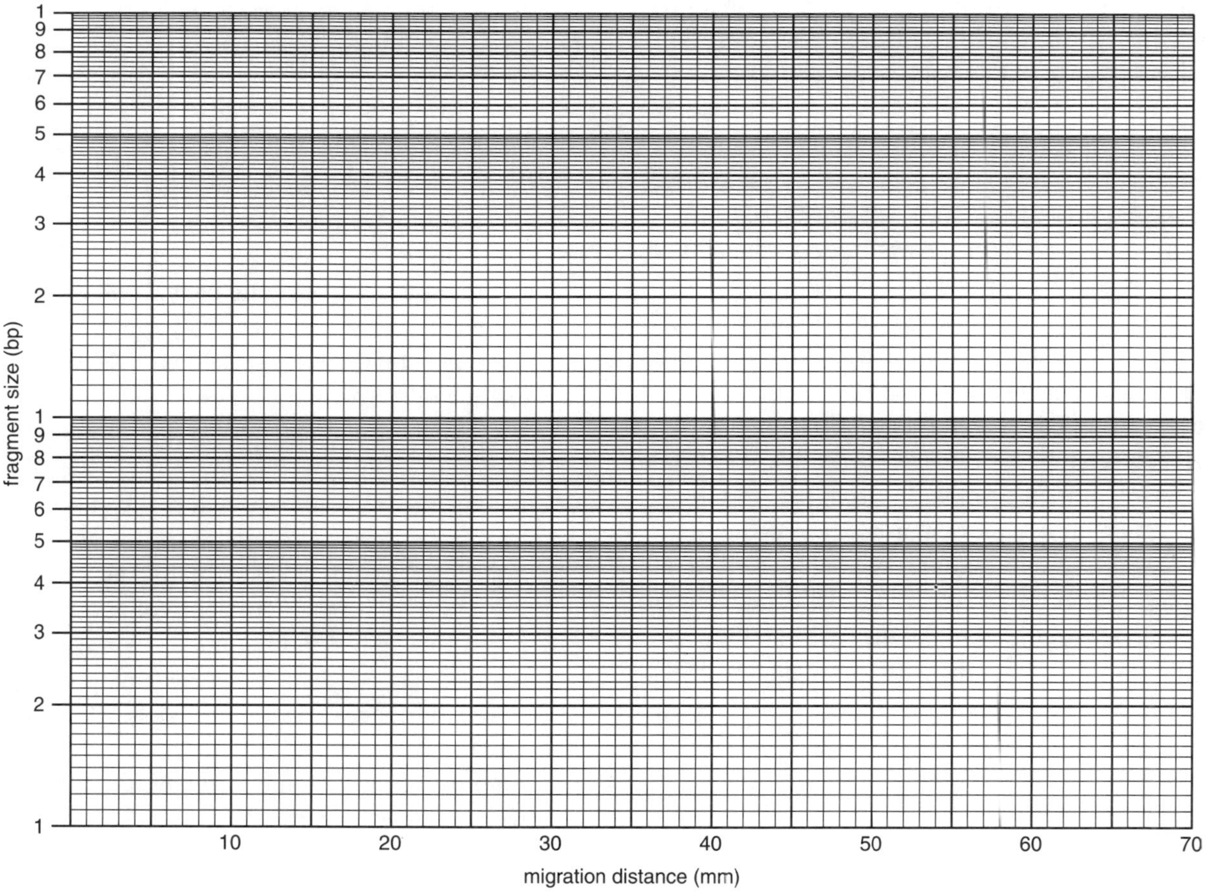


**FIGURE T13.1** Template for making a standard curve.

## ◆ Extended Laboratory Program 13

# Chapter 15 Advances in Molecular Genetics

*Restriction Digestion of Lambda DNA*

Scientists believe that restriction enzymes are thought to have evolved to protect bacteria from infection by bacteriophage. Once phage DNA enters a bacterial cell, it is recognized as foreign. An endonuclease (an enzyme that cuts within a DNA molecule) cuts the phage DNA into pieces, stopping the infection. The DNA of the host bacterium is protected by attached methyl groups and is not a substrate for the endonuclease. These endonucleases that are responsible for restricting the growth of bacteriophage are called restriction enzymes. The useful feature of restriction enzymes is that they read DNA sequences and cut only in specific positions. There are many different restriction enzymes, each of which recognizes a different DNA sequence. Together they provide scientists with a collection of molecular scissors that can be used to cut a given DNA sequence in almost any desired position. The use of restriction enzymes to manipulate DNA provides the basis for gene cloning.

In this investigation, you will gain experience analyzing DNA fragments produced by the activity of restriction enzymes. You will use agarose gel electrophoresis to sort the restriction fragments by size. An inverse relationship exists between the distance a DNA fragment migrates in the gel and its size (number of base pairs). The resulting patterns of fragments, called bands, can be analyzed to provide information about the sizes of the fragments and their relative positions along the DNA molecule. In this experiment, you will be told the sizes for each fragment in a particular restriction digest (reaction). From these data, you can construct a graph that relates the size of the DNA fragments to the distance migrated in the gel. You can then use the graph to estimate the sizes of fragments in other restriction digests.

**Materials (per team of 5)**

5 lab aprons
5 pairs safety goggles
5 pairs plastic gloves
1 set of three labeled microcentrifuge tubes containing restriction digests
1 0.8% agarose gel
1 capillary pipet or micropipet (10 µL)
350 mL TBE running buffer
1 horizontal dual gel electrophoresis chamber (for two groups)
1 DC three-cell power supply (for three groups)
5 metric rulers
5 calculators

**SAFETY** Put on your safety goggles, lab apron, and gloves; tie back long hair.

**PROCEDURE**

Before you start the procedure for running a gel, familiarize yourself with the parts of the electrophoresis apparatus. Depending on the type of equipment in the lab, your teacher may give you instructions for electrophoresis that differ from those presented here.

**Part A Casting an Agarose Gel**

Part A includes the following steps:

- Seal ends of casting tray
- Insert gel comb
- Pour gel
- Remove gel comb and unseal ends of casting tray

1. Snap the end dams onto each end of the gel casting tray. The end dams seal the ends of the tray, eliminating the need to tape the ends.
2. Using either the 8-well or the 16-well side, insert the gel comb into the slots near the end of the tray.
3. Pour approximately 25 mL melted agarose into the tray until it reaches a depth of about 3 mm.
4. Allow the gel to solidify for approximately 20 minutes. Do not disturb the gel tray or comb. When the agarose has solidified, it will turn opaque.
5. After the gel has solidified, remove the end dams from the tray by carefully sliding them out.

**Part B Loading and Running the Gel**

Part B includes the following steps:

- Position gel in chamber
- Load samples
- Add running buffer
- Electrophorese samples

6. Place the tray containing the gel in the center of the chamber. Position the well side of the gel near the black (negative) electrode.
7. Add approximately 350 mL 1X TBE running buffer to the chamber. To do this, carefully pour the buffer out of a beaker into the chamber. When the level of the buffer in the chamber reaches the gel, add buffer slowly until the buffer level is about 2 mm above the top of the gel.
8. Use the capillary pipet (or micropipet) to load 10 µL from each sample tube into a separate sample well. Rinse the pipet with buffer between additions. Do not pierce the bottom of the sample well with the pipet tip.
   - Well 1: Lambda DNA *Eco*RI digest
   - Well 2: Lambda DNA *Hind*III digest
   - Well 3: Lambda DNA *Eco*RI/*Hind*III double digest
9. Make sure the cover is dry, and then slide it onto the electrophoresis chamber. Wipe off any spills on the apparatus before proceeding.
10. Make sure the patch cords attached to the cover, as well as the female jacks on the power supply, are completely dry. Then connect the red patch cord to the red electrode terminal on the power supply. Connect the black patch cord to the black electrode terminal on the power supply.
11. Plug in the power supply, and set it to 125 volts.
12. Turn on the power supply. The red power light will illuminate, and bubbles will form along the platinum electrodes.
13. Observe the migration of the samples through the gel toward the red electrode. Turn the power off when the blue loading dye nears the end of the gel. Unplug the power supply.
14. Wait about 10 seconds; then disconnect the patch cords from the power supply, and remove the cover from the chamber.

**Part C Processing the Gel**

Part C includes the following steps:

- Remove gel
- Transfer to staining tray
- Stain gel
- Destain gel
- Store gel

15. Put a notch in the left side of the gel to ensure that the gel lanes can be correctly identified after the gel has been removed from the unit.
16. Lift the tray containing the gel out of the chamber, and gently place the gel in the staining tray.
17. Pour approximately 100 mL warmed dilute stain into the staining tray until the stain barely covers the gel.
18. Cover the tray, and let the gel stain for about 30 minutes.
19. After staining, decant the used stain. Make sure the gel remains flat and does not move up against the corner. Decant the stain directly to a sink drain, and flush with water.
20. Add warm water to the staining tray. To accelerate destaining, gently rock the tray. Destain until the bands are distinct with little background color. This will require 20–30 minutes. Note: The gel can be destained overnight in water.
21. View the gel against a light background, such as white paper, or on a light box. Gels can be stored in self-sealing plastic bags. For long-term storage, add several drops of dilute stain to the bag to prevent the DNA bands from fading. If fading does occur, the gel can be restained by using the above procedure.
22. Wash hands thoroughly before leaving the laboratory.

**Analysis**

1. What is the function of the loading dye?
2. Recall that DNA fragments migrate in inverse relation to their base-pair size. Use data from the Lambda DNA *Eco*RI/*Hind*III double digest to construct a standard curve that relates distance migrated to base-pair size. Note: The smaller DNA fragments appear fainter, and some of the smallest will likely be missing from your banding pattern.

   **a.** Orient your gel with wells at the top, and locate the Lambda DNA *Eco*RI/*Hind*III gel lane.

   **b.** Working from top to bottom, match each band with its fragment size listed in Table 13.1.

   **c.** Carefully measure the distance (in millimeters) that each fragment migrated from the sample well. Measure from the front edge of the well to the leading edge of each band. Enter the distance migrated by each band into the Migration Distance column in the table.

**d.** Set up semilog graph paper with the distance migrated as the $x$ (arithmetic) axis and the log of the base-pair length as the $y$ (log) axis. The first cycle of the $y$-axis should begin with 100 bp and end with 1,000 bp, counting in 100-bp increments. The second cycle begins with 1,000 bp, counting in 1,000-bp increments. Restriction fragments larger than 10,000 bp will not fit on your graph. They do not fall within the resolving range of your gel and will not give reliable data anyway.

**e.** Plot the distance migrated versus the base-pair length for each *Eco*RI/*Hind*III fragment.

**f.** Connect the data points.

**TABLE 13.1 LAMBDA *Eco*RI / *Hind* III MOLECULAR SIZE STANDARD**

| MIGRATION DISTANCE (mm) | FRAGMENT SIZE (bp) |
|---|---|
| | 21,226 |
| | appear as one band { 5,148 / 4,973 |
| | 4,268 |
| | 3,530 |
| | 2,037 |
| | 1,894 |
| | 1,584 |
| | 1,375 |
| | 947 |
| | 831 |
| | 564 |
| | 125 |

**3.** Next, use your standard curve to estimate the sizes of the unknown fragments in the *Eco*RI and *Hind*III gel lanes:

**a.** Measure the distance migrated (in millimeters) by each *Eco*RI and *Hind*III fragment. As before, measure from the front edge of the sample well to the leading edge of the band. Enter these distances into the appropriate Migration Distance columns in Table 13.2

**b.** Locate on the $x$-axis the distance migrated by each *Eco*RI and *Hind*III fragment. Then use a ruler to draw a vertical line from this point to its intersection with the line on the standard curve. Now extend a horizontal line from this point to the $y$-axis; this gives the bp size of the fragment. Enter the result in the appropriate Fragment Size column in Table 13.2.

**4.** Your teacher will supply you with the actual sizes of the *Eco*RI and *Hind*III fragments. Compare them with your estimated sizes. For which fragment sizes was your standard curve most accurate? For which fragment sizes was it least accurate? What does this tell you about the resolving power of agarose gel electrophoresis?

**TABLE 13.2 SIZE AND MIGRATION DATA FOR LAMBDA *Eco*RI AND *Hind* III DIGESTS**

| *Eco*RI DIGEST | | *Hind* III DIGEST | |
|---|---|---|---|
| MIGRATION DISTANCE (mm) | FRAGMENT SIZE (bp) | MIGRATION DISTANCE (mm) | FRAGMENT SIZE (bp) |
| | | | |
| | | | |
| | | | |
| | | | |
| | | | |
| | | | |

## ◆ Extended Laboratory Program 14

# Chapter 16 Population Genetics

*Variation within a Species*

**Teacher's Notes**

The idea that variations exist among individuals of a species is fundamental to the concept of natural selection. Beginning biology students are often surprised that different kinds of organisms have very similar patterns of variation. Speculating on the advantages and disadvantages of the variations gives students a chance to predict survival value.

Have students work in pairs to complete this investigation.

**Time**

One period is needed for the collection of data. Organization and evaluation of data can be done outside the class period.

**Materials (per class of 30—teams of 2)**

30 lab aprons
30 pairs safety goggles
30 pairs plastic gloves
15 petri dish halves
45 pieces graph paper
30 metric rulers
15 30-cm pieces string
450 soaked pea or bean seeds or peas in pods
200 grasshoppers of the same species

**Safety**

Students should wear lab aprons, safety goggles, and plastic gloves when handling preserved specimens. Grasshoppers obtained from Ward's Natural Science Establishment are preserved in 50% isopropyl alcohol. Both the alcohol and any residual liquid are flammable. Be sure there are no flames or other sources of ignition in the lab. If grasshoppers are obtained from other suppliers, they may be preserved in formaldehyde. See the Safety Guidelines in the Teacher's Annotated Edition for cautions about formaldehyde.

**Preparations**

Soak bean or pea seeds overnight. Fresh peas in pods do not need to be soaked. Soaking the seeds facilitates peeling the outer covering. Peeling the seed coat and separating the cotyledons allow more precise measurements.

Each student team needs 10–15 grasshoppers. Order preserved specimens from your biological supplier. You can substitute peanuts in shells or leaves from one species of tree for the grasshoppers, but students may find the grasshoppers more interesting.

Try to standardize the eye-width measuring techniques by demonstrating the method that all students are to use. The variation in eye width may be partly caused by differences in the height in the bridge of the nose.

The table and graph of eye widths are examples of the organization of data.

**Analysis**

1. All graphs should show a bell-shaped distribution. The range of variation will vary, and the graphs will look different because of diverse numerical values that may be used on the axes.
2. The longest seed would have the most stored food.
3. The volume or weight of the seed would be a better measure of stored food.
4. Because the seed contains food for the undeveloped plant's early growth, the larger the seed, the greater the food reserve for the embryo and, therefore, the greater its chance for survival under adverse conditions.
5. Answers will vary. Some advantages might be greater ability to escape enemies, to reach food, and to cover territory, all of which would improve the chance of meeting a mate and reproducing. A disadvantage might be that a long jump would exceed the grasshopper's range of vision and, hence, might take it toward danger rather than away from it.
6. The general shape of the graphs would be similar, but the doctor's graph would have many more points on it.

## ◆ Extended Laboratory Program 14

# Chapter 16 Population Genetics

*Variation within a Species*

You probably have heard the expression "as much alike as two peas in a pod." Just how much alike are two peas in a pod or two beans in a pod or leg lengths of grasshoppers or heights of people? In a flock of sheep, all the individuals look very much alike and it is not easy for us to tell one from another. Lambs, however, seem to have no difficulty in finding their own mothers when feeding time comes. There are differences in the mother sheep that the lambs use to identify them. We do not notice these differences by simply looking at them.

The differences that distinguish individuals within a species are known as variations. It is possible to describe these differences in pictures, words, or measurements. Probably the best description is made on the basis of measurements.

In this investigation, you will look for differences in a number of living things to see whether there is any pattern to these variations. In addition, you will be asked to reflect on the survival advantages of some of the variations that you measure.

**Materials (per team of 2)**

2 lab aprons
2 pairs safety goggles
2 pairs plastic gloves
petri dish or similar container
3 pieces graph paper
2 metric rulers
30-cm piece of string
30 soaked pea or bean seeds or peas in pods if available
10–15 grasshoppers of the same species

**PROCEDURE**

**Part A Collecting Data**

1. Peel the outer covering (seed coat) off the seed so that it may be easily divided in half. Lay one of the halves of the seed on the metric ruler, and determine to the nearest half or whole millimeter the longest dimension of the seed. Discard both halves. Repeat with other seeds until you have measurements from about 25 seeds. Record all measurements in a table in your logbook. Be sure to record the number of seeds of the same length. Pool your data with other students so that you have a record of at least 100 different seed measurements.

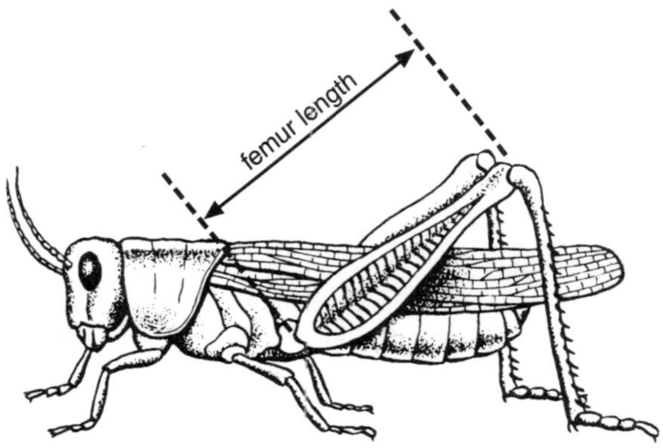


**FIGURE 14.1** The measurement to be made of the femur of a grasshopper.

**SAFETY** Put on your safety goggles, lab apron, and gloves.

2. Measure the length of the femur of the jumping leg from several grasshoppers. It is not necessary to remove the leg. Figure 14.1 will help you locate the femur. Record your measurements in your logbook. Pool your data with the entire class so that you have the measurements of all the grasshoppers.
3. Tie a knot about 5 cm from the end of a piece of string to mark a point of reference. While holding the string across the bridge of your partner's nose, position the knot in the string exactly at the outer corner of one eye. Stretch the string tightly, and mark on the string with your thumbnail the point of the outer corner of the other eye. Measure to the nearest millimeter the distance between the knot and your thumbnail. This measurement represents the distance from the outer corner of one eye to the outer corner of the other eye. Now have your partner measure this distance on you. Record these measurements in your logbook. Pool your data with the other members of your class.

**TABLE 14.1 SAMPLE DATA TABLE FOR STUDENT EYE WIDTHS**

| EYE MEASURES FROM SMALLEST TO LARGEST → | | | | | | | | | |
|---|---|---|---|---|---|---|---|---|---|
| Eye widths (mm) | | | | | | | | | |
| Number of students | | | | | | | | | |

## Part B Organizing Data

4. Arrange your three sets of measurements (bean size, leg length, and eye width) in a manner that will show the number of like measurements (see Table 14.1).
5. Prepare a graph of each set of data to show the distribution of the variations. Put the range of measurements on the horizontal axis and the number of individuals having this measurement on the vertical axis. Draw a smooth line (one without angles) connecting or passing near the dots plotted on each graph (see Figure 14.2). Put each set of measurements on a separate graph.
6. Calculate the average length in each set of measurements. Mark this length on each graph by finding the average value on the horizontal axis and making a vertical line to indicate the position of the average length.
7. Wash your hands thoroughly before leaving the laboratory.

## Analysis

1. In what ways are the three graphs alike? How do they differ?
2. The two large halves of the seed are modified leaves containing stored food that will be used by the young plant as it grows. Which of the seeds that you measured would have the most stored food?
3. What kind of measurement other than length would give you a better answer to question 2?
4. How does large seed size help the young plant to survive?
5. Assuming that a grasshopper with longer femurs can jump farther, how might leg length be a survival factor in the life of a grasshopper? List several advantages of longer femurs, but keep in mind that your conclusions are based on assumption. Also, think of a possible disadvantage of having long femurs.
6. An optometrist (a doctor who examines eyes and fits glasses) measures the width of the eyes of the people who come for new glasses. If all the thousands of measurements of eye width that any optometrist has made were plotted on a graph, how would the general shape of that graph compare with the one you made?

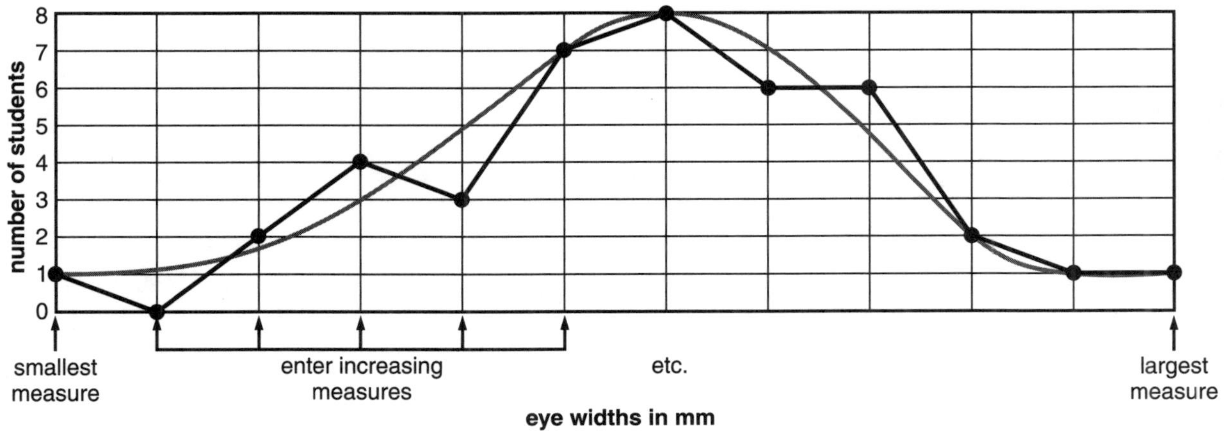


**FIGURE 14.2** Hypothetical graph of student eye widths showing straight lines connecting plotted data.

## ◆ Extended Laboratory Program 15

# Chapter 18 Diversity and Variation

*Variation in the Human Species*

**Teacher's Notes**

In this investigation, students have an opportunity to examine some human traits and realize that these traits are controlled by genes. After completing the activity, many students will go home and investigate some of these traits in their own families. Some students may note that many of these traits do not follow Mendelian inheritance. Explain that this will not affect the results of the investigation.

**Time**

One period should be sufficient.

**Materials (per class of 30—working individually)**

paper and pencils

**Preparations**

Help students determine the traits for their earlobes, thumbs, and middigits by using Figure 15.1 as a guide. Middigital hair occurs on the back side of the fingers on the middle section of any digit. Any hair here—on any finger of either hand—is considered positive for middigital hair.

Any eye color except blue is considered pigmented. This includes eyes that are combinations of both blue and other colors. Blue eyes are considered to be nonpigmented.

Visual inspection can establish hair form except when students have had their hair chemically straightened or curled. Ask students to identify the natural status of their hair.

If some students do not know their blood types, randomly assign them to a blood-type group. Do not type students' blood yourself or allow students to type one another's.

**Analysis**

1. Answers will vary. A seventh distinguishing trait might be sex, height, hair color, or the like.
2. Differences would include type of thumb (hitchhiker's), hair form (straight), blood type (A), and presence of middigital hair.
3. Blood type (AB), thumb type (straight), and eye pigment (none).
4. Answers will vary. Students may cite the homogeneous nature of their community.
5. A or AB blood type, straight hair, hitchhiker's thumb, nonblue eyes, free earlobes. Yes, assuming that each of the five traits (excluding blood type) is determined by a single gene and that the attached earlobes, blue eyes, straight thumbs, and nonstraight hair are recessive alleles.

## ◆ Extended Laboratory Program 15

## Chapter 18 Diversity and Variation

*Variation in the Human Species*

Variability is a basic characteristic of living things. The process of natural selection acts on genetic variations in a species as it adapts to changes in its environment.

Humans show variation in many traits that can be detected easily by observation. Some of these traits may persist partly because natural selection is not involved. That is, some traits such as eye color apparently have no effect on our survival.

This exercise enables you to examine variation in humans. On the basis of just six observable traits, you will find considerable variability among your class members.

**Materials (per student)**

paper and pencil

**PROCEDURE**

1. The six traits you will be examining are free versus attached earlobes, nonblue eyes (pigmented) versus blue eyes (nonpigmented), hitchhiker's thumb versus straight thumb, nonstraight hair versus straight hair, middigital hair on the second section of fingers versus none, and blood type. Three of these traits are shown in Figure 15.1. Determine your own characteristics (phenotypes) for the six traits. If you do not know your blood type (A, B, AB, or O), your teacher will assign you one.
2. Look at the genetic wheel in Figure 15.2. Start in the middle with the trait of earlobe attachment. Determine whether you are on the right or the left side of the vertical line.
3. Move to the next circle on the wheel. Decide which compartment you should select for eye color. Continue moving trait by trait until you reach the outer edge of the wheel.
4. Now read the number assigned to your particular combination of traits. Record this number in your logbook.
5. Make a list of the numbers your classmates have determined for themselves. Are any of the numbers used more than once?

**Analysis**

1. Does anyone in the class have the same phenotypes as you? That is, is anyone like you with respect to all six traits? If so, describe a seventh trait that will distinguish you from that person.

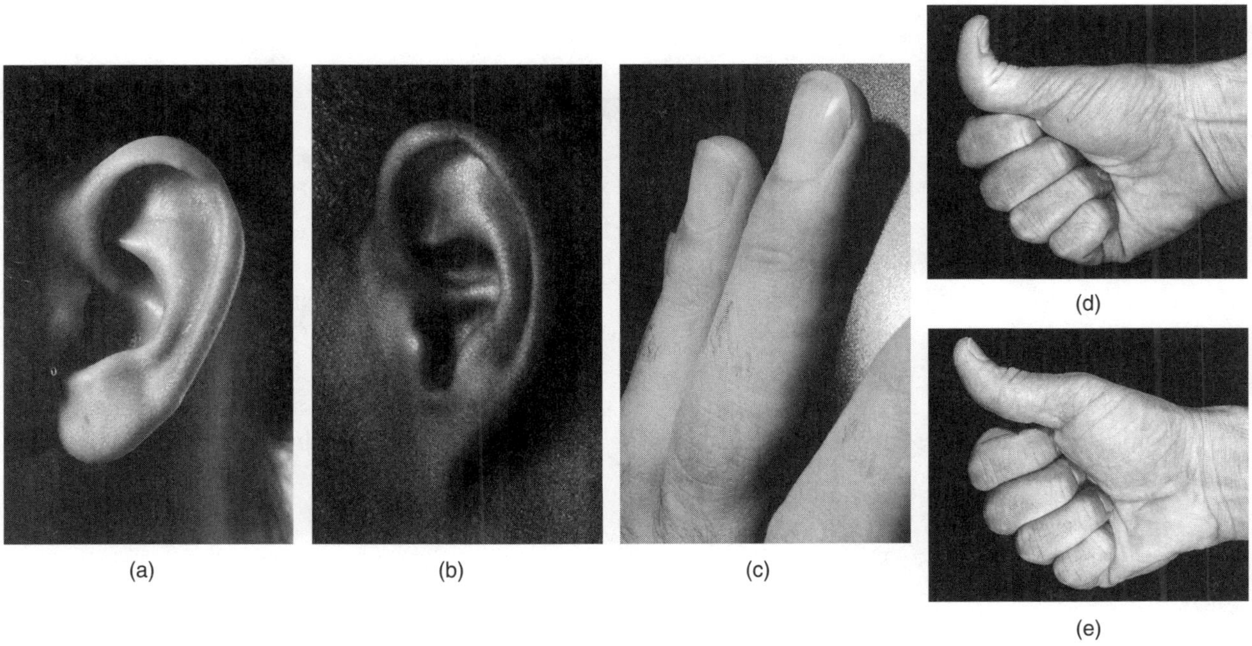

**FIGURE 15.1** Free (a) and attached (b) earlobes; middigital hair (c); hitchhiker's thumb (d) and straight thumb (e)

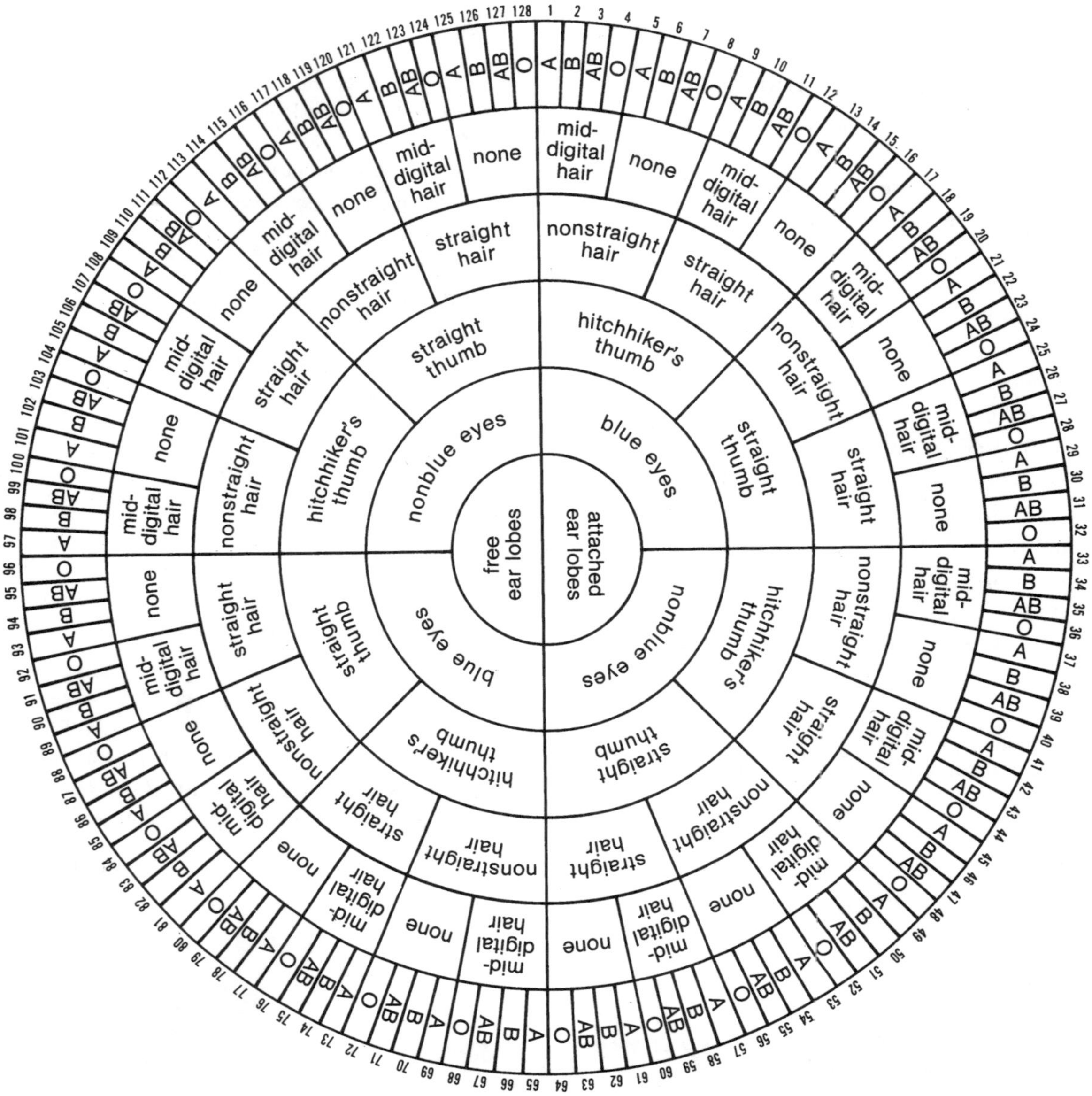


**FIGURE 15.2** The genetic wheel for five pairs of traits (phenotypes) and a sixth trait (blood group) with four phenotypes

2. By which traits would a person with the number 9 differ from a person with the number 24?
3. By which traits would a person with the number 23 differ from a person with the number 40?
4. Account for any clustering of individuals around a particular set of phenotypes.
5. Assume that two male individuals and two female individuals having the four numbers 18, 20, 22, and 24 are stranded on an uninhabited island. They start a new, isolated population. What phenotypes would be absent in the original population? Can you assume that phenotypes missing from the original population also will be missing from subsequent generations? Explain.

 

# ◆ Extended Laboratory Program 16

## Chapter 19 Changes in Species

*Dinosaur Locomotion*

**Teacher's Notes**

This investigation allows students to make measurements, perform some simple calculations, and reach conclusions about animals' walking and running speeds. Theoretically, each lab team should report exactly the same results, but errors in measurement make it unlikely that this will happen. The activity provides an opportunity for you to impress on students the importance of making careful and accurate measurements

In Part A, students are challenged to estimate walking speed. The mathematical relationship between leg length and stride length is expressed as follows: speed (in kilometers per hour) = 2.82 (stride in meters)$^{1.67}$ × (leg length in meters)$^{-1.17}$. You have the option of having students perform the calculations or allowing them to obtain their results directly from the graph.

This activity has a number of possible extensions. For example, students may take measurements from various breeds of dogs and determine their various walking speeds. They may also investigate for which animals the relationship between leg bone ratios and sprint speed holds true. (It works for mammals, birds, and many reptiles but not for turtles!)

**Time**

One class period should be sufficient.

**Materials (per class of 30—teams of 2)**

30 metric rulers
15 calculators

**Preparations**

The data is this investigation rely on length measurements made on drawings of animal footprints and bones. In some cases, errors in measurement of just 1 or 2 millimeters can have a significant effect on the final result. To minimize the effects of error in length measurement, you can enlarge the figures used in this activity on a copy machine. Students will find it easier and more accurate to work with larger drawings.

**Analysis**

1. In the imaginary race, the tyrannosaur has the greatest walking speed, about 13 kmph (or 8 mph). The brachiosaur and moa have the slowest walking speeds, about 4 kmph (or 2.5 mph).
2. There is no apparent relationship between an animal's walking speed and its size. The brachiosaur is much larger than the moa, yet they have similar walking speeds.
3. Students will suggest any number of studies, but they should mention trying the analysis on animals living today, including humans.
4. The Gallimimus, with a top speed of about 96 kmph (60 mph), wins the race. The tyrannosaur, at 51 kmph (32 mph), comes in second.
5. A number of answers are possible. Students should recognize that animals that depend on speed to catch prey or to escape from predators will be built for speed. Other animals use different strategies, such as large size or camouflage, to escape predation.
6. Students may answer by describing measurement errors, but the major assumption is that the graph relating speed to leg-bone ratios is based on analysis of mammals. It is not necessarily true that this relationship follows the same mathematics for dinosaurs. For example, there are significant differences between the cardiovascular systems of the two types of animals.

## ◆ Extended Laboratory Program 16

# Chapter 19 Changes in Species

*Dinosaur Locomotion*

Most of what we know about the appearance of dinosaurs comes from the study of their fossilized bones. However, close inspection of these skeletal remains can reveal a great deal more than just the animal's superficial appearance. For example, the skull alone can indicate a great deal about the animal's diet, bite strength, and relative intelligence.

After the first dinosaur skeletons were reconstructed in the mid-1800s, it was obvious that they represented a new class of animals that were long extinct. It was Sir Richard Owen who coined the term *dinosaur*, meaning "terrible lizard," in 1841. Meanwhile, an American, Edward Hitchcock, who was president of Amherst College, was studying dinosaur footprints. He noticed that they displayed characteristics found in the footprints of birds. So he called his animals reptile-birds.

Today paleontologists agree that dinosaurs possessed many avian features, but they are not about to give up the name *dinosaur* in favor of *reptile-bird*. This investigation focuses on the leg bones of various extinct animals, including many dinosaurs. You can see for yourself the relationship between dinosaurs and birds by comparing the drawings of the dinosaur leg bones in this activity with leg bones from a chicken. By making measurements and performing a few simple calculations, you can even deduce how quickly a dinosaur could walk or run.

**Materials (per team of 2)**

2 metric rulers
1 calculator

**PROCEDURE**

**Part A Estimating Walking Speed**

An efficient way to carry out this activity is to have one member of the team measure the leg lengths while the other member measures the stride lengths.

1. Set up a data table in your logbook as depicted in Table 16.1.
2. For each animal depicted in Figure 16.2, measure the leg length (in centimeters) and record it in the appropriate column in your data table. The leg length is defined as the distance from the point at which the femur attaches to the hip socket straight down to the toe.
3. In the next column in your data table, record the corresponding actual leg length (in meters), as provided in the drawing
4. Divide the actual leg length by the drawing leg length to obtain a scale ratio of meters to centimeters. Record the scale ratio in the appropriate column of your data table.
5. For each animal depicted in Figure 16.2, measure the stride length (in centimeters) and record it in the appropriate column of your data table. The stride length is defined as the distance covered by one foot in a complete stride. Be

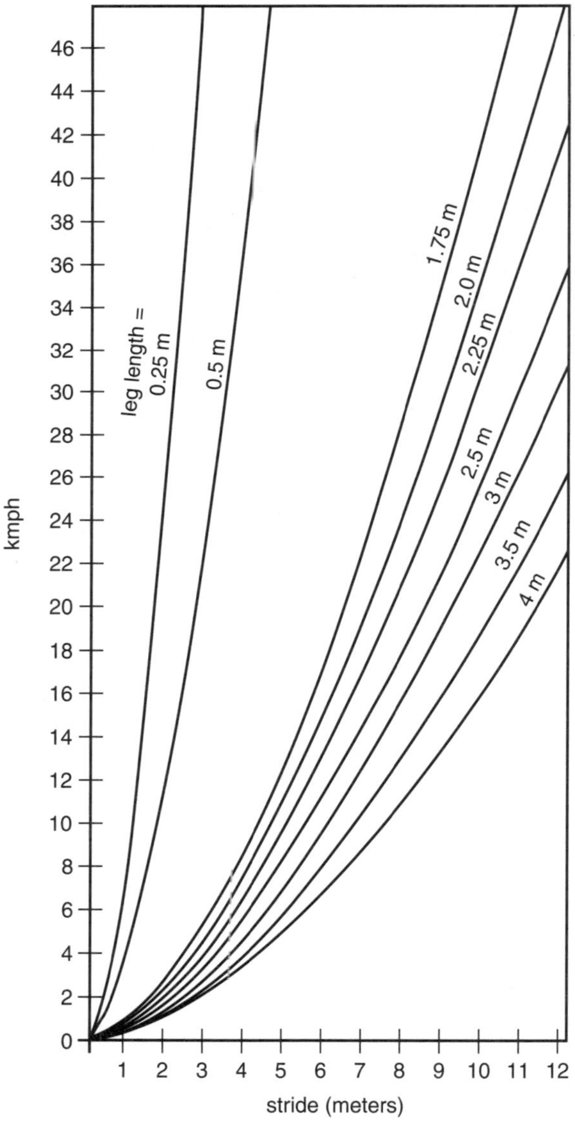


**FIGURE 16.1** Walking speed as a function of leg length

| TABLE 16.1 LEG AND STRIDE LENGTH DATA TO ESTIMATE WALKING SPEED | | | | | | |
|---|---|---|---|---|---|---|
| ANIMAL | LEG LENGTH, DRAWING (cm) | LEG LENGTH, ACTUAL (m) | SCALE RATIO (m to cm) | STRIDE LENGTH, DRAWING (cm) | STRIDE LENGTH, ACTUAL (m) | WALKING SPEED (km/h) |
| Tyrannosaur | | | | | | |
| Mammoth | | | | | | |
| Moa | | | | | | |
| Dimetrodon | | | | | | |
| Brachiosaur | | | | | | |

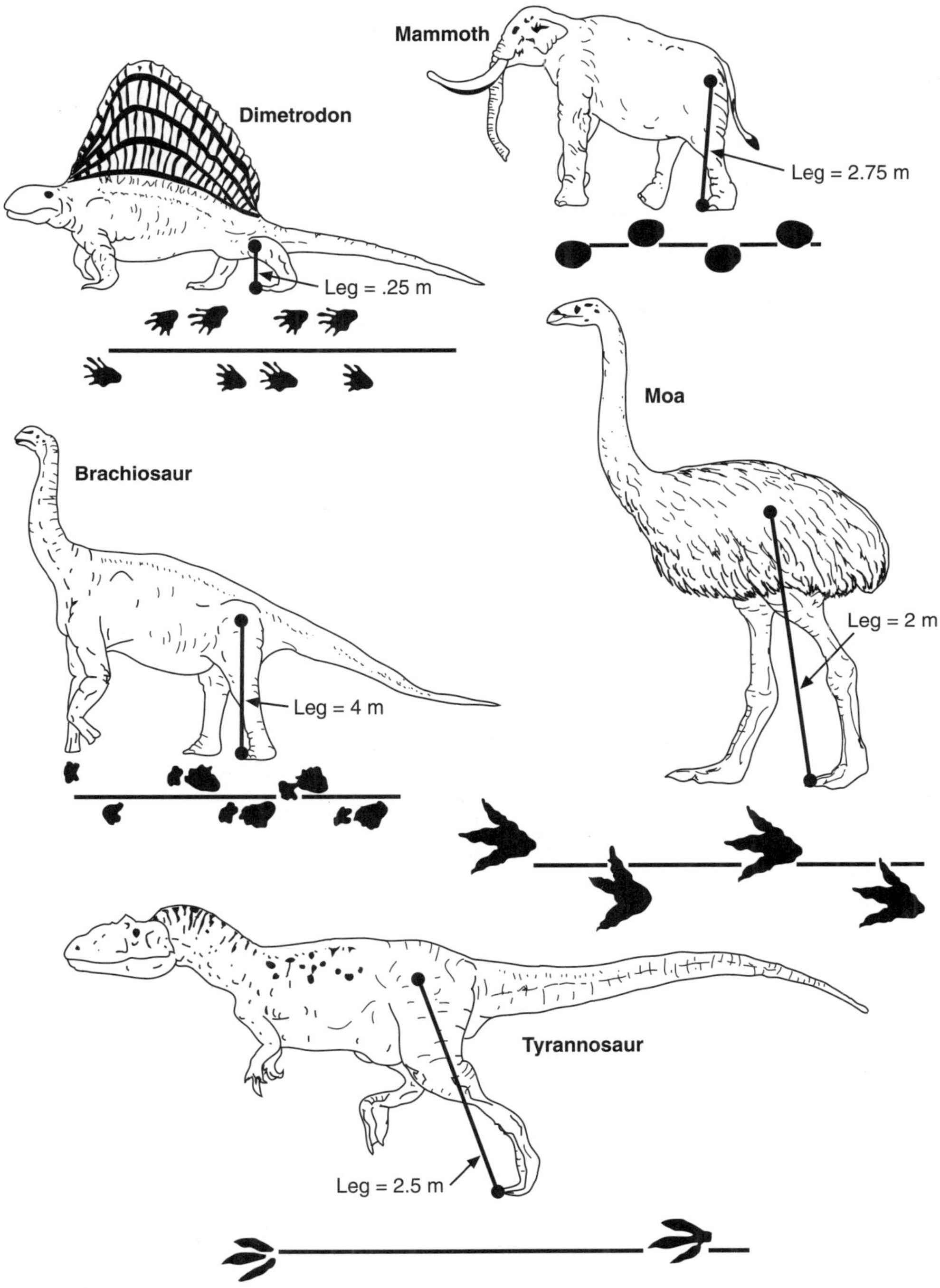


**FIGURE 16.2** Extinct animals and their footprints

sure you are measuring the distance covered by the same foot in the drawing.

6. For each animal, determine the actual stride length (in meters). To do this, multiply the stride length measured from the drawing (in centimeters) by the scale ratio (meters to centimeters). Enter these values in the appropriate column of your data table.
7. Use the graph in Figure 16.1, on page 104, to estimate the walking speed of each animal, and record the values in the last column of your data table. To do this, first find the leg-length line that is closest to the value calculated for the animal in question. Next, find its intersection with the corresponding stride length. Read across to the $y$-axis to get the speed value.

**Part B Estimating Sprint Speed**

In this activity, you will hold a hypothetical race between several different species of dinosaurs. Can you predict who will win?

1. Set up a data table in your logbook as depicted in Table 16.2.
2. For each dinosaur leg depicted in Figure 16.3, measure and sum the lengths of the shin, ankle, and foot. Enter the values in the appropriate column of your data table.

TABLE 16.2 LEG BONE LENGTH DATA TO ESTIMATE SPRINT SPEED

| CONTESTANT | SHIN + ANKLE + FOOT LENGTHS (cm) | THIGH LENGTH (cm) | SHIN + ANKLE + FOOT / THIGH | SPEED (mph) |
|---|---|---|---|---|
| Brontosaur | | | | |
| Gallimimus (ostrich dinosaur) | | | | |
| Monoclonius (horned dinosaur) | | | | |
| Stegosaur | | | | |
| Tyrannosaur | | | | |

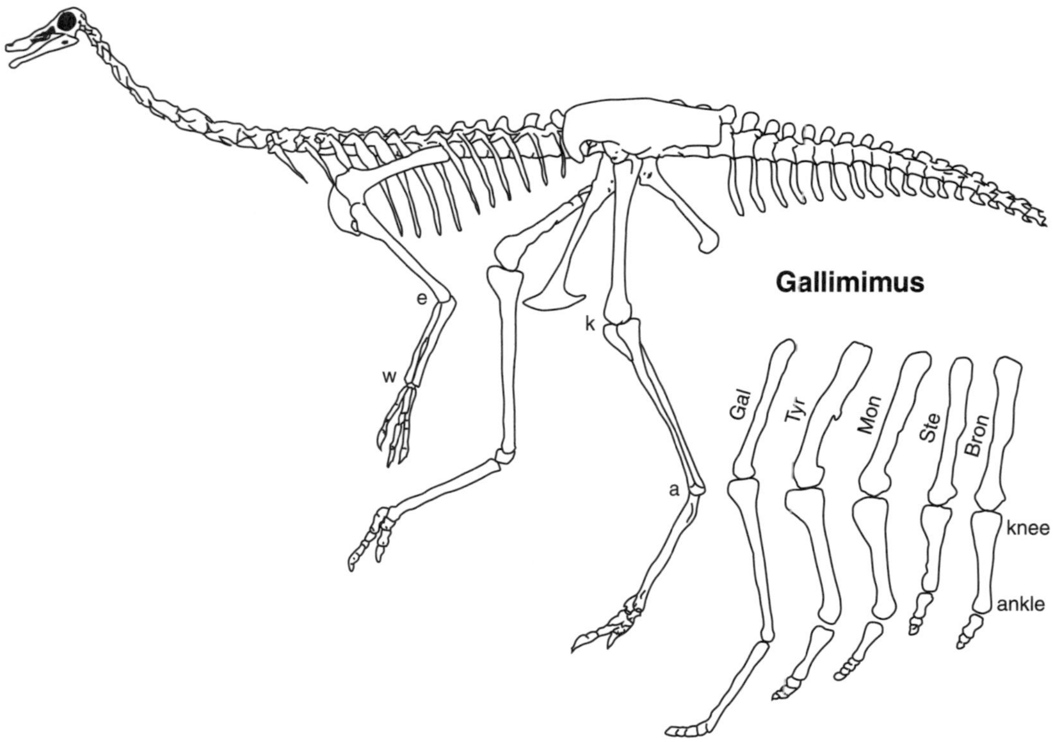


**FIGURE 16.3** Leg bones from various dinosaurs.

 

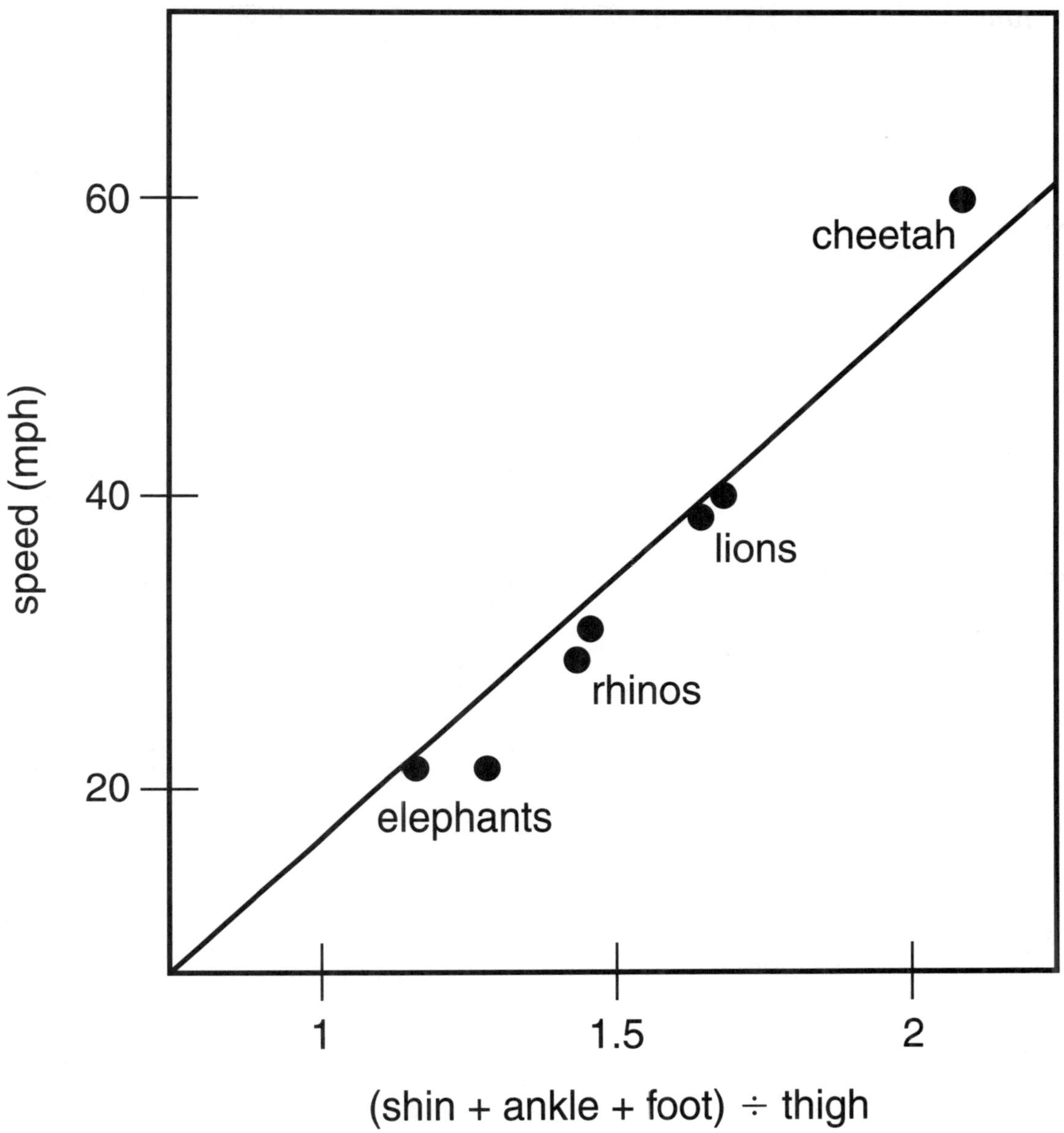


**FIGURE 16.4** Determination of sprint speed as a function of leg-bone measurements

3. Measure and record in your data table the length of the thigh bone for each dinosaur.
4. For each dinosaur, divide the sum of the shin + ankle + foot by the value of the thigh bone length. Record these results in the appropriate column of your data table.
5. Use the graph in Figure 16.4 to estimate the sprint speed for each dinosaur, and record the values in the last column of your data table.

**Analysis**

1. Which animal had the greatest walking speed? the slowest?
2. Is there a relationship between an animal's walking speed and its size?
3. Suggest some other studies that use this approach.
4. Which dinosaur won the race? Did you successfully predict the winner?
5. Describe a relationship between sprint speed and a dinosaur's lifestyle.
6. What assumption is being made in estimating dinosaur sprint speeds?

# Extended Laboratory Program 17

## Chapter 22 Behavior

*Born to It*

### Teacher's Notes

This investigation introduces the concept of biological taxes. The first objective is to have the students recognize and describe innate behaviors and to distinguish them from learned behaviors. Another objective is to classify taxis behaviors by name and to describe them as either positive or negative. Finally, students will learn to think of innate behaviors in terms of their adaptive value to the organism.

### Time

One class period is sufficient for this investigation.

### Materials (per class of 30—teams of 3)

30 lab aprons
30 pairs safety goggles
30 pairs plastic gloves

**Part A**

10 mealworms
10 sheets white paper
20 wooden strips, approximately 1 × 2 × 25 cm
10 protractors
10 rulers, 30 cm

**Part B**

10 mealworms
10 wooden boards, approximately 30 × 30 cm
fine-grade sandpaper, sufficient to cover boards
glue or thumbtacks for attaching the sandpaper
30 to 40 wooden blocks
10 protractors
10 stopwatches or watches with second hands

### Safety

These activities involve no hazards to humans. However, students should be cautioned about appropriate and gentle handling of the insects.

### Preparations

Mealworms can be maintained for at least a year in an opaque container, such as a plastic refrigerator container painted black. Punch 30–40 holes in the lid with a small nail or heavy sewing needle. Fill the container with dry, uncooked Cream of Wheat, leaving a 2.5-cm space at the top. Fold paper towels to provide four loosely stacked layers, separated by air spaces, on top of the cereal. Place the larvae in the container, and close the cover. Sprinkle a few drops of water on the paper towels each week (not enough to wet the cereal), and replace the paper as it is chewed up. Replenish the cereal at intervals of 2–3 months. A slice of apple or potato also can serve as a source of moisture. Take larvae as needed from the paper; avoid digging up the cereal.

Dim the lights in the classroom during the mealworm activities. Mealworms are negatively phototactic, and their reaction to light may interfere with the effects that you wish to observe. Explain clearly to the students how the distance and angle measurements are to be made.

To make the gravity boards for Part B, cover a piece of wood (about 30 × 30 cm) with fine sandpaper. Use glue or thumbtacks to attach the sandpaper to the boards. Draw lines across each board about 2.5 cm from each of two opposite ends. The blocks used to prop up the gravity boards can be cut at various angles or heights, or they can be stacked to various heights.

The activity can be extended by pitting taxic responses against each other. For example, students can investigate what mealworms will do when light is directed up the slope of the gravity board. Phototaxis can be made quantitative by timing the progress of the worms across a horizontal gravity board in response to various light intensities.

### Analysis

1. Answers will vary. If the worm has equal contact on both sides, it will continue to move straight ahead.
2. Answers will vary. The worm may turn to one side if that side is darker or lower.
3. Mealworms are positively thigmotactic.
4. Positive thigmotaxis leads mealworms to enter small crevices where they are safe from predators and where they may find bits of food such as seeds, flour, or plant debris that have accumulated in these narrow places.
5. Mealworms descend a steeper slope more quickly.
6. Mealworms are positively geotactic—they move downward.
7. Crawling downward brings mealworms into grass or soil, where they are less visible to predators and where they may find seeds and other food that has fallen to the ground. Also, if they happen to find a stored food area (box of cereal or squirrel's hoard), they will burrow down into the food.

## ◆ Extended Laboratory Program 17

# Chapter 22 Behavior

*Born to It*

A newborn baby knows to cry when it needs something. Crying, which is not learned, is an example of innate, or instinctive, behavior. Is your behavior mostly innate or learned? The behavior of simpler animals is mostly innate. Large intelligent animals, such as vertebrates, can learn many new behaviors.

In this activity, you will investigate the innate behavior of mealworms (beetle larvae). These animals have been kept in a simple environment and have not been trained in any way; therefore, you may assume that the behavior you observe is innate.

You will observe movement in a certain direction in response to a stimulus. This movement is called a taxis. For example, moving toward or away from light is called phototaxis. Movement in response to gravity is geotaxis. Movement in response to touch is thigmotaxis. Taxis can be positive or negative. Movement toward light is positive phototaxis. Upward movement (against gravity) is negative geotaxis.

**Materials (per team of 3)**

3 lab aprons
3 pairs safety goggles
3 pairs plastic gloves

**Part A**

1 mealworm
1 sheet of white paper
2 wooden strips, approximately 1 × 2 × 25 cm
1 protractor
1 metric ruler, 30 cm

**Part B**

1 mealworm
1 wooden board, approximately 30 × 30 cm
fine-grade sandpaper (sufficient to cover board)
glue or thumbtacks
wood blocks
protractor
stopwatch or watch with a second hand

**SAFETY** Put on your safety goggles, lab apron, and gloves.

### PROCEDURE

**Part A Response to Touch**

1. Place the mealworm on the piece of white paper.
2. Place a strip of wood in the worm's path so that the worm is facing it head-on.
3. How does the worm change direction when it reaches the wood? Does it turn around and go back, continue to move along the edge of the wood, or just stop? Record your observations in your logbook.
4. When the worm reaches the end of the wood, does it turn toward the wood, turn away from it, or continue moving in a straight line? Record your observations in your logbook.
5. Which way do you think the worm will turn after crawling between two strips of wood so that it is touched on each side? Record your prediction in your logbook.
6. Carefully place two strips of wood parallel and close together on the paper so that the worm crawls between them, touching them both. Align the ends of the strips so that the worm will reach the end of both at the same time. Do not squeeze the worm between the strips.
7. Answer Analysis questions 1 and 2.
8. Set up a data table like Table 17.1 in your logbook.
9. Repeat the procedure you used in step 2, but this time lay the ruler on top of the wooden strip so that you can measure the distance the worm must crawl along the strip before reaching the end. Before the worm reaches the end of the strip, trace the position of the end of the strip on the paper.
10. Mark on the paper to show the direction in which the worm travels after reaching the end of the strip. Use a protractor to measure the angle between the strip and the worm's path

TABLE 17.1 ANGLE AT WHICH MEALWORM TURNED AFTER FOLLOWING A STRAIGHT BARRIER

| LENGTH OF BARRIER (mm) | ANGLE OF TURN (DEGREES) |
|---|---|
| | |
| | |
| | |
| | |
| | |

after reaching the end of the strip. Record this angle and the distance traveled along the strip.

11. Repeat steps 9 and 10 four times. Each time, vary the distance from the end of the strip to the point at which the worm first encounters the strip.
12. Answer Analysis questions 3 and 4.

**Part B Response to Gravity**

13. Set up a data table like Table 17.2 in your logbook.
14. Lay a gravity board on your table. Place the mealworm at the center of one of the lines. Record the time in your logbook.
15. Record in your data table the time it takes the worm to crawl to the other line. If the worm does not reach the second line within 5 minutes, record this fact and continue with step 16.
16. Gently take the worm off the board. Place a block under one ruled end of the board. Measure the angle between the board and the table with a protractor. Record this angle in your data table.

TABLE 17.2 TIME REQUIRED FOR MEALWORM TO CRAWL DOWN BOARDS OF VARYING STEEPNESS

| SLOPE OF BOARD (DEGREES) | TIME (MINUTES:SECONDS) |
|---|---|
| | |
| | |
| | |
| | |
| | |

17. Place the worm on the line at the high end of the board, and record the time. In your data table, record the time it takes the worm to reach the lower line. If the worm has not reached the line in 5 minutes, record this fact and continue with step 18.
18. Change the slope, and repeat the measurement. Your teacher may ask you to repeat the measurement at each slope or to combine your data with another team's results to find an average time for each angle.

**Analysis**

1. Which direction did the worm take after reaching the end of the "alley" between the two strips? Was your prediction correct?
2. What are some things that might affect how sharply the worm turns after being stimulated by touch?
3. Do your results support or oppose the idea that mealworms are thigmotactic? Would you describe them as positively or negatively thigmotactic?
4. Explain your answer to question 3 on the basis of how mealworms live. How might this behavior help them to survive?
5. How did the mealworms respond to the increasing slope of the gravity board?
6. Would it be correct to call the mealworms positively or negatively geotactic?
7. What might be the adaptive value of this behavior?

## ◆ Extended Laboratory Program 18

# Chapter 23 Immune Systems

*Testing Koch's Postulates*

**Teacher's Notes**

This experiment uses the infection of apples by the fungus *Rhizopus stolonifera* to illustrate the concept of disease causation and specifically Koch's postulates. Cultures of *Rhizopus stolonifera* are available from Ward's Natural Science Establishment (catalog #85 W 6070). Many other fungi (including *Monilinia fructicola*, *Pencillium expansum*, *Nectria galigena*, and *Botrytis cinerea*) can be substituted for *Rhizopus stolonifera* in this experiment.

If you wish, the investigation may be expanded to include additional fungi and fruits. Other possible topics for study include how the fungus gets its food, what conditions promote infection, and whether the fungus secretes enzymes.

**Time**

Allow one class period plus time for making daily observations; also allow about 20 minutes two weeks later for final observations and analysis of results.

**Materials (per class of 30—teams of 3)**

30 lab aprons
30 pairs safety goggles
30 pairs plastic gloves
20 healthy apples
10 wounded and diseased apples
10 wounded apples inoculated with sterile distilled water
10 petri dishes containing a *Rhizopus stolonifera* culture growing on malt agar
40 Sabouraud dextrose agar plates
sterile distilled water
40 sterile transfer pipets
10 glass-marking pencils
10 lab burners
10 scalpels
10 pairs forceps
10 rolls tape
10 compound microscopes
20 microscope slides
20 coverslips

**Safety**

Students having allergies to mold should wear a mask or not participate in this laboratory.

Caution students to be careful using a lab burner. A clean flame burns a blue color and can be difficult to see. Be sure that students with long hair tie it back. Loose clothing must be kept away from the flame. Remind students of the location of the fire extinguisher.

The refrigerator used to store fungal cultures and agar plates must be reserved for biological work and labeled as such; no food or drink should ever be stored in that refrigerator.

**Preparations**

The *Rhizopus stolonifera* cultures should be seven days old when the experiment begins. The diseased and mock-infected apples should also be prepared a week in advance. Use sterile forceps or a probe to make a small hole in the side of the apple. Use sterile forceps to place a small quantity of fungus in the wound. The mock infection simply has a drop of sterile water placed in the wound. For an alternative control, do not wound the apple but inoculate it with fungus. Place a piece of tape over the wound site to keep the inoculum in place. After a few days, when the infection is under way, the tape may be removed. Keep the apples and inoculated agar plates at room temperature.

**Analysis**

1. The infected apple should show obvious signs of distress (discoloration, texture, and odor) emanating from the site of inoculation. The control apple may show a small lesion at the wound site but will appear much healthier than the infected apple.
2. Students may conclude that the fungus causes the disease, though that is a premature conclusion. It is possible that the symptoms displayed by the infected apple are due to the presence of a secondary pathogen that was able to infect the weakened fruit.
3. The control apple is uninfected because it was not exposed to fungus. Although it was wounded, the procedure was performed sterilely and there was no opportunity for fungus to enter the wound site.
4. The experiment provides strong evidence that the fungus causes the disease. Each infected apple contained the fungus, but each control apple did not. Furthermore, when fresh apples were inoculated with the fungus, they got the disease, whereas the mock-infected apples remained healthy.
5. The data support Koch's postulates. First, the fungus was always present with the disease. Second, the fungus was isolated and cultured in the lab. Third, the fungus caused the disease when inoculated into healthy apples. Fourth, the fungus was reisolated from the sick apple.

# ◆ Extended Laboratory Program 18

## Chapter 23 Immune Systems

### *Testing Koch's Postulates*

Robert Koch, who discovered the cause of tuberculosis, set forth criteria that are used to establish whether or not an organism causes a specific disease. First, the suspected pathogen (disease-causing organism) must be present in every case of the disease. Second, the pathogen must be isolated and cultured in the laboratory. Third, the pathogen must cause the disease when injected into animals. Fourth, the pathogen must be reisolated from the sick lab animal.

In this experiment, you will examine apples that have been infected by a fungus. Often plants that have been weakened by disease will become infected with other secondary organisms. This can complicate the isolation of the original pathogen. You will use Koch's postulates to determine whether the fungus *Rhizopus stolonifera* causes the apple disease.

**Materials (per team of 3)**

3 lab aprons
3 pairs safety goggles
3 pairs plastic gloves
2 healthy apples
1 wounded and diseased apple
1 wounded apple inoculated with sterile distilled water
1 petri dish containing a *Rhizopus stolonifera* culture growing on malt agar
4 fresh Sabouraud dextrose agar plates
sterile distilled water
4 sterile transfer pipets
1 glass–marking pencil
1 lab burner
1 scalpel
1 pair of forceps
1 roll of tape
1 compound microscope
2 microscope slides
2 coverslips

**SAFETY** Put on your safety goggles, lab apron, and gloves; tie back long hair.

**PROCEDURE**

1. Examine each apple carefully, and record your observations in your logbook.
2. Use a scalpel to cut the infected and control apples in half through the inoculation point, and answer Analysis questions 1–3.
   CAUTION: Scalpels are sharp; handle carefully.
3. Use sterile technique to culture the fungus.
   **a.** Briefly heat the scalpel blade in the burner flame.
   CAUTION: Keep loose clothing and long hair away from flame.
   **b.** Let the blade cool.
   **c.** Cut out a small piece of infected tissue.
   **d.** Use heated and cooled forceps to place the tissue on the surface of a Sabouraud dextrose agar plate.
   **e.** Label the plate with sample description, team identification, and the date.
4. Follow the procedure in step 3 to attempt to culture fungus from the control apple.
5. Use sterile technique to wound and inoculate a fresh apple.
   **a.** Use sterile forceps to wound the apple.
   **b.** Using the sterile forceps, transfer a small quantity of the fungus growing on the Sabouraud dextrose agar plate into the wound.
6. Place a piece of tape over the wound area to prevent the inoculum from falling out.
   **a.** Use sterile technique to wound and perform a control inoculation on the remaining apple.
   **b.** Use sterile forceps to wound the apple.
   **c.** Use a sterile transfer pipet to insert one drop of sterile distilled water into the wound.
7. Place tape over the wound site, as in step 6.
8. Make daily observations of the inoculated apples for one week, and record them in your logbook.
9. Compare the observations of your inoculated apples with those from the starting material.
10. Repeat the process described in step 3 to culture fungus from your inoculated apples.
11. Use a microscope to examine a sample of the original fungus and a sample of what you cultured from the infected apple.
    **a.** Label one end of a microscope slide *O* for original fungus, and label the end of a second slide *R* for reisolated fungus.

**b.** Use sterile forceps to add a small quantity of the original fungus to the microscope slide labeled *O*.

**c.** Use a transfer pipet to add one drop of water to the fungus, and place a coverslip on top of the sample.

**d.** Repeat steps a–c to prepare a slide containing the reisolated fungus.

**e.** View each slide under the high-power objective.

**12.** Note any similarities and differences between the two fungal samples. Record your observations in your logbook.

**Analysis**

1. What are your observations of the infected and control apples?
2. What do you conclude from your observations?
3. Why is the control apple uninfected?
4. Does the fungus cause the disease? Explain.
5. Does the experiment satisfy Koch's postulates? Explain.

## ◆ Extended Laboratory Program 19

# Chapter 24 Ecosystem Structure and Function

*Eutrophication*

**Teacher's Notes**

This investigation gives students an opportunity to observe the effects of fertilizer on eutrophication. You can increase the complexity (and the level of student interest) by adding more variables to the design of the experiment. Include different temperatures, different pH levels, different amounts of fertilizer (varying the amount of phosphate), and low- and high-level phosphate detergents. Assign a different variable to two separate teams of students, or assign multiple sets of test tubes to each team, depending on the amount of equipment and space you have to run the investigation. You will need to place the test tubes in a window that receives indirect sunlight or place them under a grow light while waiting for the cultures to reproduce.

**Time**

Allow one class period for teams to set up their experiments and take the initial readings. For the next 1 to 2 weeks, allow a few minutes at the beginning of each class period for teams to take readings of their test tubes.

**Materials (per class of 30—teams of 2)**

30 lab aprons
30 pairs safety goggles
30 18-mm × 150-mm test tubes
dropping pipet
4 Trekker-T-4™ units
4 turbidity probes or Insta-Rate™ density-indicator strips
15 test-tube racks
8 glass-marking pencils
4 L aged tap water
4 L aged fertilizer water
euglena culture
density indicator strips (optional)

**Preparations**

Review the Trekker-T-4™ instructions to familiarize yourself with the operation of the unit and to connect the turbidity probe. Make sure students know how to use the unit to collect data. Verify that they understand the proper use and care of the probes.

When using the turbidity probe, make certain the cap is closed firmly and the photoresistor is inserted snugly so that no stray light enters the sleeve to interfere with the turbidity readings. It may be necessary to shield the top with black cardboard.

Purchase a small container of fertilizer, or ask students to bring some from home. Try to use a brand in which the phosphate level (the second number in the formula) is higher than the nitrate level (the first number) and potash level (the third number). A 16-24-16 formula worked well in the field test. Age the water for 24 hours to remove any chlorine, and then add a rounded tablespoon of fertilizer to one of the containers. Clearly label this container *Fertilizer Water.* Label the second container *Tap Water.*

The amount of time needed to achieve good results depends on the strength of the fertilizer, the intensity of the light, the temperature, and the organisms used. Euglena cultures produce a good bloom in 1 to 2 weeks. Cyanobacteria (such as *Oscillatoria*, *Anabaena*, or *Cyanophora*) may take 2 to 3 weeks. Lower temperatures and light levels will lengthen the time required to get results. If you use euglena cultures, students should store their test tubes in the light but not in direct sunlight.

If the Trekker-T-4™ is not available, Insta-Rate™ density-indicator strips may be used as an alternative means of quantitatively evaluating cell growth in this experiment. The Insta-Rate™ density-indicator strip consists of an adhesive panel that is affixed to the culture tube. The adhesive panel contains five shaded bars, which are used to observe the increasing density of the growing culture. A growth curve (bar number versus time) can be established from these observations. The panel of bars provides only a subjective determination of cell density, but its use in this experiment is adequate in quantifying relative cell-density differences among various environmental treatments. If density-indicator strips are used, have students begin with Analysis question 4.

Using density strips:

1. Affix each panel to culture tubes, with the darkest bar (5) at the top. Label each bar (0–5).
2. Make observations by looking through the culture tube at the panel of bars. (The panel bars should be away from the observer.) Use a white index card as a background when making observations.
3. Record the cell density as the number of the bar that can no longer be observed clearly. Be sure to use the same criteria for each observation.
4. The higher the number of the bar that is obscured, the higher the cell density. Thus, if bar 5 is obscured by test tube F and bar 3 by tube T after an equal elapsed time, the cell density in tube F is higher than that in tube T. A growth curve can be established by plotting bar number obscured against time.

**Analysis**

1. The purpose of this turbidity reading was to obtain baseline readings before the experiment started.
2. The purpose of this turbidity reading was to obtain baseline readings after the euglena were added and before the experiment started.
3. Euglena are growing more rapidly in the fertilizer water.
4. Students should describe a green or dark-green color for the fertilizer water.
5. Test tube T is a control.

6. Fertilizer enhances the growth and reproduction of euglena.
7. Phosphates allow some euglena to reproduce rapidly, filling the water with the organisms. Students may infer that the large numbers of euglena use up all available oxygen when photosynthesis is not occurring. This is what kills fish.
8. Fertilizers used on golf courses to keep the grass green are washed into ponds by runoff from sprinkler water, which drains into the ponds and enhances eutrophication.
9. Farm ponds can be subject to eutrophication if fertilizers are washed from the soils and drain into the ponds.

## ◆ Extended Laboratory Program 19

# Chapter 24 Ecosystem Structure and Function

*Eutrophication*

Many ponds and lakes suddenly turn green at various times of the year. Fish die, and the water smells bad. This process, known as eutrophication, occurs frequently around areas where fertilizers and/or detergents enter streams that drain into the ponds or lakes. Are the fertilizers and/or detergents responsible for this "greening" effect? This investigation will give you an opportunity to find out.

**Materials (per team of 2)**

2 lab aprons
2 pairs safety goggles
2 18-mm × 150-mm test tubes
dropping pipet
Insta-Rate™ density strips
test-tube rack
glass-marking pencil
aged tap water
aged fertilizer water
euglena culture

**SAFETY Put on your safety goggles and lab apron.**

**PROCEDURE**

1. Record all turbidity readings in your logbook.
2. Fill one test tube about three-fourths full of aged tap water. Label this test tube *T*, and place it in a test-tube rack.
3. Fill a second test tube about three-fourths full of aged fertilizer water. Label this test tube *F*, and place it in the test-tube rack.

CAUTION: The fertilizer water may contain chemicals that are irritating to the skin. Flush spills or splashes with water.

4. Use density strips to take readings of both test tubes. Record both readings in your logbook.
5. Thoroughly stir the euglena culture, and transfer one full dropping pipet of culture into each of the two labeled test tubes.
6. Repeat step 4, and record the readings in your logbook.
7. Label both test tubes with your team symbol, and store them in the location specified by your teacher.
8. Repeat step 4 every school day for 1 or 2 weeks or for the time specified by your teacher.
9. Wash your hands thoroughly before leaving the laboratory.

**Analysis**

1. What was the purpose of the turbidity reading in each test tube before you added the euglena culture?
2. What was the purpose of the turbidity reading in each test tube immediately after you added the euglena culture?
3. Account for any differences in turbidity readings between test tubes T and F.
4. Describe any visual differences between the two test tubes.
5. What is the purpose of the test tube labeled *T*?
6. What is the effect of fertilizer on the growth of euglena?
7. Describe what happens in eutrophication.
8. Why are golf-course ponds especially susceptible to eutrophication?
9. Would farm ponds be susceptible to eutrophication? Explain.

## ◆ Extended Laboratory Program 20

# Chapter 25 Change in Ecosystems

*Acid Rain in Local Environments*

**Teacher's Notes**

This investigation gives students the opportunity to investigate local environmental situations under your guidance. Only basic procedures are provided; students should be able to design and carry out further experiments on their own. Teams of four work well with the Trekker-T-4™ data logger, but each student may want to bring his or her own samples to test. As an alternative, students can obtain data by using pH paper strips.

**Time**

Once the samples have been collected, one or two class periods are needed to gather data depending on how many samples are to be tested. Another period may be required to discuss class data and the Analysis questions.

**Materials (per class of 30—teams of 4)**

30 lab aprons
30 pairs safety goggles
16 100-mL beakers
16 glass or plastic funnels
8 100-mL graduated cylinders
Trekker-T-4™ units (2 teams per system)
8 pH probes or pH paper strips
box of filter paper
4 L acidic distilled water: pH 4
4 L boiled distilled water: pH 7
4 containers of distilled water
individual student samples of:
- rainwater
- soil
- lake or pond water
- snow (if possible)

**Preparations**

If you are using the Trekker-T-4™, review the instructions to familiarize yourself with the operation of the unit and to connect the probes. Make sure students know how to use the unit to collect data. Also make sure that they understand the proper use and care of the probes.

Obtain permission to go onto property where water samples will be taken. Be alert for rain during the school year, and collect samples of the water. If you live in a cold climate, freeze some snow during the winter to use for this investigation. Students can bring their own water and soil samples to class.

Boil the distilled water to drive off the carbon dioxide and achieve a pH of approximately 7. To prepare acidic distilled water, gradually add white vinegar to distilled water. Test the pH frequently until a pH of 4 is reached. Students may wish to test the effect of water with other pH values on their soil samples. These tests can be prepared in the same manner. Students may use acids or bases other than white vinegar to adjust pH only if such solutions are one molar or less. Store in tightly stoppered, labeled bottles.

Collect the saved data for comparisons in later years. Eventually you can develop histograms of the data showing changes in the environment.

The investigation can be performed by using wide-range pH paper, but pH probes provide more accurate data.

**Analysis**

1. Many areas on the West and East Coasts, parts of the Rocky Mountains, the Southeast, and the region around Lake Superior are most susceptible.
2. The answer depends on your location.
3. Greater production of pollutants responsible for acid rain, prevailing winds, and ability to buffer acid rain cause one area to be more susceptible than another.
4. Comparisons depend on student data. Traffic flow, proximity to industries, and use of wood-burning stoves are a few of the factors that could cause local differences in pH.
5. One factor is a high degree of sulfur and nitrogen oxides in the atmosphere. These pollutants may be produced locally or may be blown in from high-producing areas in neighboring states. Another factor is the lack of buffering capacity in the environment.
6. Comparisons depend on student data. Different pH of rainwater and different buffering action of soil samples could cause differences in pH.
7. If the pH of rainwater is low, it could increase the acidity of the soil. However, if the soil is a good buffer, this would not necessarily happen. Conversely, if the pH of rainwater is high, it should not increase the acidity of the soil. However, the pH of soil is affected to a large degree by the types of plants that have grown or are growing in it. Soil under pine trees, for example, will be more acidic than soil beneath deciduous trees.
8. Answers depend on soil samples. If the pH increased (that is, if the acid solution became less acidic), the soil does have buffering capacity. If the pH remained the same or decreased, the soil does not have buffering capacity.
9. The presence of neutralizing agents gives soil buffering capacity. Lime and limestone contain calcium carbonate, a potent buffer.
10. If the rainwater has a low pH, the soil needs a good buffering capacity to remain healthy.
11. Comparisons depend on water samples. Differences may be due to the surroundings of the lake or pond.
12. Buffering agents may be affecting the rainwater before it becomes part of the pond or lake water.
13. The presence (or absence) of neutralizing agents in the soil, gravel, or bedrock surrounding the body of water will affect its pH.
14. Answers depend on student samples. Snow meltwater would be less acidic than rainwater if the

snow fell during a period when the atmosphere contained few pollutants. The usual case, especially if the snow being tested is old snow, is that the pH of snow is lower than the pH of rainwater.

**15.** All winter the pollutant loads from snowstorms accumulate in the snowpack. The acids concentrate on the surface of the snow. The first mild weather melts this surface, releasing the accumulated acids into ponds, lakes, and soil. (This first meltwater can be 5–10 times more acidic than the remaining snowpack. Because of this, severe fish kills usually occur in early spring.)

## ◆ Extended Laboratory Program 20

# Chapter 25 Change in Ecosystems

*Acid Rain in Local Environments**

The degree to which acid rain is destructive differs in the United States from region to region. Many factors influence the probability and the potential damage of acid rain. Most experts agree that the problem begins with the burning of coal, oil, and natural gas. Smokestacks as tall as 300 meters allow winds to carry pollutants far from local sources. In addition, the capacity of the soil or bedrock to buffer the effects of the acid rain varies. Limestone bedrock neutralizes the acid very well, whereas granite bedrock has no neutralizing effect at all.

You can study the extent of acid rain in your own area by starting with a map of the United States that displays several areas of greatest vulnerability. After determining the local degree of vulnerability, you can perform actual tests of your surroundings.

**Materials (per team of 4)**

- 4 lab aprons
- 4 pairs safety goggles
- 2 100-mL beakers
- 100-mL graduated cylinder
- glass or plastic funnel
- pH probe or pH paper strips
- 8 pieces filter paper
- 400 mL acidic distilled water: pH 4
- 400 mL boiled distilled water: pH 7
- container of distilled water
- individual student samples of:
  - rainwater
  - soil
  - lake or pond water
  - snow (if possible)

**SAFETY** Put on your safety goggles and lab apron.

**PROCEDURE**

1. Examine Figure 20.1, which shows the areas of the United States with the highest susceptibility to acid rain.

* This investigation was adapted from E. Johnson and R. Bybee, supervisors, "Acid Rain: Activities for Science Teachers," *The American Biology Teacher,* 45:4, April/May 1983. © National Association of Biology Teachers, Reston, Virginia. Reprinted by permission.

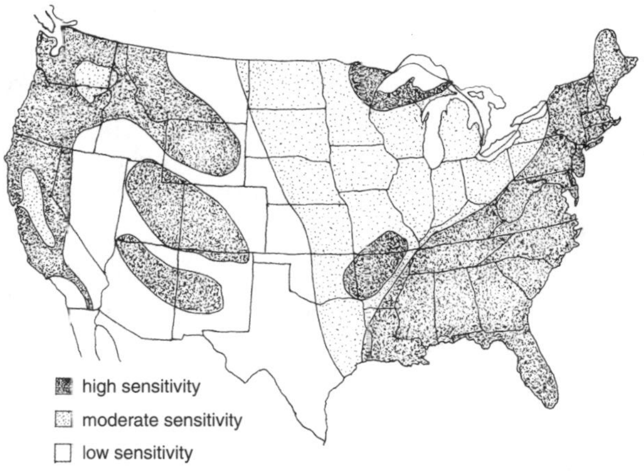


**FIGURE 20.1** Areas of the United States with high sensitivity to acid rain

2. Answer Analysis questions 1–3.
3. To measure pH during the experiment, remove the pH probe from the container of distilled water and place it in the beaker of solution being tested. If you are not using an electronic data collection system, use a pH strip. Record all pH readings in your logbook.
4. Determine and record the pH of your rainwater sample.
5. Place filter paper in a funnel, and fill the funnel with soil from your sample.
6. Hold the end of the funnel over a small beaker, and pour 100 mL of boiled distilled water (pH 7) through the soil. Dirty water should collect in the beaker.
7. Determine and record the pH of the collected water.
8. Repeat steps 5–7 with a fresh soil sample, but this time pour 100 mL of acidic distilled water (pH 4) through the sample.
9. Determine and record the pH of your pond- or lake-water sample.
10. Melt your snow in a small beaker.
11. Determine and record the pH of the meltwater.
12. If using a pH probe, place it in the container of distilled water.
13. Answer Analysis questions 4–5.
14. Wash your hands thoroughly before leaving the laboratory.

**Analysis**

1. According to the map, which parts of the country are most susceptible to acid rain?

2. According to the map, is acid rain currently destructive to your area?
3. What causes one area of the country to be more susceptible to acid rain than another area?
4. Compare the pH of your rainwater sample with that of normal rainwater (pH 5.6) and with samples from classmates who live in different neighborhoods. What are the differences in pH? What could have caused the differences?
5. What factors might cause a rainwater sample to be highly acidic?
6. How does the pH of your soil sample compare with those of your classmates? Give two reasons why the pH of soil samples might differ.
7. Consider the class data on the pH of the rainwater in your area. In what way, if any, is this rainwater affecting the acidity of your soil?
8. What is the pH of the acidic distilled water after it has passed through the soil and collected in the beaker? On the basis of these data, discuss the buffering capacity of your soil.
9. What might cause a soil to have a buffering capacity?
10. Given the pH of your rainwater, could your soil be healthy without a buffering capacity?
11. Compare the pH of your pond- or lake-water sample with the pH of healthy pond or lake water (pH around 6.5) and with samples that classmates obtained from different lakes or ponds. What are the differences in pH? What could have caused the differences?
12. If the pond or lake water is not too acidic and the rainwater is known to be acidic, what may be some factors influencing the acidity of the pond or lake water?
13. Consider the characteristics of the area around the pond or lake (for example, forest, farmland, bedrock). How may these factors influence the acidity of pond or lake water?
14. Compare the pH of your meltwater with that of your rainwater sample. Why might snow meltwater be more or less acidic than rainwater?
15. From your data, what time of year would you expect your lakes and soil to be most acidic?

## ◆ Extended Laboratory Program 21

*A Laboratory Practical Examination: Yeast Fermentation*

### Teacher's Notes

The following example of a laboratory practicum contains elements common to most BSCS investigations. A periodic check of these investigative inquiry skills will help your students place the laboratory activity in proper perspective. You may find that students who do not do well on paper-and-pencil examinations can demonstrate logical insight to solve problems in the laboratory. Students can work individually, or you may prefer to use small-group cooperative learning for the practicum.

### Time

You may need two class periods to complete the practicum.

### Materials (per setup)

lab apron
pair safety goggles
2 250-mL beakers
2 18-mm × 150-mm test tubes
graph paper
2 containers fresh yeast suspensions

### Safety

Students should submit a written procedure for your approval before they begin work.

The procedure should include information as described in the student procedure, identifying any hazards involved and the precautions that need to be taken to avoid harm.

### Preparations

Prepare two yeast suspensions. Use the same amount of yeast but different amounts of sugar. Follow the directions in the Preparation of Solutions section in the Teacher's Annotated Edition. You may wish to cut the graph paper to fit in the test tubes.

### Evaluations

1. This experiment is simple, but it requires technical skill in inverting the test tube so that no liquid is lost. Most students should be able to find the method of measurement by themselves.
2. Should they need assistance, explain the following procedure: First, fill the test tube with yeast mixture until it overflows. Second, invert the beaker over the test tube and turn the entire setup upside down. Be careful not to lose any of the solution, and make sure that no air enters the test tube. Third, measure the accumulation of $CO_2$ by examining the graph paper inside the test tube.
3. Consider the student's ability to design the experiment, the student's technical skill, and the quality of the graphs.
4. Because the amount of yeast and the concentration in the two sets is equal, students should conclude that the cause of the difference lies in the nature of the substrate.
5. Two ways to slow down fermentation are to place the system in crushed ice and to dilute the yeast suspension.
6. The student is not required to perform the identification of $CO_2$ but only to explain how the test is conducted.
7. Conclusions relate to the differences observed in the fermentation rates.
8. Because fermentation is discussed in Chapter 5, a relatively thorough knowledge and understanding of the process and related knowledge can be expected.

## ◆ Extended Laboratory Program 21

*A Laboratory Practical Examination: Yeast Fermentation*

This laboratory practicum will give you an opportunity to demonstrate your skills at designing an experiment and interpreting data. Follow the procedure step by step, and answer the questions.

**Materials**

lab apron
pair safety goggles
2 250-mL beakers
2 18-mm × 150-mm test tubes
graph paper
2 containers with fresh yeast suspensions

**PROCEDURE**

1. Examine the equipment and materials. Design an experiment that will allow you to compare the rates of fermentation between suspension A and suspension B. Write your plan out, and submit it to your teacher. Be sure your plan includes identification of the hazards involved and descriptions of the precautionary measures you will take. (See the safety information that follows.)
2. After your teacher approves the design, perform the experiment and write a report on the results.
3. Make a graph of the rate of fermentation over a period of 20 minutes for each of the yeast suspensions. What can you interpret from the graphs? The yeast concentration in A equals that in B. How would you explain the difference in fermentation rates?
4. Choose one of the yeast suspensions. Design two additional, different experiments to slow down the rate of fermentation. Write a hypothesis for each experiment.
5. Perform the experiments and record the results.

**Analysis**

1. What served as the control in these experiments?
2. What is the role of fermentation in the life of the yeast?
3. What is the gas created during this process? How can you prove this?
4. Write the conclusions you reached from your experiment.

**Safety**

Make sure you understand the hazards and precautions described on the labels and the Material Safety Data Sheets for all the chemicals you plan to use in your investigation. First, review your procedure to be sure you have included all the necessary precautions you will follow in your use of these chemicals. Second, review your procedure to be sure you plan to use the necessary precautions to prevent accidents that may be caused by the equipment you will use. Third, review your procedure to be sure your plans include the precautions you will take to control any biological hazards involved in your work. Always ask your teacher to explain any hazards that you do not understand and to help you plan the precautions you will take to prevent harm from an accident. Report all accidents, no matter how seemingly insignificant, to your teacher.

# Current Literature

# CURRENT LITERATURE

# Genetic Engineers Aim to Soup Up Crop Photosynthesis

*by Charles C. Mann*

*To improve crops' ability to turn atmospheric carbon into food, researchers hope to alter the principal enzyme or supercharge it with $CO_2$.*

Few nonbiologists may have heard of ribulose-1,5-bisphosphate carboxylase-oxygenase, the enzyme known as RuBisCO, but its importance is hard to overstate. The principal catalyst for photosynthesis, it is the basic means by which living creatures acquire the carbon necessary for life. By interacting with atmospheric carbon dioxide, RuBisCO—the world's most abundant protein—initiates the chain of biochemical reactions that creates the carbohydrates, proteins, and fats that sustain plants and other living things, ourselves included. But the enzyme also has another distinction, according to T. John Andrews, a plant physiologist at The Australian National University in Canberra: "RuBisCO is nearly the world's worst, most incompetent enzyme—it's almost certainly the most inefficient enzyme in primary metabolism that there is."

RuBisCO's ineffectiveness has been a spur to scientists since it became fully apparent in the 1970s. Indeed, the quest for a better RuBisCO is "a Holy Grail in plant biology," says George Lorimer, a biochemist at the University of Maryland, College Park, who worked with the Swedish team that mapped the enzyme's structure in 1984. "Everyone always goes in with the hope of changing the face of agriculture." Despite more than 20 years of effort, the hopes have not yet paid off. But recent advances in molecular biology—and the unexpected discovery of more efficient RuBisCO in red algae—have given new impetus to the long struggle to modify the enzyme. In what may be the most ambitious genetic-engineering project ever tried, laboratories across the world are trying to improve the RuBisCO in food crops by either replacing the existing enzyme with the red algae form or bolting on what could be thought of as molecular superchargers.

No one expects quick results—"I'm not for a second trying to minimize the task," says Andrews. The current state of the art in genetic engineering permits altering or splicing in single genes to improve a plant's resistance to pests, say, or to allow a crop to survive applications of weed-killing herbicide. To alter RuBisCO, by contrast, scientists must work with a 16-part molecule that is encoded by many genes in both the cell nucleus and the chloroplasts, where photosynthesis takes place. The enzyme also depends on a supporting cast of other enzymes, some of which will probably need to be revamped if RuBisCO is changed. But with many other avenues toward increasing crop yields seemingly blocked, RuBisCO has become an increasingly tempting target.

RuBisCO is just one actor in photosynthesis, which is a complex symphony of photochemical and enzymatic reactions. During the first, "light" stage of photosynthesis, chlorophyll—a green pigment in the chloroplasts—absorbs enough energy from sunlight to split off electrons from water molecules simultaneously releasing oxygen gas and driving the production of adenosine triphosphate (ATP), which is used in the second, "dark" stage. This step begins when RuBisCO combines with carbon dioxide to produce 3-phosphoglycerate, or PGA. Powered by energy from the ATP, a series of reactions transforms PGA into a host of starches, sugars, and other organic compounds.

Photosynthesis as a whole is not particularly efficient: a crop plant that stores as much as 1% of the total received solar energy is exceptional. As a result, the process offers many targets for bioengineers. But RuBisCO, far and away the biggest drag on the process, is the most appealing of them. First, it is torpid in the extreme—"perhaps the slowest known enzyme," William Ogren, a now-retired RuBisCO researcher from the University of Illinois, Urbana-Champaign, says with only slight exaggeration. Enzymatic rates are often on the order of 25,000 reactions per second: RuBisCO turnover in higher plants can be as little as two or three reactions per second. "Not one of evolution's finest efforts," says Ogren.

Second, RuBisCO triggers an additional reaction that interferes with the first. In 1971 Ogren and two other researchers discovered to their amazement that besides capturing and "fixing" carbon dioxide, RuBisCO catalyzes a second, opposing reaction. In what is called photorespiration, the enzyme combines with oxygen, rather than carbon dioxide, to create a compound that is subsequently converted partly into carbon dioxide. In other words, RuBisCO catalyzes one reaction that incorporates carbon into plants and another that ultimately strips them of carbon.

Reprinted with permission from "Genetic Engineers Aim to Soup Up Crop Photosynthesis" by Charles C. Mann. *Science* 283, January 15, 1999, pp. 314–316.

Typically, the RuBisCO in higher plants like rice and wheat is 100 times more likely to pick up $CO_2$ than $O_2$. But because the concentration of atmospheric $O_2$ is many times greater than that of $CO_2$ the greater affinity for $CO_2$ is largely canceled. As a result, 20% to 50% of the carbon fixed by photosynthesis is lost to photorespiration. "The oxygenation reaction is—as far as we can tell, and a lot of research over decades has gone into it—just a complete waste," says Andrews. "It doesn't do anything for the plant."

This striking inefficiency was no handicap when photosynthesis first evolved 3 billion years ago, because the atmosphere was almost devoid of oxygen. After photosynthesis filled the air with oxygen and RuBisCO's weakness was revealed, it may have been too late for evolution to fix the problem, says Murray Badger, a RuBisCO specialist at The Australian National University. "It's a somewhat general correlation that the more specific and discriminatory a reaction becomes, the slower it gets," he says. As a result, mutations that made RuBisCO target carbon dioxide better might also have made it even slower.

If genetic engineers could find a way around RuBisCO's slowness and inefficiency, they might reap a double benefit. A faster, more efficient enzyme could help crops grow and increase their biomass, letting them produce more grain at a faster rate. In addition, explains Martin Parry of the Institute of Arable Crops Research Rothamsted in Hertfordshire, Britain, RuBisCO's lethargy means that "plants need to invest incredibly heavily in it" to fix sufficient carbon. "A very large proportion of the plant's nitrogen requirements come from the need to produce the enzyme," which makes up as much as half the soluble protein in plant leaves. More efficient RuBisCO could thus lower crops' need for nitrogen, now mainly supplied by fertilizer in many countries.

### A Better RuBisCO

The discovery of photorespiration launched the effort to remodel RuBisCO. Researchers began by comparing the form of the enzyme found in higher plants with that in cyanobacteria—blue-green algae—which is even less efficient than the higher plant version. To find out why, Lorimer's group, then based at DuPont, collaborated with Carl Branden's x-ray crystallography group in Sweden to determine the molecular structures of both forms, hoping to find telling differences. "We spent years creating high-resolution structures of spinach [a model plant in RuBisCo research] and cyanobacteria," says Lorimer. But despite the finely detailed results, "the sobering reality was that you can lay down [structural maps of] these two enzymes on top of each other and you're very hard pressed to see the difference." Even if the differences could be identified, Lorimer believes, they would be so numerous and subtle "that you could not rationally reason your way to what it was that you would need to improve the enzyme."

The failure of the structures to provide a path for modifying RuBisCO dismayed many researchers; Lorimer's group disbanded. Hopes reawakened in 1992, when F. Robert Tabita and B.R. Read of Ohio State University in Columbus discovered that some diatoms and red algae have more-specific RuBisCO than that in higher plants. In 1997, a team led by Akiho Yokota, a plant molecular physiologist at the Research Institute for Innovative Technology for the Earth, in the Keihanna Society near Osaka, Japan, found red algae with RuBisCO that is about three times more efficient. "We've looked at a lot of the red algae" Tabita says, "and the trend is always the same, 2 1/2 to threefold higher than normal plants." No one yet knows why.

To try to exploit this advantage, Andrew's group is one of several that are attempting to insert RuBisCO genes from red-algal chloroplasts into the chloroplasts of higher plants, using techniques for manipulating chloroplastic DNA developed by Rutgers University biochemist Pal Maliga. "If it can be done, it would be really amazing," says Tokota, who also works at the Nara Institute of Science and Technology, in Kansai Science City, near Nara, Japan. Other groups are working on related approaches at Rothamsted, Ohio State, and the University of Nebraska, Lincoln.

"This is a little bit like transferring a V-8 engine from a big automobile into a small car," says Andrews. "It may not work." Even if the enzyme functions in its new, transgenic home, he cautions, "it's not enough simply to get the RuBisCO in there: it has to be assembled and produced in the right form, and also be connected to the regulation system that the chloroplast keeps control of RuBisCO with." Andrews hopes to see results in "about ten years."

### Supercharging Photosynthesis

While most researchers trying to modify the genetic basis of photosynthesis are focusing on RuBisCO, a few are trying another, perhaps even more ambitious strategy. Just as small engines can go faster if they are equipped with a supercharger, which force-feeds them with fuel, some plants have their own photosynthetic supercharger, known technically as the $C_4$ cycle. In $C_4$ plants, the bundle cells where photosynthesis takes place are surrounded by specialized "mesophyll" cells, which temporarily fix carbon dioxide and jam it into the bundle cells at such high concentrations that the oxygen reaction is effectively blocked. The $C_4$ cycle requires so much energy that $C_4$ plants cannot grow in dim light, but in the right, well-illuminated conditions, $C_4$ crops like sugarcane photosynthesize more efficiently than any others. About 5% of all terrestrial higher-plant species use the $C_4$ cycle; maize is economically the most important.

A joint team at Japan's National Institute of Agrobiological Resources and Nagoya University led by Nagoya microbiologist Makoto Matsuoka is now attempting to reproduce the $C_4$ cycle in rice. For the transformation to succeed, a host of altered enzymes would have to work together properly, and the plant's structure may have to be changed to create the equivalent of mesophyll cells. As a result, the project may well be the most fundamental genetic alteration that mankind has ever tried in any organism. "Don't hold your breath," Lorimer says.

Indeed, Matsuoka cautions, "I don't think we can really make a true $C_4$ [rice] plant." Rather than transferring the whole genetic structure for the $C_4$ cycle from, say maize into rice, his team is trying to identify nonfunctioning equivalents of $C_4$ type genes in rice and selectively replace them with their active counterparts from maize.

In a paper in press at *Nature Biotechnology*, Matsuoka's group reports taking a first step by replacing three silent rice genes with their more lively equivalents in maize, including the important enzyme phosphoenolpyruvate carboxylase (PEPC), which catalyzes the beginning of the C4 cycle. "We succeeded in getting [PEPC] highly expressed in a rice plant," Matsuoka says. "This is a world first." After transferring each gene to a different rice plant, the group is now crossing the results to obtain rice that produces all three enzymes.

That may not be enough to replicate $C_4$-like photosynthesis in rice, says Matsuoka. Rice has mesophyll-like cells that are not photosynthetically active, and these may have to be activated. And some of the changes may actually be deleterious—the transgenic rice with only the $C_4$ enzyme NADP-malic tends to die quickly, for example. Still, Matsuoka says, preliminary evidence suggests that active PEPC in rice cuts the destructive oxygen reaction by about a third.

Even as the work to alter photosynthesis begins to gain momentum, some critics question whether it will benefit agriculture. Since at least 1970, research has shown little correlation between crops photosynthesis rates and their yields, suggesting that improvements in RuBisCO won't automatically translate into better harvests. But according to Steven P. Long, a plant physiologist at the University of Essex in the U.K., the correlation may be hidden by the propensity of higher yielding cultivars to have bigger leaves, which increases that amount of self-shading and thus lowers the mean photosynthetic rate. When he and his colleagues temporarily boosted photosynthesis rates in wheat by flooding open fields with enough $CO_2$ to increase local atmospheric level by 50%, grain yield went up 10% to 12% in two consecutive growing seasons.

Long is more skeptical about the value of importing the $C_4$ cycle into crops like wheat and rice. Because the $C_4$ cycle imposes a high energy cost on the plants metabolism it only pays off at higher temperatures —that's why there is no winter maize crop. "You can model it fairly easily," Long says. "Below 28°C, the [standard photosynthesis] is more effective, and above 28°C, the $C_4$ is more efficient." The payoff threshold will rise even higher as the atmosphere's $CO_2$ concentration increases because of human activity. And if scientists like Andrews succeed in increasing the specificity of RuBisCO, the threshold for adding the $C_4$ cycle will go still higher—perhaps to 40°C, he says. "There'd be no point in going to $C_4$ then."

Nonetheless, Long favors working on both approaches. "These are such major steps that we don't even know how many unknowns there are in doing this. Pursuing all options is well worthwhile.

"We've now reached the limits of what we can do [with conventional breeding]," Rothamsted's Parry says. "So therefore we have to solve the next problem, which is putting in a bigger engine." It's a challenge, he observes, "with considerable practical interest."

# Coming to Grips With the Golgi

*by Carol Featherstone*

*Carol Featherstone is a freelance writer in Cambridge, U.K.*
*Research on the Golgi apparatus still shows plenty of life on the cell structure's 100th birthday, as researchers debate anew just how it transports proteins through the cell.*

PAVIA, ITALY—It's been exactly 100 years since the Italian biologist Camillo Golgi, working at the university in this medieval Lombardy town, peered through his microscope at spinal neurons and saw the stack of membranous structures that now bears his name. Over the decades since then, cell biologists have learned that this membranous network plays a key role in processing and transporting newly synthesized proteins. But a conference* recently held here to celebrate the centennial of the Golgi apparatus showed that, despite the organelle's venerable age, research on it is still lively. Indeed, new work is forcing cell biologists to reevaluate some of the key concepts produced by the past 10 to 15 years of work on the Golgi.

The challenges don't concern so much what the Golgi does, but rather how it does it. Cell biologists have established that the structure, which consists of stacks of flattened membrane sacs called cisternae surrounded by smaller membranous tubules and round sacs, or vesicles, chemically modifies and sorts two types of newly synthesized proteins—those destined to be secreted from the cell, such as hormones, blood plasma proteins, and digestive enzymes, and those that function in the cell's membranous compartments. All these proteins start their life in the protein factories of the endoplasmic reticulum (ER), a membrane network that permeates the cytoplasm. Then, they move into and through the Golgi before reaching their final destinations.

Until a year or so ago, researchers thought that proteins pass across the Golgi stack in vesicles that bud off from one cisterna and then fuse with the next. But accumulating evidence from several laboratories suggests that instead of acting as static warehouses, receiving and dispatching cargo in the shuttling vesicles, the cisternae themselves may move forward while the vesicles actually move backward to recycle components of the ER and Golgi compartments to their sites of origin.

This new view of a dynamic Golgi, known as the cisternal maturation model, "means the difference between a rigidly organized and complex [Golgi] system and a much more fluid, self-correcting, and evolving one," says Hugh Pelham of the Medical Research Council Laboratory for Molecular Biology in Cambridge, U.K. This model is already forcing researchers to rethink their notions of how new Golgi stacks are produced as the cell grows or when it needs to secrete more proteins, and also of how the Golgi forms in the two daughters when a cell divides.

But the work may have wider implications as well. "I think these principles affect the way we think about cell self-organization," Pelham adds. He's suggesting that rather than being a machine with rigid and unchanging parts, the cell instead consists of more organic components that grow, shrink, and flow as needed.

Still, the final word about how proteins are transported through the Golgi may not be in. Cell biologist Marilyn Farquhar of the University of California (UC), San Diego, cautions that "the evidence for [cisternal] maturation is not written in stone."

Indeed, if anything is constant in the Golgi field, it is change. The cisternal maturation model itself is a revival of an idea first proposed in the 1960s, but then discarded when both electron microscopic and biochemical evidence pointed to vesicles as the carriers of protein cargoes through the Golgi. "We have returned to the ideas of the '50s and '60s but now have the ability to address the questions with new insight and techniques," says Kathryn Howell, a co-organizer of the centenary meeting and a cell biologist at the University of Colorado School of Medicine, Denver.

### Traffic Patterns

The first controversy in the field arose with Golgi's description of the structure, which had its skeptics—among them his contemporary and fellow Nobel Laureate, Spanish neurobiologist Santiago Ramon y Cajal. The main problem was that Golgi could see what he called an "internal reticular apparatus" only in cells stained with a heavy metal, and this staining "was difficult and capricious," writes Eric Berger of the University of Zurich, Switzerland, editor of a book that marks the centenary of Golgi's discovery. Golgi wasn't proved right until a half-century later, when cell biologists first began using the electron microscope (EM) to examine intracellular structures and saw the

Reprinted with permission from "Coming to Grips With the Golgi" by Carol Featherstone. *Science* 282, December 18, 1998, pp. 2172–2174. ©American Association for the Advancement of Science.

* "The Golgi Complex: State of the Art 100 Years After Camillo Golgi's Discovery," 19–23 September.

 

intricate stacks of cisternae, surrounded by small vesicles, in many different cell types.

The Golgi's role in protein transport emerged in the 1960s and 1970s when George Palade, James Jamieson, and their collaborators, then at Rockefeller University in New York City, began probing its functions. By combining EM with autoradiography, which can detect the locations of radio-labeled molecules, these researchers showed that secreted proteins start their life in the ER and then enter the Golgi from "transitional vesicles" that bud from the ER near the innermost cisterna. The work also showed that the proteins exit the Golgi in vesicles that leave on the side nearest the cell membrane. The resolution of the technique was not sufficient to distinguish how the proteins move through the Golgi, however.

Meanwhile, work by several researchers had been showing that in addition to transporting the proteins, the Golgi also modifies them. Among other things, it tacks on the complex sugar chains that are attached to most secreted and transmembrane proteins. Antibodies that detect the sugar-adding enzymes further showed that some of them are present in only one or two distinct cisternae and that their location in the stack corresponds with the position of the sugars they put on in the complex chain.

Until then, cell biologists thought that proteins might be transported within cisternae moving through the cytoplasm from the interior of the cell toward the membrane—that is, through cisternal maturation. Experiments done in the early 1970s on certain unicellular algae, which secrete large scales that form a protective coating on the algal cells, supported that idea. The scales appeared to pass from one side of the Golgi to the other enclosed in a single cisterna. But the work indicating that each Golgi cisterna has unique components suggested a different picture for most secreted proteins. The idea emerged that, rather than maturing from one to another, each cisterna receives, modifies, and dispatches its protein cargo.

Then in 1981, Palade and Farquhar suggested how the proteins might be transported from one cisterna to the next—within the small vesicles that could be seen between and around the cisternae of the Golgi stack. This vesicular transport model was also consistent with the observations that proteins enter and leave the Golgi in vesicles, and the cisternal maturation model fell into disfavor. As for the algal scales, because they are much too large to fit into the vesicles, most researchers thought that their transport within cisternae might be just a peculiarity of an obscure organism.

No one could find a way to test the vesicular transport hypothesis until 1983, however, when James Rothman of New York's Memorial Sloan-Kettering Cancer Center and his colleagues devised an ingenious cell-free assay to measure transport of a cargo protein between Golgi cisternae. The cargo protein was a viral membrane protein produced when the virus takes over the cell. The researchers isolated Golgi stacks from infected cells; they then mixed these "donor" Golgi membranes carrying the viral protein with "acceptor" Golgi membranes prepared from uninfected cells, and with cytosol—the soluble components that make up the liquid phase of the cell—and an energy source.

The movement of the cargo protein from the donor membranes to the acceptors could be monitored because the acceptor membranes, but not the donors, contained a sugar-adding enzyme from the mid-Golgi cisternae. It added radioactively labeled sugars to the cargo protein, tagging it, but only when the two were in contact within the same Golgi cisterna. Rothman's results with the assay persuaded him that the Golgi normally shuttles proteins forward from one cisterna to the next, assembly-line style.

Not only did this assay mark the first time this kind of intracellular transport had been reconstructed in the test tube, says cell biologist Graham Warren of the Imperial Cancer Research Fund (ICRF) lab in London, but "it opened the way for a biochemical analysis of the proteins that are involved in this [transport] process." By staging the assay in purified fractions of cytosol instead of in the complete mixture, Rothman's team was able to identify several proteins that are essential for intra-Golgi transport. At about the same time, other researchers—including Randy Schekman of UC Berkeley, Peter Novick of Yale University, and Scott Emr of UC San Diego—were also pinning down the proteins needed for protein movements within the cell, in their case by using mutant yeasts defective in various steps of forward protein transport.

Even as this picture was developing, however, there were already signs that the vesicular transport model wasn't the full story. The findings on the algal scales were still worrisome, and the large amount of membrane that the vesicles would have to be transporting to the cell surface and other destinations also presented a problem. There had to be a way to redistribute it back through the pathway or otherwise remove it from the outer cell membrane—otherwise, the cell would expand indefinitely. But no one had a clear idea of how the cell does that.

More recent evidence suggests that this might be the job of the vesicles that are supposed to be transporting proteins forward. Among the proteins identified as important in vesicular transport is one called COPI (for coat protein I), which helps form the vesicles that bud from the cisternae by forming a coat on the flat membrane that probably makes it pinch outward. In 1994, Pierre Cosson and François Letourneur, working at the Basel Institute of Immunology in Switzerland, found that COPI binds to a peptide in the tails of certain ER proteins—a so-called retrieval signal—that is necessary for returning those proteins to the ER when they escape. In effect, COPI seemed to be specialized for leading the vesicles it coats on a backward journey to the ER, not a forward journey to the cell membrane.

The idea that COPI carries proteins to the ER gained further support when the Swiss team created mutant yeast strains that were unable to retain proteins correctly in the ER, and then traced this inability to defective COPI genes. Also, Hugh Pelham's group found that the receptor for the ER retrieval signals is concentrated in COPI-coated vesicles. "It became obvious that COPI was involved in backward transport both within and from the Golgi stack." says Pelham.

The key issue then became whether COPI-coated vesicles could be invoked in traffic in both directions, or whether Rothman had been mistaken in his original interpretation of the transport assay. Backward transport of the sugar-adding enzyme in vesicles would look the same as forward transport of the cargo protein.

Rothman did not attend the Pavia meeting and declined to be interviewed for this article, but in the 25 July 1997 issue of *Cell*, his team, with that of Lelio Orci at the University of Geneva Medical Center in Switzerland, reported evidence indicating that vesicle traffic in the Golgi might flow in both directions. They used antibodies to search cells for a secreted protein, proinsulin, which is transported backward through the Golgi, and another protein, the retrieval signal receptor, which is transported backward. They found that the two proteins are present in two different populations of COPI-coated vesicles that bud from all Golgi cisternae. That implies, the team writes, "that bidirectional transport occurs from every level of the Golgi stack."

But other researchers, some of them former students and postdocs of Rothman's, suspect that the vesicle traffic goes in just one direction: backward. For example, Joachim Ostermann, now at Vanderbilt University in Nashville, Tennessee, took a fresh look at data he produced years ago during his postdoctoral work, in which he used the Golgi transport assay. He's now noticed, he reported in Pavia, that the COPI-coated vesicles generated in the assay contained the Golgi sugar-transferring enzyme. This implies that the vesicles carry the enzyme backward to meet the cargo, rather than the other way around. After following up these findings with more experiments, Ostermann concludes, "Virtually all of the data produced from the intra-Golgi transport assay can be explained by backward transport of Golgi enzymes."

Other evidence in favor of backward transport comes from the recently completed yeast genome sequence. Researchers believe that a transport vesicle can dock to the membrane of a cisterna only if the membrane displays a protein called a SNARE. "Because the entire yeast genome has been sequenced, all the SNAREs in yeast have now been catalogued," explains Pelham, "and the only ones that are in the Golgi and essential for secretion seem to be involved in backward rather than forward vesicular traffic within the Golgi complex."

Indeed, even Schekman, who is, like Rothman, a pioneer of intracellular membrane transport research, agrees that these newer data have cast doubt on the bi-directionality of vesicular transport. "There isn't yet a smoking-pistol experiment that proves COPI vesicles are involved in forward transport through the Golgi," he says. But if proteins do not move through the Golgi in vesicles, then researchers will have to come up with another explanation. And right now, the best bet is that they are transported much as the algal scales are, by the progression of cisternae through the stack.

Some new evidence for that idea comes from Alberto Luini and his colleagues at the Consorzio Mario Negri Sud in Santa Maria Imbaro, Italy, who have been studying the transport of the large procollagen fibers needed to make connective tissue as well as bones and teeth. Like the algal scales, these fibers are much too large to enter transport vesicles, and at the meeting, Luini reported EM studies indicating they are transported by cisternal maturation across the Golgi. Also supporting the idea are EM studies from John Bergeron's team at McGill University in Quebec and others showing, in contrast to Rothman and Orci's findings about proinsulin, that secreted proteins are found only in Golgi cisternae and not in vesicles.

Because of such discrepancies, Luini, whose own work favors cisternal maturation, cautions that it is too early to rule out some role for forward vesicular transport through the Golgi. Both processes may contribute to protein transport, he says, noting that one way to find out is to compare how fast a small secreted protein and a large supramolecular structure such as a procollagen fiber move through the Golgi in the same cell. If the rates are different, he says, it may be because the transport mechanisms for the two types of protein are different.

### The Dynamic Golgi

But even if cisternal maturation accounts for only part of protein transport, there would have to be a way of constantly regenerating new Golgi cisternae, and that would require a revision of current views of how the Golgi forms. Until now, researchers thought that the structures are basically static once formed, and that the existing Golgi gives rise to the Golgi of the two daughter cells when a cell divides. For example, by tagging Golgi components with a fluorescent molecule, the ICRF's Warren found that during cell division it fragments into much smaller elements that disperse through the cytoplasm. These fragments are then passed down to the two daughters and reassembled, he postulates.

But this disappearing act looks different to cell biologist Jennifer Lippincott-Schwartz of the National Institute of Child Health and Human Development in Bethesda, Maryland, and her colleagues. When they tagged a Golgi protein with green fluorescent protein (GFP) and then treated the cells with a drug called brefeldin A, which disrupts the Golgi, they found that the fluorescence rapidly disperses throughout the ER. This indicates that the drug makes the Golgi collapse into the ER.

What's more, Lippincott-Schwartz reported at the meeting, she sees a similar, though very transient, "blink out" of the Golgi-linked GFP fluorescence in cells going through the midpoint of cell division, even without brefeldin A treatment. Like Warren, she believes that the Golgi structure fragments at the start of division. But, she proposes, all the fragments suddenly collapse into the ER, and then late in mitosis new Golgi structures begin to assemble in the daughter cells as protein-carrying vesicles bud off the ER and fuse together, eventually forming the cisternae. Such a mechanism could not only produce new Golgi structures, but also provide a way to replace them as they migrate through the stack.

Warren doesn't see this pattern, however. "We've published experiments using similar techniques, and we find that Golgi fragments persist throughout mitosis," he says.

But new data from Ben Glick at the University of Chicago add credence to Lippincott-Schwartz's story. He has been studying two yeasts, *Pichia pastoris* and *Saccharomyces cerevisiae*, both of which multiply by budding new cells off old ones, and finds that the Golgi in the yeasts do not break down during cell division. Instead, a strand of the ER enters the budding daughter cell, where it appears to give rise to a new Golgi. "If we are thinking about this correctly, then Golgi inheritance in budding yeasts is a direct consequence of ER inheritance," concludes Glick.

All in all, researchers are building a much more dynamic picture of the Golgi. Rather than being a static collection of well-defined compartments, it appears to be in constant flux, each cisterna emerging from the ER with its load of proteins and then carrying these proteins across the stack while at the same time putting the finishing touches on the proteins. That dynamism was reflected by the mood of the scientists in Pavia, who were celebrating new ideas and relishing the challenges ahead.

### Additional Reading

E. G. Berger and J. Roth, Eds., *The Golgi Apparatus* (Birkhäuser-Verlag, Basel, Switzerland, 1997).

"The Golgi Complex" (Special Issue), *Biochimica et Biophysica Acta* 1404, 1 (1998).

"Golgi Centenary Issue," *Trends in Cell Biology 8*, 1 (1998).

# Viruses Have Many Ways to Be Unwelcome Guests

*by Michael Balter*

*A recent meeting examined the lifestyles of over 40 different viruses and found that they have evolved diverse strategies for survival within their hosts.*

LYONS, FRANCE—It's not easy being a virus. Unwanted and uninvited, these tiny microbes are totally dependent on their hosts for survival. They usually show up with little baggage of their own, often wearing just a thin protein coat wrapped around a small cluster of genes. But if it seems that these lowly visitors hardly merit respect, they are nonetheless taken very seriously for the disease and misery they can cause. And despite their apparent molecular simplicity, viruses have evolved sophisticated strategies to survive and propagate within their hosts.

At a recent meeting* in this city straddling the Rhône and Saône river, some 150 virologists gathered to review current research on the lifestyles of more than 40 viruses that infect organisms from bacteria to humans. "This is the first time in my memory that there has been a meeting that set out to look at as many viruses as possible," said Timothy Greenland of the Louis Pradel Hospital in Lyons. During the 3-day meeting, researchers reviewed recent work on the many different schemes viruses use to pursue their unwelcome cohabitation with host cells, including slavish dependency, blatant molecular thievery, and occasionally haughty independence.

### Borrowers and Lenders

Although viruses form very intimate bonds with their hosts, many contribute only the bare minimum to the relationship. This is particularly true of the RNA-containing retroviruses, which reverse the normal flow of genetic information by using an enzyme called reverse transcriptase to transcribe their single-stranded RNA into double stranded DNA. Retroviruses, whose genomes are rarely longer than 10,000 nucleotides—the building blocks of RNA and DNA—can produce very few proteins of their own and must beg, borrow, and steal whatever host proteins they need to carry out their life cycle. Stephen Goff of Columbia University in New York City reviewed work that he and his collaborators, including Jeremy Luban of Columbia, have done over the past few years on the hijacking of host proteins by HIV-1, the retrovirus that causes most cases of AIDS.

Goff's group hunted for specific interactions between virus and host proteins using a technique called the yeast two-hybrid system, in which "bait" and "prey" proteins are fused to separate genetic factors that turn on detectable "reporter" genes when the proteins bind. The team demonstrated that the human enzyme cyclophilin A—which is thought to help cellular proteins fold properly—binds tightly to specific site on an HIV-1 structural protein called Gag. Goff and other researchers went on to show that the enzyme actually becomes incorporated into the virus structure and that each virus particle contains some 200 molecules of cyclophilin A. Just why HIV-1 needs this host protein is still unclear, although some researchers have speculated that cyclophilin A may help the virus assemble itself properly. But Luban and other investigators have shown that without cyclophilin A, HIV-1 cannot infect its target cells, making the protein's interaction with Gag a possible target for therapies.

Rather than just borrowing host proteins, as retroviruses do, many larger viruses have pirated host genes and made them a permanent part of their own genetic repertoire, thus acquiring the ability to produce their own versions of proteins that originally came from cells. "Viruses with large genomes have the capacity to incorporate host genes," says Robin Weiss of the Institute for Cancer Research in London. "They already have 50 or 60 genes, so they can borrow a few more." One such pilferer, described by Weiss in a talk at the meeting, is Kaposi's sarcoma-associated herpesvirus (KSHV), which was identified 3 years ago by scientists at Columbia University and which many researchers believe is responsible for a type of skin tumor often found in older European men and HIV-infected patients. KSHV's DNA-based genome, which contains at least 140,000 nucleotide base pairs, includes genes coding for myriad humanlike proteins, among them homologs of molecules that regulate cell proliferation, intracellular signaling, and immune functions.

Exactly what use the virus makes of these proteins is still under investigation, although Weiss points out that another member of the herpesvirus family, cytomegalovirus, is known to produce its own versions of cellular proteins that modulate the immune response (see Article on p. 248). "We are sure these genes do something for the survival of the



*"Strategies in Virus-Host Relationships," Lyons, France, 16–18 February.

 

virus, although they may not be necessary for basic replication." Weiss adds that various combinations of these pirated genes "may be necessary for viral growth in particular cell types under certain conditions."

Another intriguing case of gene piracy was reported at the meeting by Michael Skinner of the Institute for Animal Health in Compton, U.K. In as-yet-unpublished work, Skinner's group has identified several genes in the fowlpox virus, which infects chickens, that are homologous with genes found in a wide variety of other organisms, including yeast, roundworms, and mammals. The pirated genes include some that resemble genes coding for so-called SNAP proteins in mammals and squid proteins involved in shuttling large molecules around within the cell in bodies called vesicles. Others look like genes that code for proteins of unknown function in the roundworms, *Caenorhabditis elegans* and *Trichinella spiralis*.

Skinner told the meeting attendees that because these homologous genes are found in all strains of fowlpox so far studied, they must play an important role in the virus's interactions with his host. For example, he speculated that the viral SNAP proteins might subvert normal cellular transport processes to allow viral assembly. Alternatively, he said, they could interfere with antigen presentation, the process by which infected cells ferry proteins from a foreign invader to the cell surface to alert the immune system.

But how did all these diverse genes find their way into the fowlpox virus? Skinner hypothesized that retroviruses may have served as go-betweens. At some point in evolution, he suggested, the reverse transcriptase of retroviruses may have copied messenger RNA—an intermediary molecule between DNA and proteins—from a host organism back into the complementary DNA form. This DNA could then have been picked up by the fowlpox virus and incorporated into its genome.

### Going It Alone

Most viruses depend heavily on the host cell machinery to get their genomes replicated. Retroviruses can use cellular replication enzymes once they have created DNA copies of their genomes via reverse transcriptase, and DNA-containing viruses such as herpesviruses can use the cellular machine directly. But one group of viruses, those containing double-stranded RNA molecules, have to be more self-sufficient, bringing most of their own replication enzymes into the cell with them. The reason is simple: Host cells do not have the enzymes necessary to replicate double-stranded RNA.

A talk by B. V. Venkataram Prasad of Baylor College of Medicine in Houston illustrated the lengths to which these viruses must go to reproduce. Over the past few years, Prasad's group has been exploring the replication machinery of the rotavirus, one of the most important causes of severe diarrhea in children, causing more than a million deaths around the world each year. Earlier work by Prasad's team and other researchers had shown that rotaviruses wear three protein coats: an outer garment that is shed when they enter their target cells in the intestinal lining, and two inner layers that shelter 11 separate segments of double-stranded RNA.

Last year, using a technique called electron cryomicroscopy, which gives a high-resolution picture of frozen, well-preserved viral particles. Prasad and his coworkers demonstrated that the RNA segments are actually embedded in the innermost protein layer, which also contains the enzymes necessary for replication. Unlike most viruses, which come apart after infecting a cell and then go in search of molecular building blocks, double-stranded RNA viruses "stay intact and suck in the nutrients they need to replicate their RNA," says David Bishop of the Pasteur Institute in Paris. And remarkably, once rotavirus has replicated its double-stranded RNA, Prasad saw single-stranded messenger RNA molecules exiting from the virus via narrow channels through the protein layers. They were presumably on their way to the cell's protein-making machinery where they would direct the production of proteins needed by new virus particles.

Given the wide range of strategies viruses have evolved to adapt to life within their hosts, it is no wonder researchers have faced such daunting challenges in coming up with therapies against these unwelcome guest. Says Bishop: "The meeting gave lots of illustrations that, as far as our health and welfare are concerned, viruses are a moving target."

# New Potential for Human Embryonic Stem Cells

*by John Gearhart*

*The recent demonstration that human embryonic stem cells can be grown in culture has major implications for basic research and transplantation therapies.*

Pluripotential stem cells, present in the early stages of embryo development, can generate all of the cell types in a fetus and in the adult and are capable of self-renewal. A renewable, tissue culture source of human cells capable of differentiating into a wide variety of cell types would have broad applications in basic research and transplantation therapies. A major step in realizing this goal has now been taken with the demonstration that human embryonic stem cells can be grown in culture. These stem cells have been derived in culture from two embryonic tissues: inner cell masses of blastocysts (those cells within the conceptus that form the embryo proper) and primordial germ cells. Embryonic stem (ES) cells were first derived from the inner cell masses of mouse blastocysts in the early 1980s *(1, 2)*. More recently, primordial germ cell cultures were found to give rise to cells with characteristics of ES cells and were designated EG (embryonic germ) to distinguish their tissue of origin *(3, 4)*. ES and EG cells have now been derived from embryos of other mammals, including primates *(5–10)*. Now on page 1145 of this issue, Thomson *et al. (11)* report the derivation of ES cell lines from human blastocysts.

Pluripotential stem cells, primarily ES cells, have been used extensively in studies of embryogenesis, gene function, and development in the mouse *(11)*. ES and EG cells transferred to a mouse blastocyst can contribute substantially to all differentiated cell types in the fetus, including the germ line. Consequently, gene targeting within ES cells has enabled both whole-animal studies of gene function and the production of mouse models of human genetic diseases and abnormalities. ES and EG cells have also been used to study the differentiation of various cell types and tissues in vitro, such as neural cells *(12–16)*, hematopoietic lineages *(17–19)*, and cardiomyocytes *(20)*. ES-derived cells have been successfully transplanted into fetal and adult mice, where they have demonstrated morphological and functional integration *(19–23)*.

Thomson's group at the Wisconsin Regional Primate Research Center in Madison, in collaboration with the Departments of Obstetrics and Gynecology at the Rambam Medical Center in Haifa, Israel, and the University of Wisconsin, reports the derivation of five independent cell lines from the inner cell masses of 14 blastocysts *(11)*. The ES cell lines were continuously cultured for 5 to 6 months and expressed high levels of telomerase activity, characteristic of cells with high replicative lifespan. The cell lines had normal karyotypes (two male and three female) and expressed cell surface markers characteristic of ES cells. Four cell lines tested produced teratomas when grown in immunocompromised mice. Histology of the tumors revealed differentiated cells derived from all three embryonic germ layers (ectoderm, mesoderm, and definitive endoderm)—a result consistent with pluripotency.

This report of the derivation of ES cells from human blastocysts represents a major technical achievement with great importance for human biology. Although ES cells have been derived from several mammalian species, other species have proved refractory in yielding ES cells. It was, therefore, not a foregone conclusion that ES cells could be derived from human embryos. Earlier publications from the Thomson group reporting the derivation of ES cells from non-human primates increased the expectation that such cells could be derived from human blastocysts. In a related report, it now appears that human embryos are also amenable to EG cell derivation from primordial germ cells *(24)*.

The derivation of human ES cells now raises a whole new set of expectations. On the basis of the use and study of mouse ES cells, the research and clinical potential for human ES cells is enormous. They will be important for in vitro studies of normal human embryogenesis, abnormal development (through the generation of cell lines with targeted gene alterations and engineered chromosomes), human gene discovery, and drug and teratogen testing and as a renewable source of cells for tissue transplantation, cell replacement, and gene therapies. These latter applications could eventually preclude the direct use of fetal tissue in transplantation therapies.

It is exciting to speculate on how human ES lines could be used in tissue transplantation therapies *(25)*. Obvious clinical targets would include neurodegenerative disorders, diabetes, spinal cord injury, and hematopoietic repopulation. In addition to possibly providing large numbers of pure populations of cells for transplantation, ES cells would also lend themselves to several strategies for the prevention of immunological tissue rejection after transplantation, including (i)

Reprinted with permission from "New Potential for Human Embryonic Stem Cells" by John Gearhart. *Science* 282, November 6, 1998, pp. 1061–1062.

banking of multiple ES cell lines representing a spectrum of major histo-compatibility complex (MHC) alleles to serve as a source for MHC matching; (ii) creation of universal donor lines, in which the MHC genes could be genetically altered so rejection would not occur, an approach that has been tried with moderate success in the mouse; (iii) customization of ES cells through transgenesis and gene targeting so that a potential recipient's MHC genes are introduced into ES cells through homologous recombination; and (iv) production of ES lines containing the genome of the prospective recipient. Blastocysts obtained through nuclear transfers would be used to generate ES cells, which then could be differentiated to specific lineages for transfer to the nuclear donor (*26*). Because EG cells have been shown to reprogram adult nuclei (*27*) after cell hybrid formation, it may eventually be possible to do nuclear transfers into pluripotent stem cells, which could then be expanded and differentiated.

To realize the full potential of human pluripotent stem cells, challenging research lies ahead and several practical issues must be resolved. The conditions necessary to derive human ES cells efficiently and reliably must be defined. How did the Thomson group succeed when, on the surface, their protocol is so similar to that of other investigators? Their previous experience with nonhuman primate ES cell derivation was certainly critical for this success. Thomson *et al.* (*11*) report that of 14 inner cell masses placed in culture, five ES cell lines were established. This result is excellent, but could it be better? Are there ways of assaying blastocysts for their potential of yielding ES cells? As in the mouse, are there predisposing genes for this property? Are there other extrinsic or intrinsic factors that may lead to a greater success rate?

The conditions for directed, lineage-restricted differentiation of ES cells must be defined. Studies to date on ES cell differentiation in vitro rely primarily on the selection and enrichment of specific lineages from the many that may be present when cell differentiation is induced. Also, strategies must be developed to obtain the large numbers of pure populations of cells that would be required for engraftments. In the short term, feeder cell-independent lines will have to be derived and methods for complete cell disaggregation developed. It must also be determined if the cells are amenable to transfections, enabling selection and gene-targeting strategies.

Reports on the isolation of human pluripotent stem cells will no doubt catch the public eye, and there will be expressions of concern, rekindling the debate on human embryo research. The debate will encompass the source of the cells, human cloning potential, and the possibilities of germ line modifications. Four years ago, the Human Embryo Research Panel's report to the director of the National Institutes of Health (NIH) concluded that research deriving ES cells is acceptable as long as embryos are not created expressly for research purposes. Several issues will have to be resolved to permit the appropriate exploitation of the uniqueness and potential of these cells. Currently, as broadly written, U.S. federal law bans the use of federal funds for the derivation of these cells [Public Law 105-78, Section 513(a)]. To date, research in this area has been sponsored through private and corporate funding, with hospital and academic institutional internal review board approval and informed patient consent. It is not clear whether NIH funding necessary to realize the biomedical potential of the cells will be available to support studies using the derived ES cells. Federal legislation and funding policies should be reexamined in light of the biomedical potential of human ES cells, now made more imminent by the Thomson *et al.* report. Federal guidelines must be established so that ES cell research can be funded after appropriate peer review and oversight.

### References

1. M. J. Evans and M. H. Kaufman, *Nature* 292, 154 (1981).
2. G. R. Martin, *Proc. Natl. Acad. Sci.* U.S.A. 78, 7634 (1981).
3. Y. Matsui *et al.*, *Nature* 353, 750 (1991).
4. J. L Resnick, L. S. Bixler, L. Cheng, P. J. Donovan, *ibid.* 359, 550 (1992).
5. J. A. Thomson *et al.*, *Proc. Natl. Acad. Sci.* U.S.A. 92, 7844 (1995).
6. J. A. Thomson *et al.*, *Biol. Reprod.* 55, 254 (1996).
7. J. A. Thomson and V. S. Marshall, *Curr. Top. Dev. Biol.* 38, 133 (1998).
8. T. Doetschman, P. Williams, N. Maeda, *Dev. Biol.* 127, 224 (1988).
9. H. Shim *et al.*, *Biol. Reprod.* 57, 1089 (1997).
10. J. A. Piedrahita *et al.*, *ibid.* 58, 1321 (1998).
11. J. A. Thomson *et al.*, *Science* 282, 1145 (1998).
12. E. J. Robertson, Ed., *Teratocarcinoma and Embryonic Stem Cells. A Practical Approach* (IRL Press, Oxford, 1987).
13. G. Bain, D. Kitchens, M. Yao, J. E. Huettner, D. I. Gottlieb, *Dev. Biol.* 168, 342 (1995).
14. M. F. Finley, N. Kulkarni, J. E. Huettner, *J. Neurosci.* 16, 1056 (1996).
15. A. Fraichard et al., *J. Cell Sci.* 108, 3181 (1995).
16. C. Strubing *et al.*, *Mech. Dev.* 53, 275 (1995).
17. M. Li, L. Pevney, R. Lovell-Badge, A. G. Smith, *Curr. Biol. 8*, 971 (1998).
18. M. V. Wiles and G. Keller, *Development* 111, 259 (1991).
19. M. Kennedy *et al.*, *Nature* 386, 488 (1997).
20. N. Hole, G. J. Graham, U. Menzel, J. D. Ansell, *Blood 88*, 1266 (1996).
21. M. G. Klug, M. H. Soonpaa, G. Y. Koh, L. J. Field, *J. Clin. Invest* 98, 216 (1996).
22. T. Deacon, J. Dinsmore, L. C. Costantini, J. Ratliff, O. Isacson, *Exp. Neurol.* 149, 28 (1998).
23. O. Brustle et al., *Proc. Natl. Acad. Sci. U.S.A.* 94, 14809 (1997).
24. M. J. Shamblott *et al.*, *ibid.*, in press.
25. A. G. Smith, *Curr. Biol.*, in press.
26. O. Solter, *Nature* 394, 315 (1998).
27. M. Tada, T. Tada, L. Lefebvre, S. C. Barton, M. A. Surani, *EMBO J.* 16, 6510 (1997).

# After Dolly, a Pharming Frenzy

*by Elizabeth Pennisi*

*New successes in using DNA from fetal cells to clone transgenic animals have boosted a new biotech business, but low success rates still need to be improved.*

BOSTON—Births are usually announced on a newspaper's society or personal pages, not on the front page. But that convention didn't apply to Dolly and Polly and—just last week—George and Charlie. These white-faced lambs and Holstein calves made headlines as the products of cloning technologies that have generated fascination and fear—a reaction fanned this month by the improbable claims of a physicist who says he plans to clone adult humans within 2 years (*Science*, 16 January, p. 315). But the technologies have done more than spawn an ethical debate about the prospects for human cloning: They have also galvanized efforts to create transgenic livestock that will act as living factories, producing pharmaceutical products in their milk for treating human diseases and, perhaps, organs for transplantation.

That was always the main intention of Dolly's creators, Ian Wilmut, Keith Campbell, and their colleagues, at the Roslin Institute and PPL Therapeutics in Roslin, Scotland. But in the year since the announcement of Dolly's birth, a dozen other groups have been adapting the technique used by the team in Scotland. Some want to clone animals bearing working copies of transplanted genes. Although key problems remain to be solved, these efforts—many of which were reported here at last week's annual meeting of the International Embryo Transfer Society—have already resulted in the birth of sheep containing a human clotting factor gene and calves containing foreign marker genes. Experiments in which the nuclei of pig cells have been fused with cow eggs have also given tantalizing results.

This work is invigorating the "pharming" industry: Underwriting the cloning frenzy are biotech and pharmaceutical companies eager to cash in on its potential for creating transgenic livestock. "There is a huge industry that is organizing itself around [the new cloning] technology," says James Robl, a developmental biologist of the University of Massachusetts, Amherst.

There is, however, a crucial difference between these experiments and the original Dolly breakthrough—a distinction that has sometimes been lost in the public discussion of the implications of these new results. Dolly was cloned by taking nuclei from adult mammary gland cells, starving them of nutrients to reset their cell cycles, then fusing them with sheep eggs whose own nuclei had been removed. But this procedure was very inefficient—producing only one success out of the 277 eggs that took up the new DNA. The later experiments all use nuclei from fetal cells, which have proved more efficient at generating viable offspring than adult cells. Indeed, so far the Dolly experiment has not been exactly replicated, and some scientists have even questioned whether Dolly is in fact the clone of an adult.

Animal geneticists have jumped on the technology because it potentially offers a far more efficient way to produce transgenic animals than previous techniques, which involve the injection of foreign DNA into newly fertilized eggs. The success of an egg injection is not known until after the offspring is born. For example, using egg injection, PPL Therapeutics took years to develop a flock of 600 transgenic sheep, as only about 4% of the lambs carried the desired gene.

In contrast, nuclear transfer technology allows researchers to select as nucleus donors only those cells that express the transplanted gene. Moreover, in theory, those cells could provide as many clones as needed in a single generation. "In one fell swoop, you get what you want," says PPL research director Alan Colman. Indeed, Will Eyestone of PPL's Blacksburg, Virginia, facility told last week's meeting that egg injection "may well become old-fashioned."

### From Sheep to Cows to Pigs

Campbell, who recently moved from the Roslin Institute to PPL's labs 300 meters down the road, Wilmut, and their colleagues were the first to announce that they had been able to produce transgenic animals with cloning technology. They reported in December that they had produced three cloned sheep, two of which are still alive, carrying the human factor IX clotting protein (*Science*, 19 December 1997, p. 2130).

Now, Advanced Cell Technology has achieved in cows what the team in Scotland did with sheep: Robl and his colleague Steven Stice announced at last week's meeting the birth of two calves carrying a foreign gene. To produce these transgenic animals, the researchers first grew bovine fetal fibroblast cells in the laboratory and then added an antibiotic resistance "marker" gene. Only the cells that took up the gene survived exposure to an antibiotic

---



 

added to the culture dishes. The researchers then fused nuclei from the survivors with enucleated cow eggs, employing a variation on the technique used by Wilmut's group. About 40% of the resulting embryos continued to develop once inside foster mothers, and two calves—George and Charlie—were born in mid-January. A third has been born since the announcement, and more are on the way. "[They are] the first transgenic cloned calves, and that's great," says Campbell of PPL, which is also doing nuclear transfer work in cattle. The three calves show "the phenomenon and the technology are not restricted to one species," adds nuclear transfer pioneer Kenneth Bondioli of Alexion Inc., a biotech company in New Haven, Connecticut.

That demonstration has been eagerly awaited. Transgenic cows, which produce 9000 liters of milk per year, should be better factories for therapeutic proteins than sheep or goats. "Milk is cheap, and we have an incredible dairy infrastructure, points out Carol Ziomek, an embryologist with Genzyme Transgenics in Framingham, Massachusetts.

Indeed, that potential has already spurred a gold rush. In October 1997, Genzyme Transgenics awarded Advanced Cell Technology a 5-year, $10 million contract to develop transgenic cows that will produce albumin, a human blood protein used in fluids for treating people who have suffered large blood losses. And earlier this month, Pharming Holding, N.V. in Leiden, the Netherlands, formed an alliance with ABS Global, an animal breeding company in DeForest, Wisconsin, and its spin-off company, Infigen Inc., to develop transgenic cattle that produce the human blood proteins fibrinogen, factor IX, and factor VIII in their milk.

Other efforts are aimed at expanding the utility of pigs, particularly in biomedicine. A few companies and research groups hope to use pig organs or tissue to help meet the large unfilled demand for transplant organs. The goal is to genetically modify the animals' tissues so they are less readily rejected. Also, because a pig's physiology is more like a human's than is a mouse's, some animal scientists argue that pigs could be good models for studying human diseases if their genetic makeup could he modified so that they develop appropriate symptoms.

But this work has been lagging, partly because researchers have had trouble getting pig oocytes to start dividing after the nuclear transfers. Moreover, researchers are still working out a suitable way to keep new embryos alive until they can be placed into female pigs for continued development.

At the meeting, several teams reported progress solving these problems. At the University of Missouri, Columbia, Randall Prather has worked out a new way to activate cell division using a chemical called thimerosal as the initial trigger. And reproductive physiologist Neal First's group at the University of Wisconsin, Madison, offered a more radical potent solution: Avoid the hard-to-activate pig egg altogether by transferring nuclei from adult pigs into bovine oocytes. "Instead of using a pig oocyte, perhaps you could use sheep or cow oocyte," Robl suggests. It is unclear, however, whether such cross-species embryos would ever come to term.

### Reducing the Body Count

In spite of the rapid advances in nuclear transfer since Dolly's debut, some big obstacles still remain. At each step along the way some—often many—individuals don't survive. That low efficiency doomed an earlier version of nuclear transfer when it made its commercial debut a decade ago. At that time, several companies, including Granada Inc., based in Houston, were going great guns using nuclei from very early embryos to clone hundreds of calves to make large herds of genetically superior beef cattle. But by 1991, Granada had shut its doors. "We couldn't make as many calves as we wanted to," recalls Bondioli, who worked there. And too often, calves were oversized and unhealthy, with lungs that were not fully developed at birth.

Researchers see the same trends in the few cows and sheep produced by the newer cloning procedures. Large numbers of deaths occur around the time of birth. For example, PPL and Roslin lost eight of 11 lambs in their first experiment with transgenic clones. But it's not the nuclear transfer procedure itself that's at fault, says Robl. Animals produced by in vitro fertilization and other procedures involving the manipulation of embryos have similar problems, albeit at a lower frequency.

"Something that you do to the embryo . . . leads to a problem 9 months later," says George Seidel Jr., a physiologist at Colorado State University in Fort Collins. His data and other observations suggest that in problem calves the placenta does not function as it should. As a result, cloned calves have too little oxygen and low concentrations of certain growth factors in their blood.

While some researchers are experimenting with different nutrient solutions or making other subtle changes in their nuclear transfer techniques to make embryos and newborns thrive, others are frantically trying to hone the genetic manipulation techniques. Researchers currently have no control over where the foreign genes end up in the chromosomes or how many copies of the gene become part of that cell's genetic repertoire.

Developing that control would enable them to knock out specific genes, say the one encoding the pig protein that elicits a strong, immediate rejection response to pig organ transplants. "The Holy Grail for many is finding a way of getting targeted disruption of genes in livestock as we have in mice," explains Colman, who is confident that even this tough molecular biology problem will be solved quickly. "I expect we'll have targeting solved by next year," he predicts.

Such confidence is required in this fast-moving field, in which progress generally comes through trial and error. Understanding how it all works, say these scientists, will come later. "[There] clearly is at this point in time a pushing forward of the technology," says Alexion's Bondioli. "Have we learned any more biology? Probably not. But [we] have opened up a means to study [it]."

# Lab-grown Organs Begin to Take Shape

*by Dan Ferber*

*Dan Ferber is a writer in Urbana, Illinois.*
*With the need for transplant organs growing, researchers are making progress toward developing them, using cultured cells and special polymers.*

Call it the seaweed that's changing medicine. On a balmy summer afternoon in 1986, surgeon Joseph Vacanti of Harvard Medical School in Boston was sitting on a stone breakwater near his Cape Cod vacation house, watching his four children play on the beach. He and biomedical engineer Robert Langer of the Massachusetts Institute of Technology (MIT) had been trying for more than a year to devise new ways to grow thick layers of tissues in the laboratory—a first step toward their long-term goal of growing replacements for damaged tissues and organs.

But even though they were using the latest in cell-friendly, biodegradable polymers as scaffolds to support the growing tissue, the thickest slices they could grow were thinner than a dime—not much use for building complex three-dimensional organs like livers, kidneys, or hearts. The problem, Vacanti realized, was that as the tissues thickened, the interior cells couldn't take in enough nutrients and oxygen or get rid of sufficient carbon dioxide to continue growing.

Then, as Vacanti gazed into the water, inspiration struck. He spotted a seaweed waving its branches, silently soaking up nutrients from the water around it. He immediately made the connection: Branching is nature's way of maximizing surface area to supply thick tissues with nutrients, and polymer materials that branch, rather than being completely solid, would be porous enough to support growing tissue in the lab. Vacanti raced up the road to a pay phone to call Langer. "He asked if we could design [biodegradable] polymers that had a branching structure," Langer recalls. "I said, 'Well, we could probably do that,' and we tried and we did."

Thirteen years later, branched biodegradable plastics and related sponge-shaped plastics undergird tissues growing in dozens of laboratories around the world. Some of the simpler of these tissues, including skin and cartilage, have already made it to the clinic or are on their way. But, fueled by recent advances in polymer chemistry in the design of the bioreactors that incubate the tissues, and in the understanding of basic cell and tissue biology, researchers are also beginning to grow organs with more complex architectures.

Two months ago, for example, a team at Harvard Medical School reported in *Nature Biotechnology* that it had used tissue engineering to produce new urinary bladders that appeared to work normally in dogs. And on page 489, Langer, anesthesiologist and biomedical engineer Laura Niklason of Duke University in Durham, North Carolina, and their colleagues report growing functioning pig arteries. Less advanced, but showing progress, are efforts to engineer tissues to fill in for failing hearts, livers, and kidneys —organs for which the demand for transplants far outstrips supply.

"I'm very excited about it all," says biomedical engineer Michael Sefton of the University of Toronto. "Things are going much better than I expected." Indeed, he has organized more than 25 leading tissue engineers into an informal international network called the Life Initiative. Their goal? To raise more than $1 billion from industry, government, and private foundations for a decade-long effort to grow livers, kidneys, and hearts (*Science*, 12 June 1998, p. 1681).

### Tissue in Three Dimensions

The first bioengineered internal organ to reach the clinic may be the bladder, produced by surgeon Anthony Atala of Harvard Medical School and his colleagues. The team faced several challenges during its 9-year quest to grow a working urinary bladder in the laboratory, Atala says. First, the researchers had to isolate the necessary cells and coax them to grow in culture dishes. They had little trouble in harvesting from dog bladders the smooth muscle cells that would form the bladder's outer surface and getting them to grow. But growing the specialized epithelial cells called urothelial cells that would line the inside surface of the organ was another matter.

The main problem was that in culture, the urothelial cells, taken from the same bladder samples, tended to revert to less specialized, primitive forms that would function poorly, allowing urine to leak into bladder tissue, which could cause scarring and pain. But with the right culture conditions and combination of growth factors, the team was eventually able to steer urothelial cells toward their mature state. "It's important to get the cells to differentiate and to stop differentiating: to grow and to stop growing," Atala says.

The researchers also had to find the right polymer scaffolding on which to grow the cells. The polymers had to be elastic enough to give the cells a lifelike mechanical environment, sufficiently porous to allow the fluids bathing the growing tissues to deliver nutrients and flush away cellular wastes, and capable of degrading as the tissue developed—but not so fast that it dissolved before the cells had time to grow into it. Eventually, the Harvard team settled on a branched, porous version of a polymer called polyglycolic acid that Langer had produced. They set the polymer in a bladder-shaped mold and coated it with a second polymer called polylactide-coglycolide.

In the next step, the researchers applied the urothelial cells to the inner surface of the polymer bladders and the smooth muscle cells to the outer surface, and then nurtured the synthetic organs for 7 days in a sterile nutrient broth. At that point, they surgically replaced the bladders of six beagles with the lab-grown bladders. Within 3 months, the polymers had disappeared from the tissues, the dogs' bodies had supplied the bladders with blood vessels, and the neobladders held just as much urine as normal bladders and maintained a normal bladder shape. In addition, the bladders worked and even developed innervation. "The dogs empty by themselves," Atala says. Indeed, says bioengineer Antonios Mikos of Rice University, the Harvard group's success at engineering the bladder "proves the feasibility of engineering other organs using co-cultures of cells and synthetic, biodegradable organ scaffolds."

The Harvard group has since gone on to grow human bladders in the laboratory and is now seeking regulatory approval to start clinical trials. Lab-grown bladders could serve as replacement organs in some of the hundreds of thousands of people whose bladders have been damaged by accidents, chronic infections, bladder cancer, or congenital birth defects, Atala says.

### Building Blood Vessels

The work of the Niklason-Langer team in creating artificial arteries looks equally promising. Blood vessels have been engineered before using other techniques but with only mixed success. For example, cell biologist François Auger and colleagues at Laval University in Québec City, Québec, reported in the January 1998 issue of *The FASEB Journal* that they had assembled blood vessels from smooth muscle cells and endothelial cells from umbilical cords, and fibroblasts from adult human skin.

They did this by growing the smooth muscle and fibroblast cells as sheets, without a polymer support, in separate culture flasks, then wrapping them, in successive layers, around a pipette, "like a cinnamon bun," as Auger puts it. After lining these artificial blood vessels with endothelial cells, which form the internal surface of natural blood vessels, the Laval team showed that they held up under arterial pressures and could be stitched into dogs. But blood leaked between the sheets of cells, and many of the vessels plugged up with blood clots within a week. Auger is pleased, though, with these still-early results. "We wanted to see if the plumbing held and it did," he says.

Rather than stacking sheets of tissue, the MIT team grew its blood vessels on tubes of polyglycolic acid, under conditions Niklason designed to mimic those encountered by a newly formed artery in the body. First, co-author Jinming Gao, now of Case Western Reserve University in Cleveland, partially hydrolyzed the polymer with sodium hydroxide. This creates water loving hydroxyl groups that enable more cells to attach. Then Niklason drizzled smooth muscle cells from cows onto the tube-shaped scaffold and inserted a pliable piece of silicon into the interior. The tubing was hooked up to a pump, which pulsed 165 times a minute to mimic the pulsing pressure on a developing embryonic artery. After 8 weeks, Niklason coated the inside of the vessels with endothelial cells. "To me the most novel thing [about this study] is the idea of using a bioreactor that beats like a heart." Langer says.

The pulsing made the tissue stronger because, Niklason says, it increased the cells' production of collagen, a tough protein found in connective tissue. In any event, the pulsed blood vessels had walls that were twice as thick as nonpulsed vessels. They could also withstand sutures without tearing, although not as well as natural arteries, and they contracted in response to the same chemical signals that normally spur arterial contractions. The pulsed vessels also worked in animals. When the researchers took cells from tiny arterial biopsies of miniature pigs, grew vessels, and used them to replace an artery in the same animals' legs, the engineered arteries lasted more than 3 weeks without clogging, whereas arteries grown without pulsing clogged.

The results are "really astounding," says cardiovascular surgeon Timothy Gardner of the University of Pennsylvania School of Medicine, who is chair of the surgery council of the American Heart Association. "If this works," he adds, "it will be a major development in cardiovascular surgery." In many of the nearly 600,000 coronary bypass operations performed each year in the United States, surgeons can use blood vessels from the patient's own leg or chest to replace clogged coronary arteries. But in some cases, as when the individual has already undergone one such operation and doesn't have any suitable blood vessels left, this may not be possible. And small-diameter synthetic arteries haven't worked because they tend to clog up.

### Organizing Organs

Although engineered bladders and arteries are a significant accomplishment, making these hollow structures, formed by relatively thin layers of cells, is easier than achieving another goal the tissue engineers have set for themselves: producing large internal organs like the liver and kidney that need complex networks of arteries, veins, and capillaries to carry blood to and from every cell. "Large, vascularized tissue is the Holy Grail of tissue engineering," Vacanti says.

There has been some progress, however. In two recent papers, one published in the *Annals of the New York Academy of Sciences* in December 1997 and the other in the July 1998 *Annals of Surgery*, Vacanti, bioengineer Linda Griffith of MIT, and their colleagues reported taking a step toward producing a synthetic liver. "We are fabricating tissue-engineered structures that will have their own blood supply," Vacanti says.

To do this Vacanti, Griffith, and colleagues first created a plastic scaffold, made from polylactide and polyglycolide polymers, that contains channels that allow fluid to flow and capillaries to develop inside the tissue. The trick to making the scaffold was a method called three-dimensional printing (3DP), which was originally designed to make intricate molds for metal engine parts. By depositing successive thin layers of polymer, the technique creates a complex internal structure, including channels as narrow as 300 micrometers. When the researchers seeded the matrix with rat liver and endothelial cells and gently pumped growth medium through it for 5 weeks, the cells not only survived, but they rearranged themselves into microscopic structures resembling those in liver tissue. What's more, the liver cells grown inside the matrix produced the protein albumin just as they do in a real liver.

But despite the successes so far, researchers will need to surmount some major challenges before tissue-engineered organs are routinely available. Besides a network of capillaries, arteries, and veins, most engineered organs will also need to be wired with nerves that regulate their function. Tissue-engineered organs "will do an OK job, but ultimately you'll need fine tuning," says biomedical engineer Christine Schmidt of the University of Texas, Austin, who is designing new ways to regrow severed nerves.

Tissue engineers also need a reliable source of cells. Today, many of them are growing tissues using cells from an animal's or a patient's own body, but that can take weeks —time that some patients don't have. And engineered tissue from outside sources, although easier to come by, will run into the same problems that plague organ transplants. Unless scientists can prevent it, the immune system will reject the tissue as foreign.

Then comes the challenge of reliably manufacturing the tissue-engineered products, says Gail Naughton, president and chief operating officer of Advanced Tissue Sciences, a biotech firm in La Jolla, California, that is doing tissue engineering. That means that researchers will need to develop ways to test the products for sterility, mechanical strength, and function; ways to freeze the tissue and store it; and ways to scale up manufacturing to meet the tremendous demand. That goal has already been met for some skin and cartilage products. And researchers are optimistic that it can be achieved for internal organs as well. "I think that anyone in tissue engineering, myself included, believes that any tissue or organ can be grown outside the body," Naughton says.

# EVOLUTION OF SEX

**Current Literature 7 ◆ Chapter 12 Reproduction**

## Why Sex? Putting Theory to the Test

*by Bernice Wuethrich*

*After decades of theorizing about the evolutionary advantages of sex, biologists are at last beginning to test their ideas in the real world.*

"Let copulation thrive," exhorted Shakespeare's King Lear, and it has. Today, across the tree of life, sex reigns—many unicellular and just about all multicellular organisms do it. Yet how sex began and why it thrived remain a mystery. After all, asexual organisms were here first, and new asexual species continue to arise, if only to go extinct in fairly short order. Why did sex overtake asexual reproduction some billion or more years ago, and why does it continue to upstage asexuals? What gives sex its edge?

Biologists have come up with a profusion of theories since first posing these questions a century ago. Most ideas explore some version of the notion that sex is maintained because it enhances the rate of evolution by natural selection, says evolutionary biologist Graham Bell at McGill University in Montreal, but there are dozens of variations on that general idea (see Review on p. 1986). Most of them fall into two camps: that sex brings beneficial mutations together into a single winning combination that can spread through the population, or that sex purges the genome of harmful mutations. But the devil is in the data—or lack thereof. "I emphasize experimental problems, because we have tons of theories, and some are completely crazy," says Alexey Kondrashov, an evolutionary geneticist at Cornell University.

Sex is a paradox in part because if nature puts a premium on genetic fidelity, asexual reproduction should come out ahead. It transmits, intact, a single parental genome that is by definition successful. Sexual reproduction, on the other hand, involves extensive makeovers of the genome. The production of gametes requires recombination, in which the two copies of each chromosome pair up and exchange DNA. Fertilization, in which genes from different parents fuse, creates yet more genetic combinations. All this shuffling is more likely to break up combinations of good genes than to create them—yet nature keeps reshuffling the deck.

This paradox is compounded by the cost of sex—which is primarily the cost of producing a male. Imagine 1 million sexually reproducing snails and a single asexual female mutant. Say that she has two daughters, who (on average) have another two daughters, and so on. Meanwhile, the sexually reproducing females would be diligently producing a female and a male—who would not directly produce any progeny. Soon, the few sexual organisms would be lost in a sea of asexuals and find it all but impossible to locate a mate. All else being equal, the asexual clone would entirely replace its sexual counterparts in only about 52 generations, says evolutionary biologist Curtis Lively of Indiana University in Bloomington. Yet this happens rarely, if ever. Despite the cost, sexual species persist, while most asexuals quickly go extinct.

In recent years, evolutionary biologists have begun to find ingenious ways to test their explanations for the strange success of sex. They're observing populations in the wild and in the lab for evidence that rare packages of good genes really do offer an advantage. They're also counting mutation rates in organisms from water fleas to humans to see whether sex might play a role in eliminating harmful mutations. Although they haven't solved the mystery of sex yet, they are tackling what Kondrashov calls the limiting factor of "extremely lousy experimental data."

### On the Trail of the Red Queen

One theory put to the test in recent years is the Red Queen hypothesis, a variation of the idea that sex serves to assemble beneficial mutations and so creates a well adapted lineage in the face of a rapidly changing environment. In the case of the Red Queen, the good mutations are those that allow hosts to resist parasites. Because parasites adapt to the most common host genotype, evolution will favor hosts with rare combinations of resistant genes. Thus the Red Queen predicts that selection will favor the ability to generate diversity and rare genotypes—exactly the abilities conferred by sex and recombination.

Developed in the 1960s and 70s, the theory has been difficult to test, in part because it's hard to identify and track specific resistance alleles in sexually reproducing organisms, says Lively. Therefore, most of the evidence has been quite indirect, such as a 1987 study by Bell and Austin Burt of Imperial College in Silwood Park, U.K., showing that animals with longer generation times have more recombination. This corroborates the Red Queen, says Bell, because short-lived parasites easily adapt to long-lived animals—which therefore need extra recombination to counter the parasites.

---

Since 1985, Lively has been seeking more satisfying proof by scrutinizing teeming populations of a freshwater snail, *Potamopyrgus antipodarum*, which is found in both sexual and asexual variants in a cluster of 65 mountain lakes in New Zealand. *P. antipodarum* is plagued by a vicious parasite, *Microphallus*, a trematode that renders the snail sterile by eating its gonads. Lively found that lakes with few parasites tended to have mostly asexual snail populations, or clones, while lakes with more parasites tended to have mostly sexual snails. "This pattern suggests that parasites prevent the clones from eliminating sexual populations," he says.

In the August issue of *Evolution*, Lively and Indiana colleague Mark Dybdahl report more direct evidence for the Red Queen hypothesis from a 5-year study of the snails. They focused on clonal lineages, using identifiable genetic markers to finger each lineage; because each clone had a particular level of parasite resistance, they could track resistance without identifying specific genes. They couldn't use sexual organisms, because recombination would separate markers and resistance genes. But they reasoned that according to the Red Queen, selection pressure should act on these diverse clones just as it does on individuals genotypes and sparking an evolutionary race between snail and parasite. "We looked for rare advantage and a signature of coevolution," says Lively.

Specifically, they sought an oscillation in the frequency of host and parasite genotypes. As the rare host thrives and becomes more common, the parasite evolves to attack it and drive its numbers down. Then a new, resistant genotype surges ahead before the parasite evolves to hold it in check. Both species evolve as fast as they can, but neither gets far ahead, hence the theory's name, after the Red Queen's remark to Alice in Wonderland: "It takes all the running you can do, to keep in the same place."

Using markers for specific enzymes, the team identified rare clones (less than 5% of a lake's population) and common ones (more than 20%) and exposed both sets to parasite eggs. Three of four common clones were 100% infected, whereas rare clones were 50% infected. They also saw the telltale oscillation. Over 5 years, an overinfected common clone was driven down and replaced by what was initially a rare clone, which itself was driven down and replaced.

The two findings strongly indicate that parasites can grant an advantage to the rare genotypes efficiently produced by sex, and can introduce oscillatory dynamics, two key predictions of the Red Queen theory, says Lively. "Whether they maintain sex we can't say from this experiment," he acknowledges. "However, [the results] are consistent with one hypothesis and—more importantly—inconsistent with others." For example, the theory that sex purges bad mutations can't easily explain the pattern of asexual and sexual snails in the lakes, Lively says.

Other researchers also find Lively's correlations intriguing but not conclusive. "You can say [Lively's results] are consistent with the parasite hypothesis," says evolutionary geneticist Brian Charlesworth at the University of Edinburgh in Scotland, "but not that they prove it."

### Purging Bad Genes

Kondrashov is among those skeptical of the Red Queen; he instead favors the notion that sex removes harmful mutations. Removing bad mutations is a far more common challenge for living things than is coping with the fast-changing pressure of an evolving parasite, he says. "I do not believe that everything in the world lives under rapidly changing selection—while everything suffers from bad mutations."

Underpinning this theory are two facts of life: Mutations happen, and most are bad. Meanwhile, sex produces loser and winner offspring by re-sorting the mutations. When two parents with different harmful mutations mate, sex will produce some genetic scapegoats with plenty of bad mutations, and some winners with only a few. Selection will stop the losers dead in their tracks, getting rid of several harmful mutations in one fell swoop.

When a clone reproduces, however, its offspring inherit all of its bad genes and may pick up another through a new mutation. Without sex, mutations continue to accumulate in individuals and in the population. If the mutations interact synergistically, with each new mutation triggering an ever-bigger reduction in fitness (another assumption now under experimental investigation), at some point one more mutation will spell the death of all the clonal individuals. So sexual organisms outdo their asexual competitors because sex brings together, then purges, bad mutations, while the population as a whole is maintained with organisms carrying fewer mutations. And the higher the mutation rate, the greater the advantage of sexual reproduction.

That prediction opens the way to testing this theory. To overcome the efficiency advantage of asexuals, the rate must be on the order of one harmful mutation per individual genome per generation, says Kondrashov. "Fifty percent of modern evolutionary genetic theory may depend on the deleterious mutation rate," he says. "If it's high we can explain sex, recombination, diploidy, aging, and sexual selection."

Other researchers have embraced the mutational theory of sex, in part because unlike so many others it is testable. "This theory sticks its neck out," says Laurence Hurst, an evolutionary geneticist at the University of Bath in England. "You measure the parameters and if you don't find them, the theory is wrong."

To measure those parameters, researchers raise populations of organisms ranging from water fleas to worms and typically allow only one individual per generation to breed, so that if that individual has picked up a mutation, it won't be eliminated by natural selection. After every 10 generations or so, researchers test the lineages' fitness and translate any fitness decline into the deleterious mutation rate.

But the experiments are tricky, and problems can crop up if selection isn't adequately limited.

Additionally, mutations of very small effect may be undetectable in the lab but important in nature, where the numbers are larger and time is longer. Counting mutations "isn't counting beans," says geneticist James Crow of the University of Wisconsin, Madison.

So far, the results are disconcertingly mixed. "The state of the whole field is very much in doubt right now," says evolutionary geneticist David Houle at the University of Toronto. Benchmark studies by Terumi Mukai and Crow in the 1970s established a deleterious mutation rate of close to one per generation in the fruit fly *Drosophila*, just enough to explain sex to Kondrashov's theory. But later reanalyses of that work put the rate considerably lower. Recent worm experiments have yielded rates as low as 0.005, and recent rates in flies have ranged from just about nil to one.

Now a few scientists are bypassing the difficulties of population genetics experiments and instead simply counting mutations in sequenced stretches of DNA. They compare DNA sequences in noncoding regions in closely related species to derive a genomewide mutation rate. Then they estimate how much of genome is functional, or subject to selection, and apply the mutation rate to the functional DNA. Beneficial mutations are thought to be so rare that they aren't considered.

One such experiment, by Michael Nachman at the University of Arizona, Tucson, assumes that 5% of the human genome is subject to selection and concludes that each human infant is born with about six mildly deleterious mutations. If a higher proportion of the genome is functional—as some scientist suspect—then the rate would be even higher. Either way, it supports the mutational hypothesis for the maintenance of sex. But researchers agree that more work, in more organisms, is needed. Only the molecular method will vindicate or doom the theory, says geneticist Peter Keightley at the University of Edinburgh, who is now counting mutations in the worm *Caenorhabditis elegans*.

While scientists scrounge for data to support one or the other of the warring theories of sex, other researchers are considering merging the two schools of thought—that sex both collects beneficial mutations and purges bad ones. "My view is they're both going on," says McGill's Bell. "Something as complex, onerous, and laborious as sexuality is probably only going to be maintained if it's doing something very important."

For example, in the "ruby in the rubbish" model, the ruby—a good mutation in an asexual organism—is buried in rubbish—a glut of bad mutations that are constantly being eliminated by selection. Thus the harmful mutations drag the good ones down with them, slowing the rate of evolution relative to sexual populations that can unhitch good genes from bad ones during recombination.

But the evidence for such theories is also very indirect, and testing them is even more of a headache than testing the old theories. "We're in a world where it's easy to say such synergism is likely and harder to say how to go about falsifying it," says Bath's Hurst. For now, biologists can offer plenty of reasons why sex is good for you, but they have a ways to go before they can prove their point.

# Triple–repeat Transcripts: A Role for RNA in Disease

*by Robert H. Singer*

*Diseases caused by the expansion of a repeated sequence of three nucleotides within a gene have been discovered only within the last decade.*

A set of baffling human diseases —including myotonic dystrophy and Huntington's disease—are caused by expansion of a repeated sequence of three nucleotides within almost a dozen genes identified to date (*1*). With each generation, these repeats are replicated and the sequence gets longer, eventually compromising the function of the gene. The effect is dominant—only one of the two alleles of the gene need be expanded to result in the full pathology. Furthermore, the severity of the disease can be proportional to the length of the repeat expansion.

Unlike most genetic diseases, which are a result of a mutation that impairs or eliminates a gene product, the triplet-repeat diseases are likely caused by a "gain of function," in which a new function arises from the genetic defect. The new function is most easily understood when the expansion of the triplet CAG, which encodes the amino acid glutamine, occurs within the coding frame of a gene and creates a new protein with a polyglutamine tract. Huntington's disease is one example of such an expansion and is typical in that it exhibits a central nervous system pathology, as do all the polyglutamine diseases. Normal individuals can withstand a few repeats of glutamine at this position in their genes. As the polyglutamine expands, however, it disrupts the protein and affects cellular functions, possibly due to the high charge density of the expanding repeat. The new, gain-of-function protein can wreak havoc on cellular processes such as nuclear export, RNA and DNA binding, or membrane transport.

Expansions of these triplet repeats can also occur outside the coding region, but in these instances new proteins are not produced. Within this group of disorders myotonic dystrophy is a particular curiosity. In this disease, the expansion occurs in the untranslated region of the transcript, after protein coding has occurred, and can increase the mRNA by 6 kilobases or more. As this expansion gets progressively larger in one of the alleles, the resulting pathology of the disease becomes proportionately more severe. This behavior begs for a new model of molecular cytopathology. Such a model is provided by Philips, Timchenko, and Cooper on page 737 of this issue. They propose that the gain of function in myotonic dystrophy is at least in part a result of disrupted activity of a CUG-binding protein induced by the repeats in the RNA, which prevents it from doing its normal job of splicing a certain family of genes.

The new proposal is not the first; various models have been readily forthcoming since the first description of these diseases. None has been sufficient to explain the molecular etiology. The expansion most likely occurs initially in the germ cells or early embryos, where the DNA polymerase may "stutter" on the repeats and, in doing so, expand them. Some models rely on DNA-based mechanisms to explain the pathology of these diseases—changes in chromatin organization because of nucleosome positioning, stalling of the RNA polymerase, or suppression of other genes nearby. But these models cannot explain the trans-dominant effect of the allele, the effect of one abnormal gene on the functioning of the whole cell. Another DNA mechanism is possible. The disrupted expression of one allele could result in haploinsufficiency of the protein (not enough protein is produced because in effect there is only a haploid dose of the gene); in the case of mytonic dystrophy the protein kinase (DMPK) synthesized from the remaining allele would be present in the cell in half the normal amount. It is difficult, however, to imagine how such extreme variability in the disease—from mild throughout our adulthood to fatal early in life—could result from a protein that varies from a half to full dose. In addition, no point mutations in this gene reproduce the disease, indicating that the pathology is not primarily due to inadequacies of the protein product. [Data from animals models that resemble myotonic dystrophy suggest that a disruption of one or both *DM* genes can cause some muscle pathology (*2*), but gene dosage alone cannot explain the entire effect.]

The most likely explanation for the pathology of myotonic dystrophy is that the expansion in the untranslated region of the DMPK mRNA is like the expansion in the coding region: It leads mainly to a gain of function. Transcripts from the expanded gene accumulate in the nucleus of both cultured cells and biopsied tissues from patients with myotonic dystrophy (*3*). In the nucleus, these transcripts cannot make the appropriate cytoplasmic

---

Reprinted with permission from "Triple–repeat Transcripts: A Role for RNA in Disease" by Robert H. Singer. *Science* 280, May 1, 1998, pp. 696–697. 

 

protein, and therefore their sequestration results in a haplo-insufficiency. But in addition, these repeat expansions could also cause a new problem: They can build up within the nucleus, possibly to toxic levels, exerting a trans-dominant effect on cellular processes. Their influence would be particularly acute in differentiated cells, such as muscle and nerve, which not only express high levels of this transcript, but which no longer divide and cannot therefore dispose of this accumulated RNA during the breakdown of the nucleus during mitosis. If the toxicity results from the burden of these excess CUG repeats, the correlation between disease severity and repeat length is easy to explain —more CUG repeats cause a more severe gain-of-function phenotype.

But how might the repeats result in a gain of function? One hypothesis proposes that the repeats bind to a protein required for normal cellular function *(4)*, and in fact they are in the nucleus in an insoluble form, not easily extracted by reagents that dissolve nuclear components such as DNA or DNA bound proteins *(5)*. And the new results by Philips *et al.* relate protein binding to the repeat to a specific gain of function. The authors implicate a CUG-binding protein (CUG-BP) in the specific splicing of a family of genes containing a few CUG repeats near an exon; one member of this category is the cardiac form of troponin T *(6)*.

Two patients with myotonic dystrophy were analyzed, one with congenital disease and the other homozygous for the expansion. Cells from both patients contained increased amounts of an aberrant transcript of cardiac troponin T (cTNT) that inappropriately included exon 5, normally spliced out during embryogenesis. A constitutively spliced minigene containing a genomic fragment containing exon 5 of the human cardiac TNT was transfected into normal or DM primary muscle cells. Transcripts containing exon 5 were increased in DM cells relative to normal cells, and this increase was abolished when the CUG was mutated to CAG. The inclusion could be mimicked in normal cells by expressing the CUG-BP, or by transfecting increasing amounts of repeats, along with the minigene. Further evidence that the CUG-BP is the trans-acting factor comes from its in vitro binding to CUG motifs located within a 34-nucleotide region downstream of the alternative exon. Interestingly, an expansion in an intron may also disrupt splicing in the newest addition to the repeat diseases, Friedric's ataxia *(7)*. It is not known whether RNA-binding proteins may be implicated in this or other triplet-repeat diseases.

All of this suggests that something is going on in the nucleus of cells from these patients with myotonic dystrophy. Possibly, this nuclear CUG-BP may be titrated out by the excess nuclear CUGs of the expansion, resulting in defective regulation of those genes that depend on the CUG-BP for the proper processing. Yet, because increased CUG-BP is associated with the disease, it is difficult to reconcile this mechanism with the trans-dominant model in which repeat length is correlated with disease severity. The CUG-binding protein accumulates in the nucleus of affected cells from patients *(8)*, and this may be correlated in some as yet unknown way with the repeat expansion. Additional modifications to the CUG-BP, or its isoforms, such as phosphorylation may interfere with its intrinsic splicing activity. The CUG-BP is present in most cells of the organism, consistent with the multisystemic nature of the disease —both neuromuscular pathology and myocardiopathy are well described in this disease. The results predict that exons that contain this common CUG-repeat motif required for splicing would be present in genes expressed in nervous tissue as well.

Still unexplained is how the mechanism may relate to disease pathology and whether it is a cause or an effect. Although there is nuclear build-up of CUG repeat transcripts, the CUG-BP should colocalize with these foci of transcript aggregation *(4)*—which might be expected of a CUG binding protein. Intranuclear inclusions of protein, however, are found in brain neurons of patients with Huntington's disease and in a transgenic animal model of the same disease, also caused by a repeated polyglutamate tract *(9)*. A common feature of triplet-repeat diseases may be protein or protein-RNA complexes in the nucleus. All of these considerations solidify the gain-of-function model, but still leave room for many more questions to be answered.

### References

1. P. S. Reddy and D. E. Housman, *Curr. Opin. Cell Biol.* 9, 364 (1997).
2. S. Reddy *et al.*, *Nature Genet.* 13, 325 (1996).
3. K. Taneja, M. McCurrach, M. Schalling, D. E. Housman, R. H. Singer, *J. Cell Biol.* 128, 995 (1995).
4. L. T. Timchenko et al., *Nucleic Acids Res.* 24, 447 (1996).
5. B. M. Davis, M. E. McCurrach, K. L. Taneja, R. H. Singer, D. E. Housman, *Proc. Natl. Acad. Sci. U.S.A.* 94, 7388 (1997).
6. A. V. Philips, L. T. Timchenko, T A. Cooper, *Science* 280, 737 (1998).
7. V. Campuzano *et al.*, *ibid.* 271, 1423 (1996).
8. R. Roberts *et al.*, Proc. *Natl. Acad. Sci. U.S.A.* 94, 13221 (1997).
9. B. W. Davies *et al.*, *Lancet* 351, 131 (1998).

# Medical Consequences of Antibiotic Use in Agriculture

*by Wolfgang Witte*

*Overuse of antibiotics, especially in animal husbandry, has increased the prevalence of antibiotic–resistant bacteria. International cooperation is needed to address the problem.*

During the past decade, bacteria that cause human disease have developed resistance to many of the antibiotics commonly used for treatment. All of the pathogens usually found in hospitals are affected, as well as mycobacteria, pneumococci, and *Enterobacteriaceae*. Indeed, infections with antibiotic-resistant staphylococci and enterococci that cannot be treated by previously successful regimens have made headline news.

Resistance to antimicrobial drugs can arise either from new mutations in the bacterial genome or through the acquisition of genes coding for resistance. These genetic changes alter the defensive functions of the bacteria by changing the target of the drug, by detoxifying or ejecting the antibiotic, or by routing metabolic pathways around the disrupted point *(1)*.

Evolution of resistance to antibiotics is facilitated by the presence of resistance genes on transferable genetic elements and by the use of antibiotics in a way that allows them to act as selective agents *(2)*. Hospitals—with their concentrated combination of bacteria adapted to this environment, patients prone to infections, and antibiotic use—offer a prime opportunity for development and transfer of antibiotic resistance *(3)*.

Another arena for the development of an antibiotic resistance is found in animal husbandry in which antibacterials are used for prophylaxis, chemotherapy, and growth promotion. Animals receiving antibiotics in their feed gain 4 to 5% more body weight than animals that do not receive antibiotics *(4)*. More antibiotics are used in this manner than in medical applications: In Denmark in 1994, 24 kg of the glycopeptide vancomycin were used for human therapy, whereas 24,000 kg of the similar glycopeptide avoparcin were used in animal feed. From 1992 to 1996, Australia imported an average of 582 kg of vancomycin per year for medical purposes and 62,642 kg of avoparcin per year for animal husbandry. Vancomycin and avoparcin have the same mode of action; resistance to one can confer resistance to the other.

Antibiotic resistance that arises in animal husbandry affects such zoonotic pathogens as *Salmonella* serovars and *Campylobacter* spp., both of which are associated with diarrheal diseases, and human and animal commensals such as *Escherichia coli* and enterococci. Because the human and animal microbial ecosystems are inextricably intertwined, microbial antibiotic resistance readily crosses boundaries. Antibiotics given to animals and closely related compounds used in human therapy have been exerting selective pressure on their target baceria for decades.

In 1969, the Swann Committee of the United Kingdom concluded that antibiotics used in human chemotherapy or those that promote cross resistance should not be used as growth promoters in animals *(5)*. Since then, there has been continuous debate about the extent to which bacterial antibiotic use in food animals promotes resistance in bacteria that infect humans. Improved analytical techniques have provided circumstantial evidence that such resistance is indeed exacerbated by antibiotics in animal feed.

**Spread of Resistance**

*E. coli*, a commensal of the human and animal gut flora, easily disseminates drug resistance genes, as demonstrated by the spread of antibiotic resistance associated with antibiotics in animal feed *(6, 7)*. In the former East Germany nourseothricin was used as a growth promoter from 1983 to 1990, replacing the similar use of oxytetracycline. Resistance to nourseothricin in *Enterobacteriaceae* from humans and animals was negligible in 1983. Two years later, resistance (by means of the transposon-encoded streptothricin acetyltransferase gene) was found in *E. coli* from the gut of pigs and in meat products. By 1990, resistance to nourseothricin had spread to *E. coli* from the gut flora of pig farmers, their families, citizens from municipal communities, and patients suffering from urinary tract infections. The spread among humans occurred without apparent selective pressure. In 1987, the same resistance determinant was detected in other enteric pathogens, including *Shigella*, an organism found only in humans *(7)*.

Antibiotic use in animals also has resulted in resistance among nontyphoid *Salmonella* serovars. The resistant bacteria are transmitted to humans in food or through contact with animals. Resistance in *Salmonella* limits the therapeutic options available to veterinarians and physicians in the treatment of certain human cases of salmonellosis. *Salmonella typhimurium*

Reprinted with permission from "Medical Consequences of Antibiotic Use in Agriculture" by Wolfgang Witte. *Science* 279, February 13, 1998, pp. 996–997.

strain DT 104, which is resistant to ampicillin, tetracycline, streptomycin, chloramphenicol, and sulfonamides, has been identified in many places, including the United Kingdom, Europe, and the United States. The recent development of fluoroquinolone resistance is of special concern *(8)*. Remaining therapeutic options include only third-generation cephalosporines.

Fluoroquinolone use in poultry husbandry has promoted the evolution of fluoroquinolone-resistant *Campylobacter jejuni (9)*, which have been detected in meat products and in infected human patients. Development of quinolone resistance in both *Salmonellae* and *Campylobacter* spp. is due to mutations in the bacterial enzymes DNA gyrase and topoisomerase IV. *Campylobacter* spp. exist as a variety of strains in their animal hosts, making it impossible to link a specific antibiotic-resistant *Campylobacter* strain with a particular use of fluoroquinolones. Before these compounds were used in poultry farming, resistant *Campylobacter* were unknown in humans with no previous quinolone exposure *(9)*.

Enterococcal infections cause problems for hospitalized patients with impaired immune systems and can be treated with glycopeptide antibiotics. Resistance to glycopeptides, which is most apparent in countries that use large amounts of vancomycin, now contributes to increased morbidity and mortality in these patients, as therapeutic alternatives are limited.

Addition of avoparcin to animal feed has encouraged the presence of genes conferring resistance to related glycopeptides. Enterococci from animals can reach humans through the food chain *(10)*. The *vanA* gene cluster, encoding glycopeptide resistance, integrated into conjugative plasmids *(11)*, is disseminated among human and animal enterococcal strains.

The potential dissemination of these resistance genes to *Enterococcus faecalis* and other pathogens threatens human health. Transfer of glycopeptide antibiotic resistance to *Staphylococcus aureus*, which is already resistant to other antibiotics *(3)* and for which glycopeptides are the drugs of last resort, would be disastrous.

For infections by glycopeptide-resistant *E. faecium*, streptogramins are a potential treatment. However, streptogramin resistance has been found in bacteria isolated from both patients and animals. The resistance is due to the *sat*A gene, which codes for streptogramine A acetyltransferase. In Germany, although streptogramines have not been used in human chemotherapy, resistance has nonetheless appeared, probably driven by the use of the related antibiotic virginiamycin in animal feed for the past 20 years.

These observations suggest that antibiotic use in animal husbandry is a driving force for the development of antibiotic resistance in certain pathogenic bacterial species. However, some claim that assessing the risk incurred by the use of antibiotics in animal husbandry must include documentation of cases in which treatment of a human infection failed because of antibiotic resistance of proven animal origin. Unfortunately, once a resistance gene has become widely disseminated, it is difficult to trace it back to its origin. Prospective studies beginning with the introduction of a new antibiotic or specific labeling of a particular resistance gene might be useful in proving the causative connection.

**Global Prevention and Regulation**

A workshop sponsored by the World Health Organization (WHO) on the medical impact of the use of antimicrobial drugs in food animals reinforced the recommendations of the Swann committee *(12, 13)*. That the Swann committee's resolution needs repetition after 28 years indicates that we have not seen sufficient adherence to the principles stated. The growth-promoting effect of antibiotics is especially advantageous when primary animal performance is low *(14)*. Therefore, procedural modifications can decrease the use of antibiotics without sacrificing production. In Sweden, where the use of antimicrobials for growth was prohibited in 1986, improved hygiene has recouped the productivity losses *(4)*.

Worldwide differences in the use and licensing of antibiotics are large. Although the regulation of growth promoters in European Union countries incorporates the Swann committee's recommendations and the use of avoparcin was banned in 1997, tylosin and virginiamycin are still in use. Both drugs cause cross resistance to other compounds of the same class of antibiotics used in human therapy. Other growth promoters in the European Union are quinoxalines, ionophores, peptide antibiotics, phospholipols, and oligosaccharides.

| | |
|---|---|
| ***Enterobacteriaceae:*** | Normal enteric bacteria including *E. coli*, conditional pathogens causing urinary tract infections, wound infections, septicemia. Related pathogens, including *Salmonella*, *Shigella*, and enterotoxin-forming *E. coli*. |
| **Mycobacteria:** | Includes the cause of lung tuberculosis. Other species have significance for patients with HIV. |
| **Pneumococci:** | The main cause of acquired pneumonia. |
| **Staphylococci:** | Widespread among humans. Conditional pathogen in the hospital setting. |
| **Enterococci:** | Commensals of the intestinal flora. Pathogenic in immunocompromised patients. |
| ***Camphylobacter coli; Camphylobacter jejuni*:** | Intestinal bacteria of animals, cause of diarrheal infections in humans. |

In the countries of the developing world, which are responsible for about 25% of world meat production, policies regulating veterinary use of antibiotics are poorly developed or absent. In China, raw mycelia are used as animal growth promoters. In Russia, chloramphenicol is still in veterinary use. In Southeast Asia, use of antimicrobials in shrimp farming is unregulated.

The problems caused by inappropriate use of antibiotics reach beyond the country of origin. Meat products are traded worldwide, and evolving bacterial populations do not respect geographical boundaries. Management of antimicrobial resistance requires worldwide coordination. Pharmaceutical industries and national and international licensing authorities can support improved surveillance for antimicrobial resistance.

Current surveillance mechanisms are unable to link antimicrobial consumption to development of resistance. An analysis of this linkage is essential for arriving at the most effective response strategy. Surveillance should also include analysis of unrelated resistance genes sharing the same plasmid or transposon.

In the future, it seems desirable to refrain from using any antimicrobials for the promotion of animal growth. As exemplified by the use of virginiamycin in animal feed and the subsequent emergence of enterococci resistant to antibiotics, the use of any antimicrobial can lead to unexpected consequences that limit medical choices.

Antibiotics as promoters of animal growth can be phased out gradually. Similar benefits can be generated by improving other aspects of animal care, such as hygiene. In the long run, an industrial investment in alternatives to antimicrobials for animal growth promotion should pay off in more efficient production of food animals as well as protection of the fragile resources that are critical to successful management of human infectious disease.

### References and Notes

1. J. Davies, *Science* 264, 375 (1994).

2. S. B. Levy, *Trends Microbiol.* 2, 341 (1994).

3. F. C. Tenover and J. E. McGowan, *Am. J. M. Sci.* 311, 9 (1996).

4. *Antimicrobial Feed Additives* (Government Official Reports, no. 132, Stockholm, Sweden, 1997).

5. *Report of Joint Committee on the Use of Antibiotics in Animal Husbandry and Veterinary Medicine* (Swann Committee, Her Majesty's Stationery Offfice, London, September 1969).

6. S. Levy, *N. Engl. J. Med.* 295, 583 (1976).

7. H. Tschäpe, *FEMS Microbiol. Lett.* 15, 23 (1994).

8. J. Trelfall *et al.*, *Lancet* 347, 1053 (1996).

9. P. N. Gaunt and L. J. V. Piddock, *J. Antimicrob. Chemother.* 37, 747 (1996).

10. W. Witte and I. Klare, *Microb. Drug Res.* 1, 259 (1995).

11. M. Arthur, P. Reynolds, P. Courvalin, *Trends Microbiol.* 4, 401 (1996).

12. Report of the WHO meeting on the medical impact of the use of antimicrobial drugs in food animals, Berlin, 13 to 17 October 1997. Available www.who.ch/programmes/emc/zoo/oct97.pdf

13. WHO Scientific working group on monitoring a management of bacterial resistance to antimicrobial agents, WHO/CDS/BVI/95.7 (Geneva, Switzerland, 1994).

14. S. Thomke and K. Elwinger, *Report to the Commission on Antimicrobial Feed Additives* (Swedish University of Agriculture, Uppsala, 1997).

# The Human Genome Project: Reaching the Finish Line

*by R. Waterston and J. E. Sulston*

*Scientists involved with the Human Genome Project are anxious to begin unravelling its secrets. With the end in sight, they are redoubling their efforts to finish the project ahead of schedule.*

The human DNA sequence will be the central organizing principle for human genetics in the next century. It will be an essential reference for all biologists, and it is therefore vital that a high standard of accuracy, continuity, and accessibility be achieved as the sequence is determined. Because of the intrinsic excitement of unraveling the human genetic code and its potential to benefit humanity, there is a great temptation to acquire a view of the human genome as fast as possible. New initiatives that accelerate and enhance the program are to be welcomed and integrated into the emerging product, but none must divert us from the central aim of producing the ultimate, complete reference sequence.

Notable milestones in the International Human Genome Project include a comprehensive human genetic map *(1)*, on which landmark-based physical maps of the genome have been built *(2)*, and a human transcript map, also placed on the genetic framework, which provides any user with the positions of more than 30,000 human genes *(3)*. These resources have had an immediate effect in accelerating the identification of human disease genes. Since 1990, roughly 200 disease-associated genes have been identified by strategies that have benefitted directly from map or sequence information provided by the Human Genome Project *(4, 5)*. Such studies are leading to advances in the molecular understanding of disease, the development of powerful new diagnostic tests, and increased hopes for preventive strategies, new treatments, and even cures.

The systematic sequencing of the genomes of such model organisms as *Haemophilus influenzae* [1.8 megabases (Mb)] *(6)* and *Saccharomyces cerevisiae* (12.5 Mb) *(7)* have already given us more than a glimpse of the future. The project to sequence the 100-Mb genome of the nematode *Caenorhabditis elegans*, expected to be completed at the end of this year, will reveal all the genes and other information encoded in the DNA of a multicellular organism. The sequences of these genomes, combined with the ever-growing amounts of sequence data from a wide variety of other organisms, have opened up new lines of communication among biologists for an explosion of discoveries. Suddenly, a gene discovered in humans can be understood through its relatives in other organisms *(8)*.

When the *C. elegans* project began in 1990, there was little consensus about how to sequence genomic DNA on this scale. In fact, most scientists felt that revolutionary technology would be required before sequencing of the human genome could be undertaken. Instead, the continual evolution of sequencing methods and strategies used for model organisms has brought the human genome within reach. As a result, there has been a remarkable convergence by the scientific community on one general process (9), outlined as follows: A clone representing the region under study is selected from the map for subcloning and sequencing. Individual sequences of 400 to 1000 bases are generated by gel-based separation of dideoxy-terminated products produced by in vitro DNA synthesis, and the sequencing ladder is detected with fluorescent labels. Sequences are generated first from random fragments of the DNA to be sequenced (the "shotgun" phase). After automatic assembly of these sequence reads, based on sequence overlaps, additional data are collected in a directed fashion. In this "finishing" phase, ambiguities and gaps in the assembled sequence are resolved by editing; by determination of further sequence data on existing templates, using alternative sequencing chemistries, custom primer walks, long reads or reverse reads; and by generating new templates by constructing small-insert libraries of specific restriction fragments or polymerase chain reaction products that span gaps or difficult regions.

The Wellcome Trust, the National Human Genome Research Institute, the U.S. Department of Energy, and other government agencies launched a pilot phase of systematic human sequencing in late 1995 and early 1996 to test the feasibility of applying the methods and strategy of the model organism projects to human genomic DNA. In only 2 years, the world total of finished human genomic sequence has gone from 15 to 180 Mb. If assembled but as-yet-unfinished sequence is added on, then the total available in the public domain exceeds 330 Mb, or 11% of the genome *(10)*. With the increases in throughput previously planned, more than 800 Mb would be finished by 2001, and a total of 1500 Mb of finished and assembled sequence

Reprinted with permission from "The Human Genome Project: Reaching the Finish Line" by R. Waterston and J. E. Sulston. *Science* 282, October 2, 1998, pp. 53–54. ©American Association for the Advancement of Science.

(half the genome) would be expected to have appeared in the public domain. This is the result of a combined effort involving large and small groups, coordinating their efforts through a publicly available map *(11)*.

In addition to convergence on sequencing methods, the community has agreed on issues that are key to the success of the project and to coordination among groups, as follows:

(i) The release of all data is immediate, without constraints. The finished sequence of each clone is submitted to the public databases without delay and without patenting. All unfinished assembled sequence is released every night *(12)*.

(ii) All sequence is finished to high accuracy. The agreed error rate for finished sequence is less than 1 in 10,000 bases, and all sequences are fully contiguous. Without this standard of accuracy, the power and precision of computational searches for genes and sequence variations are compromised, and investigators have to waste time and money to correct and finish the sequence before it is reliable. To achieve this standard, rigorous cross-checking between laboratories has been carried out, and new strategies for gap closure have been developed *(13)*, so that even the most difficult (for example, GC-rich) regions in the human sequence can be completed. The only exceptions are large arrays of tandem repeats; as these are variable in the population and are unstable in available cloning vectors, attempts to generate perfect sequence from them are unwarranted.

(iii) All clones used for sequencing are physically mapped. This approach ensures good coordination and cooperation between laboratories. Sequencing on a clone-by-clone basis also has the advantage that sequencing problems can be isolated and solved more easily, which becomes critical as the scale of the project increases. In particular, the problem of assembling a consensus sequence from individual reads is greatly simplified.

The absence of a preexisting high-resolution clone map of the human genome has led to proposals that could avoid mapping. A whole-genome shotgun approach was proposed as an alternative by Weber and Myers in 1996 *(14)*. This strategy, as recently revived by Venter, *et al.* *(15)*, entails subcloning random fragments of total human genomic DNA directly and using very high-throughput sequencing technology to generate random sequences to provide at least 10-fold coverage of the genome. No doubt a vast amount of useful sequence information will be derived. But a major shortcoming of this strategy is that even with accurate linking of forward and reverse reads from templates of varying sizes, assembly would likely be woefully inadequate *(16)*. This is chiefly because the human genome is highly repetitive (40% or more), and many repeats exceed the length of a single sequence read *(17)*. Sequences containing very similar repeats will be automatically assembled into the same contigs, and careful reanalysis will be necessary to detect and correct such missassemblies. Even assemblies of shotgun sequence data of 100- to 200-kb bacterial artificial chromosome (BAC) clones or P1-derived artificial chromosome (PAC) clones sometimes require editing and additional work before they can be completed.

Another strategy to avoid mapping, proposed by Venter in 1996, was to sequence each end of every BAC in a genomic library *(18)*. Matching a finished sequence to one or more BAC ends permitted selection of the next clone *(18)*. However, this approach is also prone to errors caused by repeats. Since then, the acceleration in physical mapping has removed doubts about the feasibility of taking a map-based approach *(19)*.

The success of the international consortium gives us confidence that the human genome sequence can be completed with high quality by the target data of 2005. However, there is a growing urgency worldwide to see the whole human sequence by the millennium. More and more biologists in universities and industries are depending on genomic sequence information. Breakthroughs in understanding human disease have been made with finished and unfinished sequence.

The growing experience of the sequencing centers, plus advances in finishing strategies, suggest that with added funds, these groups could increase the output of finished sequence above the projected rates. It is also clear that the shotgun phase does not need to remain strictly linked with finishing and can be accelerated more readily. The required increase in capacity is being facilitated by advances in technology, with capillary sequencers from at least two commercial sources being developed.

The common view of the sequencing centers is that acceleration of the mapping and shotgun phases will provide a first draft of the entire human genome within 3 years from now, of which one-third will be finished. This would be of great value to biologists as it would provide segments of almost all genes for homology searching. Every sequence read would be positioned in the genome by the map location of the parent clone. In addition, all sequence data from other sources would readily be drawn in and mapped, including all available gene sequences, plus sequences from the whole-genome shotgun initiative *(16)* and possibly other sources. Indeed, the quality and organization of whole-genome shotgun initiatives will be considerably improved with the benefit of the clone-based sequence information. This should, in turn, further accelerate completion of the reference human sequence.

Although the acceleration of parts of the international program will require an initial outlay in expenditure, these are not new costs, but rather the earlier spending of funds already committed to human genome sequencing. More shotgun done now means less shotgun required later. Internationally, by June 2000, one-third of the genome will be covered by already-planned efforts. Shotgun coverage of the remainder would be achieved to a depth determined largely by the amount of available funding. To allow such a plan to proceed, the Wellcome Trust has recently agreed to bring forward part of its commitment to the Sanger Centre.

It is important to state explicitly that production of the first draft must not deflect us from the real goal of finishing the job, nor should it add to its overall cost. Acceleration of the intermediate stage must be carried out alongside the already planned commitment to increase the rate of finished sequence production. Highly accurate, fully contiguous sequence is the correct and most valuable product of the Human Genome Project *(20)*. It will be required to find all the genes in human DNA, and it will be critical in mouse-human comparisons to refine gene predictions and find regulatory and other functional elements. Such studies will be vital to a fuller understanding of the human genome, as we have already seen with the nematode and yeast genomes. Furthermore, in areas where nothing is known now, accurate finished sequence is essential to facilitate future discovery.

### References and Notes


1. C. Dib *et al.*, *Nature* 300, 152 (1996); J.C. Murray *et al.*, *Science* 265, 2049 (1994).
2. For whole-genome physical maps, see I.M. Chumakov *et al.*, *Nature* 377, 175 (1995); T. J. Hudson *et al.*, *Science* 270, 1945 (1995). Detailed physical maps of some individual chromosomes (21, 22, 16, 12, 3 and 7) appear in the following: M. Chumakov *et al.*, *Nature* 359, 380 (1992); F. S. Collins *et al.*, *ibid.* 377, 367 (1995); N. A. Doggett *et al.*, *ibid.*, p. 335; K. S. Krauter *et al.*, *ibid.*, p. 321; R. M. Gemmill *et al.*, *ibid.*, p. 299; G. G. Bouffard *et al.*, *Genome Res.* 7, 673 (1997).
3. A transcript map of the human genome is available at http://ncbi.nlm.nih.gov/genemap. P. Deloukas *et al.*, *Science*, in press (Oct. 23, 1998).
4. A list of inherited disease genes identified by positional cloning is available at http://www.nhgri.nih.gov/DIR/GTB/CLONE/.
5. F. S. Collins, *Nature Genet.* 9, 347 (1995).
6. The genome sequence of *H. influenzae* was determined by whole-genome random sequencing and assembly [R. D. Fleischmann *et al.*, *Science* 269, 496 (1995)].
7. A. Goffeau *et al.*, *Nature* 387 (suppl.), 1 (1997).
8. A database cross-referencing human genes and related genes in model organisms is available at http://www.ncbi.nlm.nih.gov/XREFdb and also in D. Bassett *et al.*, *Trends Genet.* 11, 372 (1996).
9. For methods, see E. Mardis and R. Wilson, in *Genome Analysis, A Laboratory Manual, Vol. 1: Analyzing DNA*, B. Birren, E. D. Green, S. Klapholz, R. M. Myers, J. Roskams, Eds. (Cold Spring Harbor Laboratory Press, Cold Spring Harbor, New York, 1997), pp. 397–453; R. Wilson *et al.*, *Nature* 368, 32 (1994).
10. These data are from the National Center for Biotechnology Information and Genome Monitoring Table at the European BioInformatic Institute. Currently, the largest sequence contig is over 2 Mb, and the largest mapped contig is 12.9 Mb (chromosome 22, available at http://webace.sanger.ac.uk/HGP/Chr22).
11. Current and planned efforts of the international genome sequencing consortium are listed in the human genome sequence index (http://ncbi.nlm.nih.gov/HUGO). Graphic displays of the information can be viewed at http://webace.sanger.ac.uk or at http://www.ncbi.nlm.hih.gov/Entrez/Genome/org.html. D. Bentley *et al.*, *Trends Genet.*, in press.
12. Participants at the first international strategy meeting on human genome sequencing unanimously endorsed the "Bermuda statement" that "all human genomic information produced at large-scale sequencing centres should be freely available and in the public domain, in order to encourage research and development and to maximize its benefit to society." For details of access, see *(8)*.
13. A. A. MacMurray, J. E. Sulston, M. A. Quail, *Genome Res.* 8, 562 (1998).
14. J. L. Weber and E. W. Myers, *ibid.* 7, 401 (1997).
15. J. C. Venter *et al.*, *Science* 280, 1540 (1998).
16. P. Green, *Genome Res.* 7, 410 (1997).
17. A. F. A. Smit, *Curr. Opin. Gen. Dev.* 6, 743 (1996).
18. J. C. Venter, H. O. Smith, L. Hood, *Nature* 381, 364 (1996).
19. Over 400 Mb of mapped clones for sequencing have been produced in our centers (see http://www.sanger.ac.uk or http://genetics.wustl.edu).
20. M. V. Olson and P. Green, *Genome Res.* 8, 414 (1998).


21. We are grateful for discussions with F. Collins, M. Morgan, E. Lander, D. Cox, R. Gibbs, and M. Olson who have expressed support. We also thank the staff of the Sanger Centre and the Genome Sequencing Center for assistance in preparing the manuscript.

# Genetic Enhancement in Humans

*by Jon W. Gordon*

*Advances in gene transfer technology have caused scientists and bioethicists to question what is meant by genetic enhancement and how it should be applied to humans.*

Dramatic advances in gene transfer technology since the early 1980s have prompted consideration of its use in humans to enhance phenotypic traits. The notion that genetic modification could confer special advantages on an individual has generated excitement. Controversial issues surround this prospect, however. A practical concern is determining how to ensure equal access to such advanced medical technologies. There has also been speculation that genetic enhancement might affect human evolution, and philosophical objections have been raised, based on the belief that to intervene in such fundamental biological processes is to "play God." Although such philosophical questions cannot be resolved through data analysis, we nevertheless have the tools in hand to objectively assess our state of progress. We can also assess the impact that promulgation of such technology might have on human evolution and formulate sensible guidelines for developing policies governing human genetic enhancements.

### Defining Genetic Enhancement

Some experts have argued that "enhancement" can have different meanings depending on the circumstances. For example, when a disease is common, the risk for developing the disorder may be considered the norm, and genetic alleviation of that risk might be regarded as a form of enhancement *(1)*. This kind of semantic gamesmanship is misleading. The obvious public concern does not relate to improvement of traits for alleviation of deficiencies or reduction of disease risk, but to augmentation of functions that without intervention would be considered entirely normal. To raise the athletic capabilities of a schoolyard basketball player to those of a professional or to confer the talents of Chopin on a typical college music professor is the sort of genetic enhancement that many find troublesome. The experts in the gene transfer field should acknowledge the distinction in order to avoid causing public distrust and undermining the deliberative process.

Another important distinction is that between genetic changes that are heritable and those that cannot be genetically transmitted. At the present time, gene transfer approaches that involve the early embryo are far more effective than somatic cell gene therapy methodologies. Embryo gene transfer affords the opportunity to transform most or all cells of the organism and thus overcomes the inefficient transformation that plagues somatic cell gene transfer protocols. Moreover, the commonly used approaches to embryo gene insertion—pronuclear microinjection *(2)* and transfection of embryonic stem cells *(3)*—are associated with stable, high expression of donor DNA. Typically, however, genetic changes introduced into the embryo extend to the gametes and are heritable.

Scenarios can be constructed wherein introduced genes could be deleted from germ cells or early embryos derived from the treated individual. For example, transferred genes could reside on artificial chromosomes that could be deleted by activating a recombinase that induced recombination of the chromosome ends *(1)*. Such approaches, however, are currently speculative. Germline gene transfer has already succeeded in several animal species. Because of this and the general belief that voluntary abstention from germline modification in humans is unlikely, a candid discussion of genetic enhancement must include the possibility that changes introduced will be transmitted to offspring.

### The State of the Art

Animal experiments thus far have attempted to improve what are intuitively regarded as "simple" traits such as growth rate or muscle mass. Efforts to genetically improve the growth of swine have involved insertion of transgenes encoding growth hormone *(4, 5)*. Nevertheless, despite the fact that growth hormone transgenics are expressed well in swine, increased growth does not occur *(4, 5)*. Although the transgenic animals fortuitously have less body fat *(5)*, these unexpected benefits cannot be extrapolated to human clinical protocols. Before a human embryo is treated with recombinant DNA, we must know exactly what we are doing.

Another spectacular failed attempt at enhancement resulted from efforts to increase muscle mass in cattle. When expressed in mice, the avian *c-ski* gene, the cellular counterpart of the retroviral *v-ski* oncogene, induced massive muscle hypertrophy *(6)*. This prompted efforts to produce cattle expressing a *c-ski* transgene. When gene transfer was accomplished, the transgenic calf initially exhibited muscle hypertrophy, but muscle degeneration and wasting soon followed. Unable to stand, the debilitated animal was killed *(7)*.

Why did these enhancement experiments fail? For clues, it is useful to compare modern-day gene transfer technology with the more traditional approach to genetic engineering: selective breeding. Selective breeding maximizes the reproductive efficiency of individuals that exhibit desired characteristics. The selection strategy is oblivious to the number of genes responsible for generating the phenotype. Swine selected for rapid growth may consume more food, produce more growth hormone, respond more briskly to endogenous growth hormone, divert proteins toward somatic growth, and possess skeletal anatomy that allows the animal to tolerate increased weight. Dozens or perhaps hundreds of genes may influence these traits, but in selective breeding, favorable alleles at all loci can simultaneously be selected. In contrast, gene transfer selects one relevant locus and attempts to improve it in isolation. It is little wonder that this approach, albeit potentially powerful and efficient, is more chancy, and has, despite more than 10 years of effort, failed to yield even one unequivocal success. Greater success has been achieved in genetic enhancement of plants, which are more easily manipulated genetically and reproductively: for example, see *(8)*.

Given the inherent limitations of the gene transfer approach to enhancement, discussion of extending such procedures to humans is scientifically unjustified. We clearly do not yet understand how to accomplish controlled genetic modification of even simple phenotypes. Where more complex traits such as intelligence are concerned, we have no idea what to do, and in fact we may never be able to use gene transfer for enhancement of such phenotypes. A useful way to appreciate the daunting task of manipulating intelligence through gene transfer is by considering the fact that a single cerebellar Purkinje cell may possess more synapses than the total number of genes in the human genome. There are tens of millions of Purkinje cells in the cerebellum, and these cells are involved in only one aspect of brain function: motor coordination. The genome only provides a blueprint for formation of the brain; the finer details of assembly and intellectual development are beyond direct genetic control and must perforce be subject to innumerable stochastic and environmental influences.

### Genetic Engineering and Human Evolution

Some have suggested that genetic enhancement and related reproductive technologies now give us the power to control human evolution. This solemn pronouncement is totally without scientific foundation. The evolution of the human species may be understood as a nonrandom change in allelic frequencies resulting from selective pressure. The change progresses over generations because individuals with specific patterns of alleles are favored reproductively. If new alleles were introduced by gene transfer, the impact on the species would be negligible. Every month worldwide approximately 11 million babies are born. The addition of one genetically modified individual could not significantly affect gene frequencies. Moreover, if the "enhanced" individual had his or her first child at the age of 20, then 2,640,000,000 unengineered children would be born during the interval between the birth and procreation of the gene recipient. Even if 1000 successful gene transfers were performed per year, a number not likely to be achieved in the foreseeable future, those newborns would constitute only 1/132,000 of all live births. Thus, any effort to enhance the human species experimentally would be swamped by the random attempts of Mother Nature.

Finally, there is no certainty that genetically enhanced individuals would have greater biological fitness, as measured by reproductive success. A genius or great athlete who has no children has no biological fitness as defined in evolutionary theory. For these reasons, neither gene transfer nor any of the other emerging reproductive technologies will ever have a significant impact on human evolution.

### Developing Policy

If we accept the notion that genetic enhancement is not practicable in the near future, what policies should we develop concerning the use of such technology? The decision to undertake any form of invasive medical intervention immediately renders the treatment subject a patient who has a right to informed consent as well as to protection from unjustifiably dangerous medical manipulation. Our inability to predict the consequences of an attempt at genetic enhancement makes informed consent impossible, and current knowledge from animal experiments tells us that embryo gene transfer is unsafe: The common approach of pronuclear microinjection is characterized by random integration of donor DNA, a lack of control of the number of gene copies inserted, significant rearrangements of host genetic material, and a 5 to 10% frequency of insertional mutagenesis *(9)*. Homologous recombination *(10)* in embryonic stem cells overcomes many of these shortcomings, but human embryonic stem cell transfection would necessarily be followed by nuclear transfer into enucleated oocytes. Because nuclear transfer in at least two animal models is associated with a low birth rate and a very high rate of late pregnancy loss or newborn death *(11)*, this procedure is also unsafe. The risks are so high and the documented efficacy is so low for gene transfer that it could not compare favorably to straightforward prenatal diagnosis even when a compelling need for therapy exists, as in cases of genetic disease. The use of gene transfer for elective purposes such as enhancement would stray far beyond the limits of acceptable medical intervention.

To attempt genetic enhancement with extant methods would clearly be medically unacceptable, but attempts to ban gene transfer legally could be a cumbersome approach to limiting its clinical use. Verification of compliance would be difficult. The diverse resources required for gene transfer necessitate that the procedure be carried out in facilities equipped for in vitro fertilization. Direct inspection would be required to uncover gene transfer procedures in such facilities. This would impose on the privacy of patients undergoing accepted assisted reproduction procedures such as sperm injection. Moreover, gene transfer can be easily concealed; in the case of pronuclear microinjection, only a few seconds are needed to complete the process. Legal restrictions can also be easily avoided by performing the procedure outside the area of jurisdiction.

Finally, and perhaps most important, broad legal restrictions incur the risk of limiting invaluable research. Exemplifying this problem is the current overly broad ban on federal funding for experiments with human embryos. The recent derivation of human embryonic stem cells from preimplantation embryos *(12)* has created important new research opportunities, accompanied by pressure to provide federal funds for the work. This pressure has led to the odd situation in which federal funds will likely be allowed for research with embryonic stem cells but not for manipulating human embryos to produce embryonic stem cell lines. If, as a society, we feel compelled to make a statement against genetic enhancement, we need not enact anticipatory legislation. Instead we can evaluate such manipulations as we would any other invasive clinical procedure. If we require that gene transfer be accompanied by informed consent, that it have a reasonable possibility of succeeding, that its cost not be excessive, that it have acceptable side effects and toxicities, that it not be accompanied by a burdensome requirement for long-term follow-up evaluation, and that it compare favorably with other treatment options, we will currently reject the procedure on all counts as medically unethical. Were entities such as the National Bioethics Advisory Commission or Congress to make such statements formally, no responsible physician would attempt genetic enhancement. Irresponsible use of technology can never be stopped, even by legislation.

Fear of genetic manipulation may encourage proposals to limit basic investigations that might ultimately lead to effective human gene transfer. History has shown that effort is far better spent in preparing society to cope with scientific advances than in attempting to restrict basic research. Gene transfer studies may never lead to successful genetic enhancement, but they are certain to provide new treatment and prevention strategies for a variety of devastating diseases. No less significant is the potential for this research to improve our understanding of the most complex and compelling phenomenon ever observed—the life process. We cannot be expected to deny ourselves this knowledge.

### References


1. G. Stock and J. Campbell, Eds., *Summary Report, Engineering the Human Germline Symposium* (University of California, Los Angeles, 1998).
2. J. W. Gordon *et al.*, *Proc. Natl. Acad. Sci. U.S.A.* 77, 7380 (1980); J. W. Gordon and F. H. Ruddle, *Science* 214, 1244 (1981); F. Costantini and F. E. Lacy, *Nature* 294, 92 (1981); R. L. Brinster *et al.*, *Cell* 27, 223 (1981).
3. M. J. Evans and M. H. Kaufman, *Nature* 292, 154 (1981); G. R. Martin, *Proc. Natl. Acad. Sci. U.S.A.* 78, 7634 (1981); S. L. Mansour, K. R. Thomas, M. R. Capecchi, *Nature* 336, 348 (1988); S. Thompson *et al.*, *Cell* 56, 313 (1989).
4. V. G. Pursel *et al.*, *Vet. Immunol. Immunopathol* 17, 303 (1987).
5. C. A. Pinkert, E. J. Galbreath, C. W. Yang, L. J. Striker, *Transgenic Res.* 3, 401 (1994); M. B. Solomon *et al.*, *J. Anim. Sci.* 72, 1242 (1994).
6. P. Sutrave, A. M. Kelly, S. H. Hughes, *Genes Dev.* 4, 1462 (1990).
7. R. A. Bowen *et al.*, *Biol. Reprod.* 50, 664 (1994).
8. K. J. Kramer and S. Muthukrishnan, *Insect. Biochem. Mol. Biol.* 27, 887 (1997).
9. R. D. Palmiter and R. L. Brinster, *Annu. Rev. Genet.* 20, 465 (1986); J. W. Gordon, *Int. Rev. Cytol.* 115, 171 (1989).
10. O. Smithies, R. G. Grett, S. S. Boggs, M. A. Koralewski, R.S. Kuchertapati, *Nature* 317, 230 (1981).
11. I. Wilmut *et al.*, *ibid.* 385, 810 (1997); T. Wakayama, A. C. F. Perry, M. Zucotti, K. R. Johnson, R. Yanagimachi, *ibid.* 394, 369 (1998).
12. M. J. Shamblott *et al.*, *Proc. Natl. Acad. Sci. U.S.A.* 95, 13726 (1998); J. A. Thomson *et al.*, *Science* 282, 114S (1998).

# Is It Time to Uproot the Tree of Life?

*by Elizabeth Pennisi*

*More genomes have only further blurred the branching pattern of the tree of life. Some blame shanghaied genes; others say the tree is wrong.*

A year ago, biologists looking over newly sequenced genomes from more than a dozen microorganisms thought these data might support the accepted plot lines of life's early history. But what they saw confounded them. Comparisons of the genomes then available not only didn't clarify the picture of how life's major groupings evolved, they confused it (*Science*, 1 May 1998, p. 672). And now, with an additional eight microbial sequences in hand, the situation has gotten even more confusing—so confusing that some biologists are ready to replace what has become the standard history with something new.

Many evolutionary biologists had thought they could roughly see the beginnings of life's three kingdoms: the ordinary bacteria; the Archaea, which are microbes best known for living in extreme environments; and the eukaryotes, which are all organisms, from yeast to people, whose cells have distinct nuclei. Comparisons of the genes encoding microbial species' ribosomal RNAs—the RNAs that make up the ribosomes, the small cellular structures where proteins are synthesized—suggested that life began with some sort of primitive bacteria. These then diverged into two branches, one leading to the modern bacteria and the other producing the Archaea and later branching again to produce the eukaryotes. When full DNA sequences opened the way to comparing other kinds of genes, researchers expected that they would simply add detail to this tree. But "nothing could be further from the truth," says Claire Fraser, head of The Institute for Genomic Research (TIGR) in Rockville, Maryland.

Instead, the comparisons have yielded many versions of the tree of life that differ from the rRNA tree and conflict with each other as well. Some put Archaea and bacteria together; others divide the Archaea into multiple groups. And microbiologists' response to this confusion is equally varied, as this year's Microbial Genomes III meeting, held in early February in Chantilly, Virginia, showed.

Some, pointing to evidence that microbes have swapped genes wantonly over evolutionary history, say that many of these genes are an unreliable guide to evolutionary history and the old RNA-based tree is still basically sound. But others think that it's time to uproot the old tree and are proposing candidates for new trees based on specific features of the genome and cell structure. And still others worry that gene swapping has turned the tree of life into a tangled briar whose lineages will be next to impossible to discern. "There's so much lateral transfer that even the concept of the tree is debatable," says Andre Goffeau, a geneticist at the Université Catholique de Louvain in Louvain la Neuve, Belgium.

## A Study in Confusion

As a case study in how new data are muddying the evolutionary picture, consider *Thermotoga*, a heat-loving microbe discovered 13 years ago on Volcano Island off Italy. The sequence of one of the microbe's rRNAs puts it on the bacterial branch of the evolutionary tree, just a bit higher than an ancient bacterium called *Aquifex*, sequenced last year by Diversa. But after TIGR scientists completed *Thermotoga's* genome 9 months ago, TIGR's Karen Nelson and her colleagues compared gene sequences from the two microbes. They found, she says, that "there's no consistent picture [of] where these two organisms fall."

As she described at the meeting, Nelson first identified 33 genes that are found in both *Thermotoga* and *Aquifex*, as well as in an additional 10 bacterial species, four Archaea, and the eukaryote yeast. She then used the base changes in the genes of the various organisms to construct separate trees reflecting the evolutionary history of each gene. "We could find only three situations that supported the branching order [derived] from the ribosomal subunit," she said. "It was impossible to say whether *Aquifex* or *Thermotoga* was more ancient."

Some evolutionary specialists suspect that such confusion is the result of rampant gene swapping. A gene acquired that way by one microbe would then look very similar to its counterpart in the donor organism, indicating their close kinship, while another gene could be very different in the two, possibly because it came from still another species. Thus, the histories inferred from the two genes would be contradictory, making the true history of the microbes difficult to discern.

This phenomenon is becoming more apparent with each new genome. The *Thermotoga* genome, for example, confirmed a suggestion made last year by Frank Robb and Dennis Maeder, microbiologists at the University of Maryland Center of

Reprinted with permission from "Is It Time to Uproot the Tree of Life?" by Elizabeth Pennisi. *Science* 284, May 21, 1999, pp. 1305–1307. ©American Association for the Advancement of Science.

 

Marine Biotechnology in Baltimore. Based on the partial *Thermotoga* genome, they proposed that the microbe has many genes in common with Archaea, and now that idea has been borne out. Because the two aren't supposed to be closely related, the likely explanation is that one somehow acquired genes from another, possibly because those genes were once part of mobile genetic elements capable of inserting into foreign genomes. Other researchers recently buttressed the case for such microbe-to-microbe transfers by finding what appears to be evidence for a recent transfer—evolutionarily speaking—within the past 100 to 1000 years.

Indeed, microbes can apparently even appropriate genes from "higher" organisms. At the microbial genome meeting, for example, Kira Makarova of the Uniformed Services University of the Health Sciences in Bethesda, Maryland, and her colleagues reported that the genome of the bacterium *Deinococcus radiodurans* contains several genes previously found only in plants. Thus it appears that these genes somehow got transferred to the microbe. And evolutionary biologist Winston Hide of the University of the Western Cape in Bellville, South Africa, reported that his team found that *Mycobacterium tuberculosis*, which causes tuberculosis, has taken on at least eight human genes. This is apparently beneficial to *M. tuberculosis*, as the genes encode proteins that help break down the hydrogen peroxide that some white blood cells use to kill foreign invaders. They could thus help the bacteria fight off host defenses.

Experts differ, however, on what such gene transfers mean to researchers trying to trace evolutionary relationships. Some, such as Goffeau, wonder whether they make tree building difficult, if not impossible. These researchers point out that the transfers may have gone on even more extensively in life's early days, when the molecular machinery for replicating and processing genetic material was still universal and organisms had not yet evolved ways of getting rid of foreign DNA. If so, tree building may be a meaningless exercise. The genomes of modern microbes may be mosaics of genes from different organisms rather than descendants of any single early form of life, and thus not even the ribosomal genes reflect true historical relationships.

Yet a few evolutionary biologists question whether the gene swapping has been as widespread as some researchers are now suggesting. TIGR's Owen White points out that the conclusion that *Deinococcus* has picked up plant genes is based primarily on researchers' inability to find those genes in any other microbes. But he cautions that the genome sequences currently available represent "a mere sliver" of all microbial genomes so that the genes may actually exist in a much broader range of organisms than researchers now realize. If so, the "plant" genes in *Deinococcus* might just be relics of genes lost from most other microbes. That might also hold for other apparent transfers.

And Peer Bork, a biochemist at the European Molecular Biology Laboratory (EMBL) in Heidelberg, Germany, argues that even if there has been a lot of lateral gene transfer, it might still be possible to detect the outlines of the tree of life. In a study published in the January issue of *Nature Genetics*, Bork and EMBL's Berend Snel and Martijn Huynen looked for the percentage of genes in common among 13 fully sequenced genomes, including those of bacteria, Archaea, and the eukaryote yeast. They found that each microbe shares between 77% and 17% of its genes with another organism. They didn't try to estimate what percentage of those genes had transferred in from elsewhere.

But based on those percentages, the researchers concluded that gene swapping would not displace any organism from its designated place in the existing RNA tree of life, with the kingdoms of the Archaea and eukaryotes branching off from a common ancestor, much as is now envisioned. That's because close relatives on that tree, such as *Escherichia coli* and *Haemophilus influenzae*, have higher percentages of genes in common than do organisms further apart, a finding that supports the microbes' current positions. "Gene transfer happens quite frequently at all levels," says Bork. "But there is still a strong [evolutionary] signal in the gene content."

### New Trees Needed?

Despite supportive data based on the relative gene contents in the genomes, some researchers are not convinced that the tree of life based on ribosomal genes is correct. Other types of gene comparisons and analyses of cell morphology make other evolutionary scenarios seem just as—if not more—likely, they suggest. Radhey Gupta, a molecular biologist at McMaster University in Hamilton, Ontario, for example, focuses on two different microbial features.

One feature involves small insertions or deletions, which he calls "indels," in the coding regions of genes. When an indel exists between two conserved stretches of coding sequence, Gupta says, it almost certainly has been retained through evolutionary time, as it's highly unlikely that two unrelated organisms would have the exact same alteration. Such indels can therefore be used to trace gene lineages. Gupta also takes into account the structure of the cell membrane. The result is a drastic reorganization of the classic tree in that Gupta finds a fundamental split among the bacteria. In addition, he divides the Archaea between the two bacterial branches instead of placing them in a separate kingdom.

Gupta begins with the supposition—that the most primitive organism resembled today's gram-positive bacteria (which are so called because they are dyed by the Gram stain). Based on the indels present or absent in other genes coding for highly conserved proteins, such as those involved in DNA synthesis, Gupta then came up with a scenario for how other organisms arose.

In his view, one group of these gram-positive bacteria eventually gave rise to the salt-tolerant Archaea and to the modern gram-positive bacteria, while some descendants of a second group eventually split into two lines, one of which became the gram-negative bacteria, while the other includes the Archaea that today thrive in methane—or acid-rich environments.

This picture is supported, Gupta says, by the presence in all the gram-negative bacteria examined of a particular 26-amino acid insert that is not found in the gram-positive organisms or in the Archaea. The insertion must have occurred, he says, after the gram-negative bacteria diverged from their gram-positive ancestors. Other support comes from the proportions of the bases guanine and cytosine in the genomes of these microbes, he notes. The salt-loving Archaea and many of their close gram-positive cousins belong on one major branch because they have a high proportion of these bases, while the other Archaea and gram-positive bacteria, with a low proportion, belong on the other.

In keeping with other recent evidence (*Science*, 13 March 1998, p. 1633), Gupta then suggests that eukaryotes arose when an archaeon fused with a gram-negative bacterium. The archaeon became the nucleus, while the gram-negative bacterium contributed most of the rest of the cell's components as indicated by the bacterial nature of the modern eukaryotic cell's membranes.

His theory got a lukewarm reception at the meeting, however. "It's an interesting hypothesis," says TIGR's Fraser. "But I don't think the story is anywhere near that simple." Nor does mathematician Hervé Philippe, who with microbiologist Patrick Forterre of the University of Paris-Sud has tentatively put forth an even more radical scenario. These researchers propose that a eukaryote-like cell, rather than a prokaryote, is the last common ancestor of all life-forms. "That's a more fruitful working hypothesis," Philippe maintains.

According to the analysis Philippe presented at the meeting, one problem with previous tree building efforts is that researchers didn't compensate sufficiently for the fact that genes can differ in their rates of evolution. (The results are also in press in *Molecular Biology and Evolution* and in the *Journal of Molecular Evolution*.) As a result, organisms that have a lot of the faster evolving genes tended to appear more ancient than they really are, confusing the trees. And Philippe says bacterial genes are the faster evolving ones, causing bacteria to fall to the tree's base.

But that's actually where eukaryotes belong, he says, basing this conclusion on an examination of the genes involved in the synthesis and use of the genetic information in the various types of organisms. He notes, for example, that many researchers think that the earliest organisms lived in an RNA world, where RNA, rather than DNA, was used to carry life's instructions. And compared to prokaryotes, eukaryotic cells have many more genes involved in processing the RNAs that may be relics of that RNA world.

In eukaryotes and Archaea, he says, the complex interactions between the proteins encoded by the genes needed to process the genetic information would have limited the ability of the genes to change. But those constraints would be less severe in bacteria, which somehow lost a few of these genes. The bacteria could have then evolved faster—and as a result they end up at the bottom of the tree when they don't deserve to be there.

This new tree of life pleases some researchers, particularly those who already have reason to suspect that the current tree might be misleading. For example, Maryland's Robb says he's often suspected that the placement of the hyperthermophiles, microbes that live at extremely high temperatures, toward the bottom of the tree might be an artifact resulting from assumptions about how fast these microbes evolve and on their initial discovery in extreme environments reminiscent of what early life might have experienced.

That placement was based in part on the finding that the DNA in the ribosomal genes of the organisms has lots of guanine and cytosine bases, an indication that their genomes had been around long enough for certain bases to become overrepresented. But it turns out that these bases tend to dominate in the genomes of organisms living in extreme environments, as they help stabilize DNA (*Science*, 8 January, pp. 155, 220). Thus, these DNAs do not necessarily tell the true story of evolutionary relatedness. "A lot of what Philippe is saying makes sense to me," Robb notes.

Nonetheless, like Gupta, Philippe has his skeptics, as many researchers think it's too early to reach any firm conclusions about early evolution from the new genomes. "We just don't have that high a resolution for ancient [events]," says EMBL's Bork. To try to resolve the issue, both he and Philippe advocate not only looking at more genomes but also working to understand better how genomes change through time.

Fraser agrees that more needs to be learned about the molecular mechanisms that underlie evolution. At this point, she says, the only thing that can be said with any certainty "is that the story is by far more complicated than is suggested by the ribosomal DNA tree."

# How the Genome Readies Itself for Evolution

*by Elizabeth Pennisi*

*Built into the genome's DNA sequences are regions that can promote rapid and extensive genetic change.*

The renowned author and cancer scientist Lewis Thomas once wrote: "The capacity to blunder slightly is the real marvel of DNA. Without this special attribute, we would still be anaerobic bacteria and there would be no music." Like many others—Nobel laureate Barbara McClintock was a notable exception—Thomas thought that genetic change, and hence the evolution of new species, results from small, random mutations in individuals genes. But a growing wealth of data, much of it presented at a recent meeting,* indicate that mainstream biologists need to consider genomes, and the kinds of evolutionary changes they undergo, in a much different light.

The work shows that the mutations leading to evolutionary change are neither as small nor as rare as many biologists have long assumed. Sometimes they involve the movements of relatively large pieces of DNA, like transposable elements, the stretches of mobile DNA originally discovered in maize by McClintock. They can even take form of wholesale shuffling or duplication of the genetic material (see p. 119). All these changes can affect the expression of genes or free up duplicated genes to evolve new functions.

What's more, these changes may not be totally random. Researchers have found, for example, that some stretches of DNA are more likely to be duplicated or moved to another place than others, depending on the nature of their sequences. They are also learning that the enzymes that copy and maintain the DNA introduce changes in some parts of the genome and not others, creating hotspots of mutation that increase the efficiency of evolution. As James Shapiro, a bacterial geneticist at the University of Chicago, puts it, "Cells engineer their own genomes."

Findings such as these are leading to what Lynn Caporale, a biotechnology consultant based in New York City, describes as a "paradigm shift." In the past, researchers assumed that genomes evolve to minimize mutation rates and prevent random genetic change. But the new findings are persuading them that the most successful genomes may be those that have evolved to be able to change quickly and substantially if necessary. Or as McClintock said in her 1983 Nobel lecture, the genome is "a highly sensitive organ of the cell, that in times of stress could initiate its own restructuring and renovation."

### Nature's Genetic Engineers

One of the oddest examples of how the genome can restructure itself comes from David Prescott, a molecular geneticist at the University of Colorado, Boulder. For the past 25 years, he has studied the genetic makeup of a group of single-celled organisms called hypotrichous ciliates, whose genomes are truly bizarre.

In addition to its largely working nucleus, called the macronucleus, which contains multiple copies of all the genes, every ciliate has one or more micronuclei, each of which carries two copies of all the genes. The micronuclear genes, which are normally inactive, are split into multiple sections, with lots of interrupting DNA, called internal eliminated sequences, between the coding regions. In three of the 10 genes analyzed thus far, the coding regions are also in the wrong order.

These micronuclear genes have to undergo a dramatic change during sexual reproduction when the micronuclei from the two partners fuse and give rise to both a new micronucleus and a new macronucleus. As the macronucleus takes shape, not only is the DNA between gene coding regions removed, but the coding regions have to be put into the correct order—"all in a matter of hours," Prescott notes.

Although having the gene-coding regions shuffled and split up in the micronucleus may seem a waste of time, Prescott thinks this arrangement helps generate new genes that can help the ciliates adapt to changes in their environment. In examining the DNA sequences of six species, he and his colleagues found that the sizes of the internal eliminated sequences vary from one to another. Sometimes they shift their positions by a few nucleotides, creating a slightly different coding region. This adds a new dimension to the changes that can occur, as now not just the original coding region but also this slightly altered one become available for shuffling and rearrangement into a new gene.

Most organisms do not go to the extremes that the ciliates do, but they have the potential to perform similar DNA rearrangements, if only on a lesser scale, because they have the necessary enzymes for cutting and rearranging DNA, as well as splicing it back together. Thus,



* "Molecular Strategies in Biological Evolution," 27–29 June, organized by the New York Academy of Sciences, in New York City.

 

Prescott concludes, the ciliates are "teaching us about what trickery DNA can perform to support evolution."

### Nonrandom Mutations

The changes in the ciliate genes appear to occur randomly, but researchers studying other species are finding, says Caporale, that the "rate of mutation is not monotonous throughout the whole genome." One striking example comes from Baldomero Olivera, a molecular biologist at the University of Utah, Salt Lake City, who has been assessing the incredible variation that exists in the toxins of predatory *Conus* snails.

With 500 species, these organisms are the most successful marine invertebrates. All use venom-laden harpoons to immobilize worms, fish, and other mollusks for food, and depending on the species, the venom contains 50 to 200 peptides. "Each *Conus* has its own distinct set" of toxin peptides, says Olivera. The genetic variation responsible for this toxin diversity, which may enable *Conus* species to identify one another, seems to be concentrated at particular hotspots in the snail DNA.

When Olivera and his colleagues began looking at the coding sequences for the large precursor proteins that break down to form the toxin peptides, they found that the first coding section, or exon, was almost identical in the pairs of *Conus* species examined. In contrast, Olivera reported at the meeting, the third exon, which is the one that codes for the peptide that becomes part of the venomous cocktail, "has just gone crazy in terms of changes." Because normally the functional parts of precursor proteins tend to change less over the course of evolution than the other parts, this finding suggests, he says, that the malleable exon somehow has a built-in tendency to mutate. The snails "are a good example of how expert nature is in using genome flexibility . . . to generate genetic variation," says Thomas Kunkel, a biochemist at the National Institute of Environmental Health Sciences in Research Triangle Park, North Carolina.

Kunkel stresses that there could be several different explanations for such mutational hot spots. For one thing, he notes, the efficacy with which the errors that creep into the genome are repaired can vary greatly. In test tube studies of DNA repair enzymes, Kunkel finds that the error rate for the repair of mistakes made during DNA replication can vary from 99% to 3%, depending on the nature of the sequence that needs repair. That may be because the sequence influences the ease with which the repair enzymes do their job.

What's more, some sequences are much more prone to error when the DNA is copied than others. In particular, DNA consisting of the same base repeated several times or simple two- and three-base repeats can be quite hard to replicate accurately. The problem is that the newly synthesized strand tends to slip relative to the strand being copied so that it may end up longer or shorter than the original, depending on whether it slips forward or backward.

### Genomic Flexibility

Such simple repeats are best known for the problems they cause. Researchers have traced the mutations causing Huntington's disease and several other hereditary disorders to large increases in the number of repeating triplets associated with the relevant genes. But the repeats also provide the genome with a means for fine-tuning gene expression, say Edward Trifonov, a biophysicist at the Weizmann Institute for Science in Rehovot, Israel, and David King, an evolutionary biologist at Southern Illinois University in Carbondale, and their colleagues. Because the number of repeats associated with genes can vary randomly among individuals, some people may have an advantage over others under certain environmental conditions if the number of repeats influences gene function. One example King cites are the genes controlling how much fat is produced in a mother's milk. Those genes sometimes contain repeats, which could help set their level of activity without changing the coding regions of the genes themselves, he suggests.

Repeats may also affect gene activity indirectly. In the fruit fly, for example, researchers have found that certain genes are less likely to be expressed when they are close to a series of repeats than when they are located elsewhere on the chromosome. Also, at least 50 genes encoding transcription factors, which control the expression of other genes, contain the expression of other genes, contain adjustable repeats. In 1994, researchers showed that a change in the number of the repeats can affect the rate at which a transcription factor is produced (*Science*, 11 February 1994, p. 808). This change could in turn affect the activity of other genes controlled by the factor, King points out.

The repeated sequences known as transposons or transposable elements, which can move from one stretch of DNA to another, provide another source of genomic flexibility. McClintock first proposed the existence of transposons in 1948 based on studies of such things as color variation in corn kernels. She noted, for example, that the colors can change from one generation to the next—far too fast for ordinary evolutionary change. This led her to propose that what she called "controlling elements" could jump about the genome each time it is replicated. If one of those elements landed in a gene, say for a corn pigment, it would interrupt the gene sequence, resulting in loss of function. Then, if it popped back out, the gene's function, and the corn kernel color, would be restored.

A great deal of work has since confirmed that idea. But transposable elements have long been considered to be rogue DNAs that seemed to land anywhere in the genome, making their effects little different from random gene mutations that are more likely than not to be detrimental to the organism's survival. "Some people see them as parasites," says Caporale.

As McClintock realized, however, these movements can not only have profound effects on the expression of genes but also provide grist for the mill of evolution. A recent study of morning glories by

Shigeru Iida, a molecular biologist at the National Institute for Basic Biology in Okazaki, Japan, and his colleagues provides an example. They traced the loss of blue or purple pigmentation in the flowers of some mutants to the insertion of transposable elements into genes needed to make the colored pigments.

As in corn, the transposons don't always stay put. Some of the mutant plants, whose flowers were originally all white, later produced speckled or streaked blossoms, a signal that, in the colored cells, the function of the altered gene had been restored because the transposon had moved. In the wild, the resulting streaked and speckled patterns could play a role in evolution, as one design may be more attractive to pollinators than another, causing plants with those patterns to become predominant, Iida notes.

Transposons likely played a pivotal role in the evolution of higher vertebrates as well. New data suggest that two enzymes key to generating the immune system's inexhaustible variety of antibodies are relics of an ancient transposon. A great deal of this variability is due to the way antibody genes are assembled: by joining three separate sequences (denoted V, J, and D), each of which comes in many variants, into a single antibody gene. Two reports, one in this week's *Nature* from David Schatz of Yale University and his colleagues and the other in *Cell* from Martin Gellert's team at the National Institute of Diabetes and Digestive and Kidney Diseases in Bethesda, Maryland, now show that the enzymes that do this assembly, Rag1 and Rag2, work just like enzymes called transposases that mobilize transposons.

The transposases recognize the ends of the transposon and cut it out of a chromosome. That freed-up piece of DNA, with the transposase still attached, can loop around itself and move to a new place in the genome. That's just what Rag1 and -2, which act together in a complex, do, Schatz says: "In a test tube, [Rag1 and -2] are indistinguishable from a transposase."

This discovery helps explain why the adaptive immune system seems to have appeared so abruptly during evolution. Unlike most vertebrate features, the immune system has no counterpart in invertebrates, says Ronald Plasterk, a molecular biologist at the Netherlands Cancer Institute in Amsterdam. The sudden introduction of a transposon into the part of the genome containing the remote predecessors of the antibody genes could have set the stage for the many gene duplications that allowed the large antibody gene complex to evolve. "[The immune system] is a wonderful example of how a mobile piece of DNA can have an astounding impact on evolution," Schatz says.

Besides altering genomes by disrupting specific genes, mobile elements may also reshape the whole architecture of the genome, providing yet another kind of genomic flexibility. Iida has found parts of foreign genes buried in the morning glory's transposons, suggesting that transposons may capture genes and move them wholesale to new parts of the genome. If this is true, then they make possible DNA shuffling that can place genes in new regulatory contexts and, possible, new roles.

And in May, Evan Eichler of Case Western Reserve University in Cleveland, Ohio, and his colleagues showed how a repeat sequence was likely responsible for the duplication of several genes' worth of DNA in humans and other primates (*Science*, 12 June, p. 1692). The researchers found copies of one piece of chromosome 16 in several other parts of the human genome, and Eichler suggests that repeated sequences at the ends of the inserted pieces of DNA were signals to enzymes that cut out that DNA as it replicated and allowed it to insert elsewhere. Sequence similarities between transposons and the repeats hint that the repeat sequence might be a remnant of an ancient mobile element. "It's no longer a matter of conjecture that transposable elements contribute to evolution," concludes Nina Fedoroff, a molecular biologist at Pennsylvania State University in University Park. "It's a fact."

These evolutionary contributions may increase when a species faces new selective pressures, says Caporale. As she puts it, "Chance favors the prepared genome." Researchers first noticed this happening in 1988 when John Cairns, then at Harvard University, showed that mutation rates in the bacterium *Escherichia coli* increased when the microbes needed to evolve new capabilities in order to survive changes in their environment. At the time, it seemed that only those genes directly involved with the adaptation changed, and this idea of adaptive or directed evolution caused quite a stir.

But then last year, molecular geneticist Susan Rosenberg at Baylor College of Medicine in Houston and her colleagues showed that mutation rates increase throughout the genome, although only in a subset of the population. Another group also found that more than just the relevant genes changed. The bacteria are "smart" about how they evolve, explains Rosenberg. In response to adverse conditions, they "turn on mechanisms of mutation that are nothing like the usual [mutation] mechanisms," she adds. They activate different repair enzymes and promote DNA shuffling, for example.

Not long ago, these ideas would have been considered heretical, but "I see more and more people looking at [evolution] this way," says Werner Arber, a Nobel laureate and microbiologist at the University of Basel in Switzerland. Whether by radically rearranging themselves, making use of mobile elements to generate variation, or causing certain stretches of DNA to mutate at high rates, genomes are showing that they can help themselves cope with a changing environment. Says Shapiro, "The capability of cells has gone far beyond what we had imagined."

# Calibrating the Mitochondrial Clock

*by Ann Gibbons*

*Mitochondrial DNA appears to mutate much faster than expected, prompting new DNA forensics procedures and raising troubling questions about the dating of evolutionary events*

In 1991, Russians exhumed a Siberian grave containing nine skeletons thought to be the remains of the last Russian tsar, Nicholas II, and his family and retinue, who were shot by firing squad in 1918. But two bodies were missing, so no one could be absolutely certain of the identity of the remains. And DNA testing done in 1992—expected to settle the issue quickly—instead raised a new mystery.

Some of the DNA from the tsar's mitochondria—cellular organelles with their own DNA—didn't quite match that of his living relatives. Forensic experts thought that most people carry only one type of mitochondrial DNA (mtDNA), but the tsar had two: The same site sometimes contained a cytosine and sometimes a thymine. His relatives had only thymine, a mismatch that fueled controversy over the authenticity of the skeletons.

The question of the tsar's bones was finally put to rest after the remains of his brother, the Grand Duke of Russia Georgij Romanov, were exhumed; the results of the DNA analysis were published in *Nature Genetics* in 1996. Like the tsar, the duke had inherited two different sequences of mtDNA from their mother, a condition known as heteroplasmy. But solving the mystery of the Romanov's [sic] remains raised another puzzle that first troubled forensics experts and is now worrying evolutionists. "How often will this heteroplasmy pop up?" wondered Thomas J. Parsons, a molecular geneticist at the Armed Forces DNA Identification Laboratory in Rockville, Maryland, who helped identify the tsar's bones.

Several new studies suggest that heteroplasmy may in fact be a frequent event. They have found that it occurs in at least 10% and probably 20% of humans, says molecular biologist Mitchell Holland, director of the Armed Forces lab. And because heteroplasmy is caused by mutations, this unexpectedly high incidence suggests that mtDNA mutates much more often than previously estimated—as much as 20-fold faster, according to two studies that are causing a stir. Other studies have not found such rapid mutation rates, however.

Resolving the issue is vital. For forensic scientists like Parsons, who use mtDNA to identify soldiers' remains and to convict or exonerate suspects, a high mutation rate might cause them to miss a match in their samples. It could also complicate the lives of evolutionary scientists who use the mtDNA mutation rate as a clock to date such key events as when human ancestors spread around the globe.

Evolutionists have assumed that the clock is constant, ticking off mutations every 6,000 to 12,000 years or so. But if the clock ticks faster or at different rates at different times, some of the spectacular results—such as dating our ancestors' first journeys into Europe at about 40,000 years ago—may be in question. "We've been treating this like a stopwatch, and I'm concerned that it's as precise as a sun dial," says Neil Howell, a geneticist at the University of Texas Medical Branch in Galveston. "I don't mean to be inflammatory, but I'm concerned that we're pushing this system more than we should."

## Counting Mutations

The small circles of DNA in mitochondria have been the favored tool for evolutionary and forensic studies since their sequence was unraveled in 1981. Unlike the DNA in the nucleus of the cell, which comes from both egg and sperm, an organism's mtDNA comes only from the mother's egg. Thus mtDNA can be used to trace maternal ancestry without the complicating effects of the mixing of genes from both parents. And every cell in the body has hundreds of energy-producing mitochondria, so it's far easier to retrieve mtDNA than nuclear DNA.

It seemed like a relatively straightforward genetic system. Researchers could count the differences in the same sequence of mtDNA in different groups of people and, assuming a constant mutation rate, calculate how long ago the populations diverged. But the case of the tsar highlights how little is known about the way mtDNA is inherited. His mother must have carried or acquired a mutation, so there were hundreds of copies of each of two kinds of mtDNA in her egg cells. She then passed some of each kind to her sons. But just how often do such mutations occur?

The most widely used mutation rate for noncoding human mtDNA relies on estimates of the date when humans and chimpanzees shared a common ancestor, taken to be 5 million years ago. That date is based on counting the mtDNA and protein differences between all the great apes and timing their divergence using dates from fossils of one great ape's ancestor. In humans, this yields a rate of about one mutation every 300 to 600 generations, or one every 6,000 to 12,000 years (assuming a generation is 20 years), says molecular anthropologist Mark Stoneking of Pennsylvania State University in University Park. Those estimates are also calibrated with other archaeological dates, but nonetheless yield wide margins of error in published dates. But a few studies have begun to suggest that the actual rates are much faster, prompting researchers to think twice about the mtDNA clock they depend upon.

For example, after working on the tsar's DNA, Parsons was surprised to find heteroplasmy popping up more frequently than expected in the families of missing soldiers. He and his colleagues in the United States and England began a systematic study of mtDNA from soldiers' families and Amish and British families. Like most such studies, this one compares so-called "noncoding" sequences of the control region of mtDNA, which do not code for gene products and therefore are thought to be free from natural selection.

The researchers sequenced 610 base pairs of the mtDNA control region in 357 individuals from 134 different families, representing 327 generational events, or times that mothers passed on mtDNA to their offspring. Evolutionary studies led them to expect about one mutation in 600 generations (one every 12,000 years). So they were "stunned" to find 10 base-pair changes, which gave them a rate of one mutation every 40 generations, or one every 800 years. The data were published last year in *Nature Genetics*, and the rate has held up as the number of families has doubled, Parsons told scientists who gathered at a recent international workshop* on the problem of mtDNA mutation rates.

Howell's team independently arrived at a similar conclusion after looking deep within the pedigree of one Australian family affected with Leber hereditary optic neuropathy, a disease caused by an mtDNA gene mutation. When the researchers analyzed mtDNA from 40 members of this family, they found that one individual carried two mutations in the control region (presumably unrelated to the disease, because it is noncoding mtDNA). That condition is known as triplasmy, because including the nonmutated sequence, he had three different mtDNA sequences in his cells.

By tracing the mutations back through the family pedigree, Howell was able to estimate that both mutations probably arose in the same woman who was born in 1861, yielding an overall divergence rate of one mutation every 25 to 40 generations. "Both of our studies came to a remarkably similar conclusion," says Howell, whose study was published in late 1996 in the *American Journal of Human Genetics*. Both also warned that phylogenetic studies have "substantially underestimated the rate of mtDNA divergence."

Several teams of evolutionists promptly went back to their labs to count mtDNA mutations in families of known pedigree. So far, Stoneking's team has sequenced segments of the control region in closely related families on the Atlantic island of Tristan da Cunha, where pedigrees trace back to five female founders in the early 19th century. But neither that study nor one of 33 Swedish families has found a higher mutation rate. "After we read Howell's study, we looked in vain for mutations in our families," says geneticist Ulf Gyllensten of Uppsala University in Sweden, whose results are in press in *Nature Genetics*. More work is under way in Polynesia, Israel, and Europe.

Troubled by the discrepancy in their results, the scientists have pooled their data with a few other studies showing heteroplasmy, hoping to glean a more accurate estimate of the overall mutation rate. According to papers in press by Parsons, and Stoneking and Gyllensten, the combined mutation rate—one mutation per 1200 years—is still higher than the one mutation per 6,000 to 12,000 years estimated by evolutionists, although not as fast as the rate observed by Parsons and Howell. "The fact that we see such relatively large differences among studies indicates that we have some unknown variable which is causing this," says Gyllensten.

Because few studies have been done, the discrepancy in rates could simply be a statistical artifact, in which case it should vanish as sample sizes grow larger, notes Eric Shoubridge, a molecular geneticist at the Montreal Neurological Institute. Another possibility is that the rate is higher in some sites of the DNA than others—so-called "hot spots." Indeed, almost all the mutations detected in Parsons and Howell's studies occur at known hot spots, says University of Munich molecular geneticist Svante Pääbo.

Also, the time span of observation plays a role. For example, because hot spots mutate so frequently, over tens of thousands of years they can revert back to their original sequences, overwriting previous mutations at that site. As a result, the long-term mutation rate would underestimate how often hot spots mutate—and the average long-term mutation rate for the entire control region would be slower than that from near-term studies of families. "The easiest explanation is that these two rates are caused by hot spots," says Pääbo.

If so, these short-term rates need not perturb long-term studies. "It may be that the faster rate works on the short time scale and that you use the phylogenetic rate for long-term events," says Shoubridge.

But Parsons doubts that hot spots account for all the mutations he has observed. He says that some of the difference between the long-term and short-term rates could be explained if the noncoding DNA in the control region is not entirely immune to selection pressure. The control region, for example, promotes replication and transcription of mtDNA, so any mutation that interferes with the efficiency of these

* "First International Workshop on Human Mitochondrial DNA," 25 –28 October 1997, Washington, D.C.

processes might be deleterious and therefore selected against, reducing the apparent mutation rate.

Regardless of the cause, evolutionists are most concerned about the effect of a faster mutation rate. For example, researchers have calculated that "mitochondrial Eve" —the woman whose mtDNA was ancestral to that in all living people—lived 100,000 to 200,000 years ago in Africa. Using the new clock, she would be a mere 6,000 years old.

No one thinks that's the case, but at what point should models switch from one mtDNA time zone to the other? "I'm worried that people who are looking at very recent events, such as the peopling of Europe, are ignoring this problem," says Laurent Excoffier, a population geneticist at the University of Geneva. Indeed, the mysterious and sudden expansion of modern humans into Europe and other parts of the globe, which other genetic evidence puts at about 40,000 years ago, may actually have happened 10,000 to 20,000 years ago—around the time of agriculture, says Excoffier. And mtDNA studies now date the peopling of the Americas at 34,000 years ago, even though the oldest noncontroversial archaeologic sites are 12,500 years old. Recalibrating the mtDNA clock would narrow the difference (*Science*, 28 February 1997, p. 1256).

But not everyone is ready to redate evolutionary history on the basis of a few studies of mutation rates in living people. "This is all a fuss about nothing," says Oxford University geneticist Martin Richards, who thinks the fast rate reaches back hundreds of years at most.

That, however, is squarely within the time frame of forensics cases. Heteroplasmy isn't always a complicating factor in such analyses. When it exists in more than one family member, the confidence in the identification gets stronger, as in the case of the tsar. But otherwise, it could let a criminal off the hook if his mtDNA differed by one nucleotide from a crime scene sample. Therefore, Parsons and Holland, in their work identifying 220 soldiers' remains from World War II to the present, now have new guidelines—adopted by the FBI as well—to account for a faster mutation rate. When a missing soldier's or criminal suspect's mtDNA comes up with a single difference from that of a relative or at a crime scene, the scientists no longer call it a "mismatch." Instead the results are considered "inconclusive." And, for now, so are some of the evolutionary results gained by using the mtDNA clock.

# Which of Our Genes Make Us Human?

*by Ann Gibbons*

*For decades scientists have known that at least 98% of human DNA is identical to that of chimpanzees. Now they have at last begun to explore which genes separate us from the apes.*

We humans like to think of ourselves as special, set apart from the rest of the animal kingdom by our ability to talk, write, build complex structures, and make moral distinctions. But when it comes to genes, humans are so similar to the two species of chimpanzee that physiologist Jared Diamond has called us "the third chimpanzee." A quarter-century of genetic studies has consistently found that for any given region of the genome, humans and chimpanzees share at least 98.5% of their DNA. This means that a very small portion of human DNA is responsible for the traits that make us human, and that a handful of genes somehow confer everything from an upright gait to the ability to recite poetry and compose music.

But what are these genes, so few in number yet so powerful in effect? Until now there has been little funding or research to track them down, and the primate genome has been almost virgin territory. "You could write everything we knew about the genetic differences in a one-sentence article," quips neuroscientist Thomas Insel, director of the Yerkes Regional Primate Research Center of Emory University in Atlanta.

Now that is changing, as a persistent band of geneticists and evolutionary biologists launches new studies to find differences in the genes, chromosomes, and biochemistry of humans and chimpanzees. Next month, researchers will report finding the first significant biochemical variation between humans and other apes: Humans lack a particular form of a ubiquitous cell surface molecule found in all other apes. Other teams are reporting newfound differences in the arrangements of DNA on the chromosomes of humans and other primates.

And new sequencing projects are starting to compare primate and human DNA base by base. Two new centers—at Yerkes and in Leipzig—opened simultaneously last year to study the molecular evolution of the great apes; next month, at an international meeting in Chicago, researchers will push for an organized, international primate genome project. Because primates are less susceptible than humans to certain diseases, including cancer and AIDS, the sequence differences are of more than evolutionary interest and are already drawing commercial attention. One Denver, Colorado, biotech company has already submitted patents on key human and chimp genes.

For the moment, no one can tie the few known molecular variations with the familiar litany of chimp-human differences, such as body hair, language, or brain size. But the leaders of genomic and evolutionary research alike say that now is the time to explore those links. "This is one of the major questions that those of us interested in our own biology would like to ask—what does that 1.5% difference look like?" says Francis Collins, director of the National Human Genome Research Institute.

### Vive La Difference

That question was first raised in print in a landmark 1975 paper by geneticist Mary-Claire King and the late biochemist Allan Wilson, both then of the University of California, Berkeley. They surveyed protein and nucleic acid studies and found that the average human protein was more than 99% identical to its chimpanzee counterpart; the coarse DNA hybridization methods of the time showed that the average nucleic acid sequence was almost as similar (*Science*, 11 April 1975, p. 107). Thus, King and Wilson concluded humans and chimpanzees were genetically as similar as sibling species of other organisms, such as fruit flies or mice.

This left a great paradox: Our DNA is almost identical to that of our chimp cousins, but we don't look or act alike. "The molecular similarity between chimpanzees and humans is extraordinary because they differ far more than many other sibling species in anatomy and way of life," the pair wrote.

What's more, much of the DNA in any organism is so-called "junk DNA" that has no apparent function, and mutations in these regions do not change the function of genes. Thus, many of the genetic differences between humans and chimps probably don't affect the organisms at all. The challenge is to find those few mutations that do make a difference—either by altering genes that code for proteins or by changing how genes are regulated, King and Wilson said.

But although many labs have since confirmed that our nuclear



 

DNA is 98% to 99% identical to that of chimpanzees, few have taken on the quest to find the differences that matter. "It was one of those fields that fell through the cracks," says Ajit Varki, a glycobiologist at the University of California, San Diego (UCSD), who has recently surveyed the known differences between humans and apes.

The tools and funds for sequencing large amounts of DNA rapidly weren't available until recently. And this line of work required a bold shift in thinking for most labs. "Most people in comparative genetics ask what's similar and conserved," says molecular evolutionist Caro-Beth Stewart at the State University of New York, Albany, who trained with Wilson. "Just a few of us have been trying to ask: What's different? What makes us human?"

## The Biochemical Trail

One way to answer that question is to start with biochemical differences, and then trace them back to their genetic origins. That approach has yielded its first big payoff, to be reported in the October issue of the *American Journal of Physical Anthropology*. After studying tissues and blood samples from the great apes and 60 humans from diverse ethnic groups, Varki and his colleagues Elaine Muchmore and Sandra Diaz at UCSD were surprised to find that human cells are missing a particular form of sialic acid, a type of sugar, found in all other mammals studied so far, including the great apes. "Now you've got something that is changing the surfaces of all cells in the body," says Varki.

The sialic acid molecule is found on the surface of every cell in the body, and previous work has shown that it can take on a surprising variety of roles. In some cases, it acts as a receptor for messages from other cells, but pathogens including those that cause cholera, influenza, and malaria also use it to gain a foothold on the cell. Chimpanzees are not as susceptible as humans to some of these pathogens, and the researchers speculate that this molecular change may be part of the reason why. There are even hints that sialic acid may be involved in cellular communication during brain development and function, says Varki.

The chimp and mammalian form of sialic acid, known as *N*-glycolylneuraminic acid (Neu5Gc), is modified from the basic form of the compound (called Neu5Ac) by the addition of an oxygen atom. But the human form is simply the basic acid, lacking the additional oxygen atom. That changes the shape of the molecule in a region that could alter how it is recognized by other molecules, whether pathogens or cellular messengers, says Varki. Other researchers are intrigued: "I think this is really nice work," says Svante Pääbo, a molecular geneticist at the University of Munich and leader of a German effort to sequence ape genomes. "It's really the first difference [in the expression of a gene product] that has come up."

In recent months, Varki's team has traced this difference back to a gene that codes for a hydroxylase enzyme in apes, which adds the extra oxygen atom. Humans are missing a 92-base pair section of this gene, according to new results from a team of Japanese researchers, led by glycobiologists Akemi Suzuki and Yasunori Kozutsumi at Tokyo Metropolitan Institute. Varki's lab has a paper in press in the *Proceedings of the National Academy of Sciences* that includes similar studies on the gene in the great apes; they are now also working on human fossils with Pääbo's lab to see if this change was recent in humans.

Yet the question still remains: Does this biochemical difference matter? No one has yet identified a specific function altered by the loss of this particular version of the molecule. "Until we know what this gene does, I remain skeptical of its importance," says Harvard University molecular anthropologist Maryellen Ruvolo. In search of clues, Kozutsumi is raising mice where the hydroxylase enzyme gene is knocked out, as it is in humans, to see if they produce the simpler form of sialic acid seen in humans—and if they have any anatomical or behavioral differences. "Maybe their mice will speak," jokes Varki, whose own lab is raising transgenic mice that overexpress hydroxylase in the brain to see if it affects anatomy or behavior.

But Ruvolo doesn't expect dramatic results: "Somehow, I can't believe that switching off the expression of this gene in adult humans is responsible for a myriad of important changes in human evolution. "Most researchers, including Varki, agree that the difference won't come down to a single genetic change. "There won't be one magic gene that makes us human,"says King, now at the University of Washington, Seattle.

## Remodeling Chromosomes

Another tactic for finding the key differences is to start with chromosomes, because even a simple karyotype—a picture of the chromosomes—reveals that other apes have 24 pairs of chromosomes while humans have 23 pairs. While 18 of the 23 pairs are virtually identical in humans and other apes, it has long been known that the remaining pairs have segments that have been reshuffled since the great apes went their separate ways (*Science*, 19 March 1982, p. 1525). In recent years, several labs in the United States and Germany have been homing in on this chromosomal remodeling.

For example, the gene mutated in a rare human disorder called adrenoleukodystrophy—made famous by the movie *Lorenzo's Oil*—turns up on the X chromosome in chimpanzees and humans. But nonfunctioning copies of it also have been found in different places in the chimpanzee and human genomes (*Science*, 12 June, p. 1693). Chromosomes 4, 9, and 12 have also been remodeled differently in chimps and humans, according to work published last month in *Genomics* by human geneticists David Nelson and Elizabeth Nickerson of Baylor College of Medicine in Houston. For example, the researchers spotted a chunk of DNA that sits on chromosome 4 in all apes and humans, but in chimpanzees it has moved to a new spot on the same chromosome and been inverted. The translocated chunk includes a gene called *AF4*, which codes for a transcription factor—and is mutated in some forms of acute leukemia in humans. Because apes are much less prone to certain cancers, including leukemia, this is an intriguing finding, raising the possibility that the inversion alters the factor's expression in chimps and so helps protect them from leukemia, says Nelson.

Still, the functional significance of this and other chromosomal differences between humans and other apes is unknown. One possibility, says Nelson, is that similar remodeling disrupted specific genes in our primate ancestors, altering human physiology or function. Because sperm and eggs can't mingle their genetic material unless the chromosomes line up properly, he adds that such rearrangements could have created a reproductive barrier between our ancestors and other primates—the first step in creating new species like our own.

But others, such as Pääbo, think that chromosomal rearrangements at influential sites are rare and so are skeptical that they play a major role in the differences between chimp and human. Pääbo and King think instead that the most promising research avenue is to identify small sequence differences that subtly change the expression of genes that regulate the timing of development, such as those that code for transcription factors that might lengthen the growth period of the brain and, hence, allow more complex brain structure in human fetuses.

The sequencing efforts that may reveal these differences are now under way. Pääbo's group in Munich and Leipzig has sequenced a 10,156-base pair segment of DNA in the X chromosome of humans and chimpanzees, confirming again that they are about 99% similar. Now they're seeking differences in the expression of the identified genes in the brain and in the immune system.

And at GenoPlex Inc., a Denver-based company founded last year by University of Colorado Health Sciences Center geneticists Jim Sikela and Tom Johnson, researchers have come up with a rapid method to find meaningful sequencing differences between humans and chimps. After sequencing a stretch of DNA in each species, they count two different types of nucleotide differences: those that change the structure and function of a protein product, and silent substitutions that don't. If the ratio of replacement to silent substitutions is high, they consider that the gene sequence is likely to have undergone a functional change that was selected for in humans.

Preliminary results suggest that they have found uniquely human genes involved in AIDS susceptibility and learning and memory, says Walter Messier, an evolutionary biologist at the company. The firm has submitted patents on novel uses of these gene sequences, which they hope may become targets for new drugs.

Much of this work is in its infancy, but researchers say they are poised on the verge of a brave new world where they will be able to identify and tinker with the DNA that makes us human—and will face new ethical dilemmas. "What happens if scientists identify a human gene that controls development of the larynx—a gene that might give chimpanzees the anatomy needed for speech?" asks Edwin McConkey, a molecular biologist at the University of Colorado, Boulder. "Can you imagine the ethical debate involved in whether or not to create transgenic chimps? It will open a real Pandora's box."

 

# BSE and Prions: Uncertainties About the Agent

*by Bruce Chesebro*

*The idea that prions are composed solely of protein was recognized by a Nobel prize. Still, some researchers caution that much remains unclear about these mysterious agents and their ability to cause disease.*

Humans and domestic animals are tragically susceptible to a family of rare, fatal brain diseases called transmissible spongiform encephalopathies (TSEs) or prion diseases. A prominent example is the recent epidemic of bovine spongiform encephalopathy (BSE) in the United Kingdom. In March 1996, a new variant of a human TSE, Creutzfeldt-Jakob Disease (nvCJD), was reported in the United Kingdom in a small group of people, all of whom were much younger than most individuals with CJD *(1)*. The unusual nature of these outbreaks prompted ready speculation that the causative agent for BSE was transmitted from cattle to humans, triggering the nvCJD cases. Although no definitive answer to this question has yet been obtained, several recent papers shed new light on this issue.

Brain damage in prion diseases is thought to occur when abnormal prion protein (PrP) molecules gain access to the brain and cause normal PrPs to take on the abnormal, disease-causing form. Raymond *et al. (2)* mimicked this process in a test tube and, in their cell-free biochemical system, analyzed interactions between normal and abnormal PrP molecules obtained from various species. Positive interactions between prions from different species, leading to formation of additional abnormal protein, usually predict the susceptibility of the different species to cross-species transmission of the prion diseases. Using abnormal, protease-resistant PrP from cattle with BSE, these workers found positive interactions with normal PrP of cattle, sheep, and mice (all of which are susceptible to experimental BSE), and negative interactions with hamster (which is resistant to BSE). Normal human PrP gave a weak, but detectable, positive interaction, a result that suggests that the biochemical potential for BSE transmission to humans exists, but that humans may be more resistant than other known susceptible species.

Two other groups have focused their efforts on nvCJD to test its possible relation to BSE. Bruce *et al. (3)* found that when brain tissue from three of the individuals with nvCJD and from BSE-infected cattle was injected into mice, all exhibited the same patterns of lesions in their brain. Furthermore, when brain tissue from eight of the BSE cases and three nvCJD cases *(3, 4)* was injected into several different mouse strains, all required similar incubation periods before the disease was evident. In a separate communication, Hill *et al.* observed that the pattern of PrP bands detected by protein immunoblot analysis in mice inoculated with six nvCJD brain extracts was similar to that observed in BSE-inoculated mice *(5)*. These similarities suggest that a similar strain of transmissible agent causes both BSE and nvCJD, but does not provide any indication of how BSE might have been transmitted from cattle to humans.

Prior to the above results, epidemiology had provided the strongest evidence that nvCJD is a form of a human prion disease related to BSE. Although the BSE epidemic itself is well past its peak, it is still uncertain whether the very low number of nvCJD cases detected since the initial report in 1996 reflect the gradual onset of a much larger incidence to be expected in the future. In this regard, data by Cousens *et al.* recently reveal an increased incidence of CJD in the United Kingdom in certain groups of farm workers exposed to cattle *(6)*. However, these patients did not exhibit the type of plaques typical of the pathology of nvCJD, and the pattern of PrP bands detected was also not the same as that of nvCJD. Thus, this increase in CJD incidence is likely due to factors other than BSE, a conclusion supported by the finding of a similar increase in CJD in dairy farmers in Italy where BSE has not been found.

Although most interesting, these recent findings fall short of proof and illustrate the extreme difficulty of establishing the susceptibility of humans to the BSE agent. The dilemma is accentuated by the fact that the causative agents of all the TSEs remain an enigma. Although the notion that "protein only" can account for the infectious agent has received considerable publicity as a result of the Nobel prize award to S. Prusiner for the discovery of prions, the fact remains that there are no definitive data on the nature of prions. Prions continue to be vaguely defined, and for the most part this

Reprinted with permission from "BSE and Prions: Uncertainties About the Agent" by Bruce Chesebro. *Science* 279, January 2, 1998, pp. 42–43. ©American Association for the Advancement of Science.

The author is in the Laboratory of Persistent Virus Diseases, Rocky Mountain Laboratories, Hamilton, MT 59840, USA. E-mail: bchesebro@nih.gov

### Is Abnormal Prion Protein the Transmissible Agent?

| | For | Against |
|---|---|---|
| **Amyloid diseases** | • Amyloid fibrils can induce polymerization of precursors | • Amyloid diseases not transmitted by amyloid fibrils |
| **Cell-free conversion** | • $PrP^{res}$ induces conversion of $PrP^{sen}$ | • Do not know if infectivity is generated |
| **Scrapie strains** | • $PrP^{res}$ structural variations | • Too many strains (20) in mice |
| **PrP null mice** | • No disease or agent replication | • PrP important as agent cofactor or receptor; involved in pathogenesis |
| **Genetics** | • TSE (both genetic and infectious)<br>• PrP mutant causes disease? | • Retroviruses (both genetic and infectious)<br>• PrP mutant = susceptibility gene? |
| **Transgenic mice** | • Overexpression of mutant PrP ($Leu^{102}$) causes brain disease | • No $PrP^{res}$ by protein immunoblot<br>• Transmission?<br>• No disease if no overexpression |
| **Irradiation** | • Target size too small | • Target size = small virus |
| **Composition** | • Infectious samples have $PrP^{res}$<br>• May be too little DNA for a virus<br>• No virus found so far | • $>10^5$ $PrP^{res}$ molecules per infectious unit<br>• Infectious samples have nucleic acids<br>• New viruses can be difficult to find |

term is used as an operational term for the transmissible agent, but without structural implications *(7)*. There are arguments both for and against the hypothesis that abnormal PrP itself is the transmissible agent (see the table above), but on either side of this controversy no argument is as yet completely convincing *(8)*.

The central unresolved issue is the difference between TSEs and nontransmissible amyloid diseases such as Alzheimer's disease, type II diabetes, and amyloidoses associated with various cancers and inflammatory diseases (see table on next page). In both groups of diseases, a normal or mutant precursor protein or peptide is incorporated into an aggregated, noncovalently linked polymeric fibril-like structure (amyloid), often with a high level of ß-sheet folding. This ordered aggregate can induce further polymerization of additional precursor protein, thus providing a possible basis for propagation or transmission. Furthermore, in the case of PrP, striking species and TSE strain specificity operates in this induced polymerization process, which might explain the species barrier effects and agent strains in vivo *(2, 8, 9)*. However, none of the non-PrP amyloid diseases has been found to be transmissible by injection or ingestion of the polymers themselves in spite of extensive attempts [see the table below and *(10)*]. Thus, there is no precedent for the transmission of TSE by the abnormal polymeric PrP. Either we are missing some vital and unique factor that distinguishes the polymers of TSE from those of other amyloid diseases, or the agent of transmission may not be the abnormal polymers themselves.

The alternative is that a virus is the transmissible agent. It is certainly conceivable that a virus might induce polymerization of a normal or mutant cell protein, as has been shown for B19 parvovirus and sickle cell hemoglobin *(8, 11)*. Although no direct evidence for a TSE virus has been obtained, searching for unknown viruses is not straightforward. Even with the powerful methods of molecular biology, without a specific probe or unique sequence, new viruses can be extremely difficult to detect. However, this should not deter future search efforts as new viruses (Borna and hepatitis), continue to be discovered.

Support for the protein-only hypothesis also comes from the apparent spontaneous generation of a neurodegenerativc disease in transgenic mice that overexpress a mutant form of PrP associated with familial TSE disease in humans *(12, 13)*. These results are inconclusive for several reasons. First, these mice differ from all known TSE models in that no abnormal protease-resistant PrP is detectable in diseased brain tissues by protein immunoblot. Second, brain disease is produced only when the mutant transgene is overexpressed, but not when it is present as a single-copy gene in the normal PrP locus. Third, transmission of this disease occurs only in transgenic mice expressing low levels of the same mutant PrP transgene, and not in normal nontransgenic mice. Thus, "transmission" could be due to transfer of cytokines or other molecules from diseased brain, which up-regulate mutant PrP transgene expression in the recipient mice. Nor have the transmitted materials been shown to have the typical properties of TSE agents, such as unusual resistance to inactivation by heat and harsh chemicals. To explain the lack of transmission to normal mice, it has been proposed that the PrP mutation ($Leu^{102}$) in the transgene causes a barrier to transmission to normal ($Pro^{102}$) mice. Although theoretically possible, it is unlikely because effective transmission can occur from humans with $Leu^{102}$ PrP humans to both monkeys and mice with $Pro^{102}$ *(14, 15)*.

## Comparison of Transmissibility of Amyloid Diseases

| Disease | Amyloid Protein | Transmissible by Amyloid | Transmissible Agent |
|---|---|---|---|
| **Alzheimer's disease** | β A4 | No | None |
| **Multiple myeloma** | IgL chain | No | None |
| **Diabetes (type 2)** | Amylin (IAPP) | No | None |
| **Familial amyloidotic polyneuropathy** | Transthyretin (TTR) | No | None |
| **Chronic dialysis** | $\beta_2$ Microglobulin | No | None |
| **Thyroid carcinoma** | Calcitonin | No | None |
| **Secondary amyloidosis** | SAA and SAP | No | *M. tuberculosis*, *M. leprae*, malaria |
| **TSE** | $PrP^{res}$ | ? | ? |

Hopefully, open-minded future research will resolve the mysteries regarding the nature of the TSE agent. Additional information on this subject must be obtained to better facilitate development of drugs. Evidence suggests that treatment with drugs that interact with the abnormal PrP can prolong survival in experimental mouse or hamster scrapie, a TSE disease originally derived from sheep. Other approaches—gene therapy with TSE-resistant PrP alleles or elimination of the PrP gene itself—are also possible, particularly for domestic animals. Clearly, we are in the very early stages of exploration of this subject. It would be tragic if the recent Nobel Prize award were to lead to complacency regarding the obstacles still remaining. It is not mere detail, but rather the central core of the problem, that remains to be solved.

An enhanced version of this commentary with links to additional resources is available for *Science* Online subscribers at www.sciencemag.org.

### References


1. R. G. Will *et al.*, *Lancet* 347, 921 (1996).
2. G. Raymond *et al.*, *Nature* 388, 285 (1997).
3. M. E. Bruce *et al.*, *ibid.* 389, 498 (1997).
4. M. E. Bruce, personal communication (1997).
5. A. F. Hill *et al.*, *Nature*, 389, 448 (1997).
6. S. N. Cousens *et al.*, *Br. Med. J.* 315, 389 (1997).
7. A. Aguzzi and C. Weissmann, *Nature* 389, 795 (1997).
8. B. Caughey and B. Chesebro, *Trends Cell Biol.* 7, 56 (1997).
9. R. A. Bessen *et al.*, *Nature* 375, 698 (1995).
10. Secondary amyloidosis can be induced by transmissible diseases such as tuberculosis, leprosy, and malaria, which all induce strong inflammatory responses, and is also associated with nontransmissible diseases with a strong inflammatory component such as rheumatoid arthritis.
11. B. Chesebro, *Nature Med.* 3, 491 (1997).
12. K. K. Hsiao *et al.*, *Science* 250, 1587 (1990).
13. G. C. Telling *et al.*, *Genes Dev.* 10, 1736 (1996).
14. P. Brown *et al.*, *Ann. Neurol.* 35, 513 (1994).
15. J. Tateishi *et al.*, *Neurology* 46, 532 (1996).

# Tweaking the Human Circadian Clock with Light

*by Dan A. Oren and Michael Terman*

*The circadian clocks of plants and animals were thought to use different timekeeping machinery. New research suggests that they may not be so different after all.*

For more than two millennia, keen observers have noted the circadian rhythms of flowering plants that bloom only in the daytime or of trees with leaves that open during the day and close at night. Such movements are regulated by a clock within each organism *(1)*, one that can be reset by light as days get shorter or longer.

Animals show strikingly similar circadian behavioral responses, driven by an internal clock and also reset (or entrained) by light *(2)*. The timekeeping similarities between plants and animals have not been thought to extend to the light receptors *(3)*. Information about chronobiological phototransduction in plants has blossomed, yet the primary visual light-sensitive system in animals—rod or cone photoreceptor molecules—may not be necessary for entrainment at all. But recently animals have started to seem more like plants. The fruit fly *Drosophila* has autonomous clocks throughout the body, each sensitive to its own intrinsic, as-yet-identified photoreceptors *(4)*, and in a report on page 396 of this issue, Campbell and Murphy show that the endogenous clock of humans can be entrained by light application to an unexpected spot, the popliteal region—that is, the back of the knees *(5)*.

The relation of light and other zeitgebers (time setters) to intrinsic biological rhythms is classically defined by phase response curves. These are constructed by exposing the organism to a zeitgeber at all phases of the endogenous rhythm and measuring the resulting effect on the phase of the cycle. With remarkable consistency, zeitgebers shift the rhythms by advancing the phase of the clock at certain parts of the circadian cycle and by delaying the phase when impinging on other parts of the cycle. The magnitude of such phase shifts is dose-dependent, the product of intensity × duration. The new phase response curves for extraretinal popliteal illumination in humans *(5)* conform to the classic pattern, with phase delays of body temperature and pineal melatonin secretion (outputs of the clock) occurring during the night, followed by phase advances early in the morning.

How is the entrainment signal carried from the knees to the endogenous clock? Presuming the conservation of chronobiological properties of plants and animals, one of us recently made a parsimonious proposal: that "humoral phototransduction" could subserve the zeitgeber effect *(6)*. This model postulates that heme moieties could serve as photoreceptors and that tetrapyrrole-based pigments, such as hemoglobin and bilirubin in mammals, act as counterparts to the tetrapyrrole-based, primary light-sensitive plant pigments of chlorophyll and phytochrome to mediate light's chronobiological effects. Hemoglobin and bilirubin could act in concert with the blood-borne nocturnal antioxidant melatonin to shift the phase of the clock. Such a model suggests a physiological function for light's known capacities to stimulate heme-based enzymes to form reactive gases, to cause dissociation of gases such as NO and CO from heme moieties, and to facilitate the destruction and excretion of unconjugated bilirubin.

The results show that popliteal illumination with a visible-spectrum light source—originally developed to treat hyperbilirubinemia (or jaundice) in neonatal babies—can shift human circadian rhythms without any transduction of light through the eye. This result is consistent with, but does not prove, the idea that blood contains chronobiological photoreceptors *(5)*. Further work will differentiate between light's effects on hemoglobin, bilirubin, the vascular tissue, and other components of the skin. Already we know from Campbell and Murphy's results that ultraviolet radiation, an obvious hazard to both eye and skin, is not required for the response.

In addition to its phase-shifting action, light exposure to the eyes immediately suppresses secretion of melatonin, a hormone that depresses body temperature, facilitates sleep onset, and—when administered in pill form at appropriate times of the day—induces phase advances or delays of the circadian system *(7)*. The inhibitory signal originates at the suprachiasmatic nuclei (SCN), whose self-sustaining oscillations underlie the mammalian circadian pacemaker. The retinohypothalamic tract, which carries information from the eyes to the SCN, thus provides phase-shifting signals and inhibits melatonin secretion. Because melatonin is secreted when the phase response curve for light is active—in the "subjective" night—it will be important to ascertain whether popliteal illumination also directly inhibits melatonin secretory activity.

Ocular light therapy has been applied successfully, for adjustment of aberrant sleep-wake cycles and alleviation of winter depression. When sleep occurs abnormally late (for example, with onset at 3 a.m. and awakening at 11 a.m.), light exposure at the end of the subjective night can elicit corrective advances, whereas early sleep (most common in the elderly) can be delayed by evening light *(8)*. Winter depression has been blamed on the delayed sunrise at northerly latitudes, which might allow the circadian pacemaker to drift later relative to habitual sleep. Although depressed patients and normal individuals show similar melatonin secretion phases, morning light preferentially alleviates symptoms —the earlier, relative to evening melatonin onset, the better, as predicted by the phase response curve *(9)*. On the other hand, evening light also has a mild benefit, which raises doubt that phase advances are essential to antidepressant action. If ocular illumination has antidepressant effects beyond its circadian action—even at midday, an immediate energizing effect in fatigued winter depressives has been noted—will popliteal illumination provide a mimic?

Campbell and Murphy posit a potential advantage of popliteal illumination over illumination of the eyes for therapy: It can be administered while patients are asleep. Outside our dark bedrooms, nature's gradual, dim dawn signal often rises during the final hours of sleep, when the propensity for phase advances is greatest. Artificial simulations of dawn shift the melatonin rhythm, expediting wake-up, increasing morning alertness, and countering winter depression *(10)*. Thus, bright light to open eyes is not a therapeutic sine qua non. The Campbell-Murphy method uses 13,000 lux of high illumination, similar to the amount used in postawakening bright light therapy (although the lux metric, adjusted for human photopic spectral sensitivity, may confuse dose specification). A comparison of popliteal illumination with dawn simulation at corresponding phases of sleep would clarify whether the two methods share parallel chronobiological and antidepressant effects. It is premature to assume therapeutic equivalence, especially since ocular light may well affect mood by trigger extracircadian mechanisms.

### References and Notes


1. *Histoire de L'Academie Royale des Sciences* (1729), p. 47.
2. C. P. Richter, *Res. Publ. Assoc. Res. Nerv. Ment Dis.* 45, 8 (1967); E. Bünning, *The Physiological Clock* (English Universities Press, London, ed. 3, 1973).
3. C. S. Pittendrigh, in *Seasonal Affective Disorders and Phototherapy*, N. E. Rosenthal and M. C. Blehar, Eds. (Guilford, New York, 1989), p. 87.
4. J. S. Takahashi, P. J. DeCoursey, L. Bauman, M. Menaker, *Nature* 308, 186 (1984); D. A. Oren *et al.*, *Depression* 1, 29 (1993); I. Provencio *et al.*, *Vision Res.* 34, 1799 (1994); J. G. Plautz, M. Kaneko, J. G. Hall, S. A. Kay, *Science* 278, 1632 (1997).
5. S. S. Campbell and P. J. Murphy, *Science* 279, 396 (1998).
6. D. A. Oren, *Neuroscientist* 2, 207 (1996); *Chronobiol. Int.* 14, 319 (1997).
7. A. J. Lewy *et al.*, *Chronobiol Int.* 9, 380 (1992); J. Arendt *et al.*, *J. Biol. Rhythms* 12, 604 (1997); C. A. Czeisler, *ibid.*, p. 618.
8. For a consensus report on bright light treatment of sleep disorders, see M. Terman *et al. Sleep*, in press.
9. A. J. Lewy, R. L. Sack, L. S. Miller, T. M. Hoban, *Science* 235, 352 (1987); C. Thompson, P. A. Childs, N. J. Martin, I. Rodin, P. J. Smythe, *Br. J. Psychiatry* 170, 431 (1997); M. Terman, in *Seasonal Affective Disorder and Beyond. Light Treatment of SAD and Non-SAD Conditions*, R. W. Lam, Ed. (American Psychiatric Press, Washington, DC, 1998, in press).
10. M. Terman and D Schlager, in *Sleep and Biological Rhythms*, J. Montplaisir and R. Godbout, Eds (Oxford Univ. Press, New York, 1990), p. 113; D. H. Avery *et al.*, *Am. J. Psychiatry* 150, 113 (1963) M. Terman, in *Sleep Science. Integrating Basic Research and Clinical Practice*, W. J. Schwartz, Ed. (Karger, Basel, 1997), p. 230.


11. The authors' research has been sponsored by a Department of Veterans Affairs Career Development Award (D.A.O.) and NIH grant MH-42931 (M.T.)

# Probing the Biology of Emotion

*by Christine Mlot*

*Christine Mlot is a science writer in Madison, Wisconsin.*
*Neuroscientists using an array of new methods to explore the anatomy and chemistry of emotion are finding that intense emotions can leave a long-lasting physical imprint in the brain.*

After he cracked the origin of species, Charles Darwin turned to another of life's mysteries: the nature of emotions. In his 1872 book *The Expression of the Emotions in Man and Animals*, Darwin took an outside-in approach, scrutinizing sulky monkeys, snarling dogs, faces of the insane, even his own wailing infant. He found that different species have common ways of expressing certain emotions, reinforcing his belief in the shared ancestry of animals.

A century later, researchers have picked up on Darwin's cross-species approach, but they've turned the study of emotions inside out. In rats, monkeys, and people, neuroscientists are looking beyond the surface expression of emotion into the pockets of brain chemistry in rats and other animals, and scanning brain activity in humans, they are beginning to map the neural circuits that send emotional messages. And although their field is in its infancy, they are poised to integrate these studies into an understanding of the biological basis of emotion. "Emotion is now tractable at a mechanistic level," says neuroscientist Richard J. Davidson of the University of Wisconsin, Madison. "It's been a huge advance, just enormous."

Increasingly, researchers are finding that intense emotions, particularly at key times in early life, can trigger not only behavioral changes but long-lasting physical changes in the brain. These persist long after the emotions themselves have passed and shape emotional responses later in life. This inside-out approach is also lending new insight into another favored Victorian notion—that of emotional temperaments. Individuals who are fearful or resilient not only have characteristic behaviors, they have distinct patterns of brain activity, too.

The new findings have sparked excitement—and more research. In the past month, scientists have gathered at three major meetings on the science of emotions.* And many young clinicians are eager to enter the field, hoping for better ways to diagnose and treat emotional disorders, which afflict an estimated 15 million adults in the United States. Meanwhile, spending by the National Institute of Mental Health (NIMH) on the neurobiological basis of emotion has been relatively modest, but it's growing, reaching \$6.3 million in fiscal year 1997. "We think the funding future is very bright," says psychiatrist Ned H. Kalin, who directs the 2-year-old Health Emotions Research Institute at the University of Wisconsin, Madison. Scientifically, "emotion is way behind in many respects," admits Antonio R. Damasio of the University of Iowa College of Medicine in Iowa City. "But it's catching up."

### The Anatomy of Emotion

Emotions were long the province of behavioral scientists, while neuroscientists typically focused on cognitive or sensory functions such as vision. Emotions were considered "too vague and difficult to quantify," says Damasio. "It's not as bad as being a sex researcher," adds Standord University psychiatrist David Speigel, who pioneered studies of stress and cancer survival, "but it comes close."

When researchers did begin to probe the neurobiology of feelings, they first focused on anatomy. They have traced emotional messages to such areas as the prefrontal cortex, just behind the forehead, and the ventral striatum, deep in the brain. But one of the most important emotional sites, as shown over the last 15 years by New York University neuroscientist Joseph LeDoux and others, is the amygdala—an almond-shaped structure in the center of the brain that is a key station in the processing of fear. Over the last several years, work in rats have identified finer and finer areas within the amygdala that are part of the neural fear circuits, says LeDoux. Researchers have shown, for example, that a cluster of neurons called the lateral nucleus brings the fear message in from the senses, and another cluster, the central nucleus, sends it out to other brain structures. Researchers are now pinpointing the neural fear connections at even finer scales.

Meanwhile, less detailed imaging studies show that when humans feel fear, the amygdala becomes active. "It's the hot area," says Yale University psychologist Elizabeth

Reprinted with permission from "Probing the Biology of Emotion" by Christine Mlot. *Science* 280, May 15, 1998, pp. 1005–1007. ©American Association for the Advancement of Science.

*"Wisconsin Symposium on Emotion," Madison, 17–18 April; Society for Neuroscience 1998 Brain Awareness Symposium, Chicago, 24 April; National Institute of Mental Health and Library of Congress, "Discovering Our Selves: The Science of Emotion," Washington, D.C., 5–6 May.

Phelps. She, LeDoux, and others report in a study in press in *Neuron* that when human subjects who have been conditioned to associate a visual cue with a shock see the telltale cue, blood flow to the amygdala increases. In another new study from Damasio's lab, patients with damaged amygdalas rated faces with negative expression to be as trustworthy as faces with positive expression. Without the amygdala to issue a warning, these patients apparently don't feel the usual wariness sparked by a stranger, Damasio said at the NIMH meeting.

Now that this wave of research has marked off some of the brain territory crucial to emotions, other researchers are studying the biochemical events that take place there. They are detailing how intense or even mundane emotional experiences leave their marks on the chemistry of the developing mammalian brain. For example, Trevor W. Robbin's group at the University of Cambridge has been comparing adult rats subjected to two different kinds of early life stress. One set of rats is stressed by being raised in solitary cages after weaning. The other group is stressed early by repeated separations from their mothers but then grow up among other rats.

The two different stressors produced almost opposite behavioral syndromes, Robbins said at the Madison meeting. Isolation-related rats appear frenzied, become overexcited in response to food cues and a new environment, and they are unusually sensitive to amphetamines—similar in some ways to the attention dysfunction seen in schizophrenia and other diseases. In contrast, maternally deprived rats seem to be less responsive to their environment and exhibit dull reactions, similar to human mood disorders, he says.

These behavioral syndromes correspond to different changes in the rat's brain chemistry. By inserting microprobes into rat brains and sampling neurotransmitters, Robbins' group found that the isolation-reared rats had higher levels of dopamine—a neurotransmitter thought to go awry in schizophrenia—in certain areas of the brain known to be involved in addiction and motor control. The maternally deprived rats, in contrast, were found in postmortem tissue analyses to have reduced levels of the mood-mediating neurotransmitter serotonin—malregulated in depressed humans—in parts of the brain that process emotions and memory, Robbins reported at the Madison meeting.

In contrast, researchers have shown that intense mothering—presumably an emotionally positive experience for the infant rat—also has a powerful effect on brain development. Graduate students have long noticed that when rat pups are handled often by people, they grow up to be relatively less anxious and more resilient. The key, Michael Meaney of McGill University in Montreal, Paul Plotsky of Emory University in Atlanta, and their colleagues found last year, is the intense attention—licking, grooming, and nursing—that the rat mothers lavish on the pups each time they are returned to the nest (*Science*, 12 September 1997, pp. 1620 and 1659.) The team found that a certain subset of "good mothers" gave even nonhandled pups this extra attention, and these pups showed similar beneficial effects.

And in the 28 April *Proceedings of the National Academy of Sciences*, researchers report neurochemical changes that correspond to these behavioral differences. Rats with especially attentive mothers have more receptors for neurotransmitters that inhibit the activity of the amygdala and fewer for corticotropin-releasing hormone, a stress hormone. Those changes in receptor numbers could explain why the adult animals display more equanimity in novel environments, says Plotsky.

Even subtle factors in a young animal's environment can color the emotional life of the adult. For example, in a study of 28 lab rhesus monkeys, Kalin and his colleagues identified a birth-order effect. Later born monkeys had lower levels of cortisol—a neurally controlled stress hormone—than first- or early-borns, the researchers report in the February, *Behavioral Neuroscience*. Noting that experienced mothers have different mothering styles, Kalin suspects that the emotional state of the mother, who might be calmer with later borns, somehow affects the offspring's hormone levels.

Although no one is ready to make a direct leap from rats or monkeys to humans, the point is clear: Emotional events in young mammals can have major, long-lasting effects on the neurochemistry of the developing brain—and therefore on mood and behavior. "We're talking about the effects of very early experience on the adult brain, when most of the very early [hormonal] stressor effects have waned," says Robbins. "They are lasting effects." Adds Plotsky, "you can do something fairly mundane in the first days of the animal's life, and somehow this changes how that animal responds to its environment for the rest of its life."

### Tracing Temperament

But it's not only environmental effects—both extreme and subtle—that color emotional responses. Studies in both animals and humans support the idea that individuals carry certain dispositions throughout their lives. For example, Kalin has found that some infant monkeys are abnormally fearful, exhibiting a startled "freeze" behavior with very little provocation and having high baseline cortisol levels. And in humans, decades of study by Harvard University psychologist Jerome Kagan and his colleagues are revealing what look like innate, lifelong temperaments. Kagan's group examined 450 baby boys and girls, first at 16 weeks, then again at 14 months, 21 months, 4 years, and 7 years, by testing their response to cues they could see, hear, and smell, such as a cotton swab dipped in alcohol.

They found that 20% of the 16-week-old infants fell into a test category Kagan calls "high reactive": The tests made them fretful and agitated. Another 35% responded with little distress and low motor activity. Over time, some of the high reactives began to respond normally, while others began to show extreme shyness. None became a bold, fearless child, says Kagan. By age 7, about one-third of the high reactives had developed extreme fears compared with 10% of the others, Kagan said at the NIMH meeting.

Brain imaging complements these behavioral studies by showing a consistent package of brain activation that dovetails with temperamental differences. In Kalin's study, the abnormally fearful rhesus monkeys also had relatively more right frontal brain activity, as recorded by electroencephalograms.

Davidson finds a similar asymmetry in people. People who are negative or depressed according to standardized psychological tests tend to show more baseline prefrontal activity on the right, he says. And the happy-go-lucky folks who are more likely to bounce back when life throws a curve ball tend to show more activity in the left prefrontal cortex.

He speculates that the prefrontal cortex modulates the emotional activity of the amygdala. People with more left prefrontal cortex activity can shut off the response to negative stimuli more quickly, he says. "Being able to shut off negative emotion once it's turned on is a skill that goes with left activation." He adds that it's not yet known whether such temperaments are inborn or a product of very early life experiences.

Indeed, Davidson and others caution that they have far to go in explaining the full biological basis of our passions. LeDoux calls the state of the science of emotions "infantile," as the only emotion for which the neural hardware and software is well understood is fear, and even that has mostly been passed in the rat. "Things have not entirely coalesced into a coherent picture," agrees NIMH director Steve Hyman. He hopes to help it develop, by pushing neuroimages to test hypotheses about the neural circuits, and by "goosing cognitive neuroscience to start considering emotion." Darwin would no doubt approve and sympathize. Understanding the origin of emotional expressions remains a great difficulty, he wrote, and "it deserves still further attention."

 

# AIDS THERAPIES



## Exploring How to Get at–and Eradicate–Hidden HIV

*by Jon Cohen*

*Highly active antiviral therapy can reduce HIV loads to undetectable levels. Thus emboldened, researchers are now asking whether HIV can be eradicated from the body.*

DEDHAM, MASSACHUSETTS—Arthur Ammann, head of the American Foundation for AIDS Research (AmFAR), placed a small placard next to a slide projector sitting in the back of a lecture hall at a small scientific meeting he co-organized 2 weeks ago. "Don't even think about it," the placard warned the three dozen scientists who attended the weekend gathering in this quaint Massachusetts town.

Ammann's sign was simply a warning to participants that the meeting was supposed to be a slide-free discussion-heavy think tank. But 3 years ago, that same sentiment—don't even think about it—would have applied equally well to the topic of the conference* itself: how to eradicate HIV from an infected person. Now, the remarkable success achieved by various combinations of anti-HIV drugs—called highly active antiretroviral therapy, HAART—has allowed researchers not only to think about whether the last bits of virus can be eliminated from the body, but also to speculate about how it might be done. "Up until now, we've been talking about an ocean of HIV in patients," said Richard Kornbluth, a researcher at the University of California, San Diego, who specializes in HIV infection of scavenger white blood cells called macrophages. "Now it's like the tide has gone out with HAART."

Despite this sea change, however, discussions at the meeting made it clear that eliminating HIV from the body may be every bit as difficult as attempting to roll back the ocean itself. The main obstacle: Even in people whose viral load is "undetectable," the virus continues to lurk in cells throughout the body. It hides out, under the immune system's radar and out of reach of antiviral drugs, in "latently" infected blood cells and in sites such as the brain, eye, and testes.

"The issue is answering the question: Is eradication possible?" said David Ho, head of the Aaron Diamond AIDS Research Center in New York City. He and others proposed several strategies, including injecting people with immune system messengers that could flush out the virus so that the patient's own immune system could pick it off and looking for ways to sneak drugs past borders such as the one that protects the brain. "Whether you succeed or not" said Ho, "you're going to learn a lot of science."

The hiding place that should be the most vulnerable to current treatments, Ho thinks, is latently infected blood cells, which have the viral genome woven into their own but do not actively produce new copies of the virus. Keeping patients on HAART to hold the virus at bay and waiting for the latently infected cells—which include macrophages and T lymphocytes—to simply die off is one approach. Using mathematical models recently developed by his team, Ho calculates that the macrophages would die off first leaving the more stubborn T cells around for—at a minimum—5 years. That's a long time to be on HAART, which requires taking dozens of pills a day and also can cause serious side effects.

Other results suggest that Ho's estimates about macrophages may be too optimistic. Donald Moiser, a mouse researcher at The Scripps Research Institute in La Jolla, California, has found that HIV infection somehow lengthens the lifespan of macrophages in mice. And at the meeting, Suzanne Crowe from Australia's Macfarlane Burnet Center for Medical Research revealed that she still finds latently infected macrophages in "undetectable" patients who have been on HAART for up to a year. "It may well be that the macrophage is one of the toughest nuts we have to crack when it comes to eradication," said Crowe.

### Battle Plans

Instead of simply waiting for the latently infected blood cells to die off, some researchers proposed a more active approach. The idea is to compel latently infected cells to produce HIVs, drawing the attention of the immune system so that it can destroy the infected cells. HAART, meanwhile, would theoretically prevent any new HIV that was "flushed out" from infecting virgin cells.

Latently infected T cells might be flushed, proposed Aaron Diamond's Martin Markowitz by injecting patients with the mouse antibody OKT3, which specifically tells T cells to copy themselves. Latently infected macrophages might be unmasked, others suggested, with immune system messengers like granulocyte-macrophage colony-stimulating factor (GM-CSF), tumor necrosis factor, and

---

Reprinted with permission from "Exploring How to Get at–and Eradicate–Hidden HIV" by Jon Cohen. *Science* 279, March 20, 1998, pp. 1854–1855. ©American Association for the Advancement of Science.

*Cellular and Systemic Reservoirs for HIV Replication Under Highly Active Antiretroviral Therapy, Massachusetts Institute of Technology's Endicott House, organized by Am FAR and Treatment Action Group, 27 February–1 March.

interleukin-12, all of which stimulate macrophage reproduction.

Harvard immunologist Abul Abbas suggested yet another strategy, aimed at the whole class of quiescent T cells that can harbor latent virus. "Identify stimuli that are particularly good at keeping these memory cells alive and then inhibit them," said Abbas. He gave the example of interleukin-15 (IL-15), which in mice blocks the "death by neglect" that memory cells usually undergo if they don't receive growth stimulation. Theoretically, then blocking an IL-15-like substance could speed the death of memory cells.

On the road to eradication, the last hairpin turn requires stopping HAART. It's possible that if there are still tiny amount of virus left, the person's immune system can then mop them up, ultimately clearing the infection. But Markowitz and Ho worry that the immune system may not do its part. In several of their patients, HAART has so successfully controlled the infection that they no longer mount strong immune responses against HIV. To try to thwart this resurgence, Ho and Markowitz conclude that it makes sense to boost these patients' immune systems with an HIV vaccine. "Before you stop the drug, wouldn't it be wise to enhance the immune response?" asked Ho.

John Croffin, a leading retrovirologist from Tufts University didn't think so, sparking the hottest debate of the meeting. Coffin argued that while such vaccinations might hamper any remaining HIV's attempt to spread, they could not prevent it. "You're not going to eradicate the virus with an immune response," Coffin contended. "So far that hasn't been good enough to remove the infection from a single individual." Ho countered with a calculation suggesting that in patients who only have latently infected cells, a few cells—50 to 75—come out of latency each day and produce HIV. Vaccinating, he proposed, would give the immune system enough punch to clear that small number of producing cells and, over time, eradicate the infection.

But several researchers suggested that complete eradication of HIV from the blood may be the wrong target in any case. Instead, they asked, why not enlist the immune system—rather than drugs—to control the virus for extended periods? The goal, Scripps's Mosier proposed, could be achieved by cycling patients with both undetectable viral loads and waning HIV immunity off therapy, and then cycling them back on when their virus reemerges. Several studies, he noted, have shown that such "treatment holidays" do not allow the virus to develop resistance to the drugs. And this repeated exposure to the virus would keep boosting immune responses to it. "It surprises me that you're even actually thinking about vaccines [to boost immunity] when you have got the natural infection of immunized people to begin with," said Mosier.

### Brain Drain

Even if HIV can be eliminated from the blood, however, it might not be possible to rid the body of it altogether because of the virus's propensity for skulking around the central nervous system. "I'm very worried that there's a viral reservoir in the brain," said Stuart Lipton, who studies HIV neuropathogenesis at Harvard's Brigham and Women's Hospital. There's no accurate way to measure levels of HIV in the brain, where it can stimulate the production of toxins that produce dementia. But it seems to be protected there by the so-called "blood-brain barrier," which keeps both drugs and immune cells that might fight the virus out of the brain.

Anti-HIV drugs that do reach the brain are a double-edged sword, warned immunologist Angela McLean of Oxford University in the United Kingdom. Only low concentrations of the drugs likely make it past the blood-brain barrier, she noted, and "suboptimal" levels of anti-HIV drugs allow the virus to develop resistance to the agents. "The last thing you want is intermediate drug levels," cautioned McLean.

To make matters worse, the brain is but one "sanctuary site" for HIV. Other areas of the body that have similar barriers include the testes, the eye, the thymus, and the spinal cord. And even less is known about HAART's impact on HIV in these sites. "The sanctuary issue is still up in the air," said Ho.

The think tank participants proposed a practical way to learn much more about the impact of anti-HIV drugs on these remote compartments: autopsies of relatively healthy, HIV-infected people. Although many autopsies have been done on people who have died from AIDS, by definition they have failed treatment and their bodies hold no clues to how HAART affects viral levels in the brain and other hard-to-access sites. Retrovirologist Ashely Haase of the University of Minnesota Medical School in Minneapolis suggested that researchers make a concerted effort to find HIV-infected people who have died from car accidents and other traumas. "One thing that might come out of this meeting is the need to develop such a tissue bank," said Haase.

In the end, researchers could not agree whether eradication is possible, but they clearly are nowhere done thinking about it.

 

# A Species' Fate, By the Numbers

*by Charles C. Mann and Mark L. Plummer*

*Mann and Plummer are the authors of Noah's Choice.*
*A popular approach for predicting a population's survival is coming under scrutiny now that its use in critical decisions on endangered species is on the rise.*

SAN DIEGO—When the National Marine Fisheries Service (NMFS) announced on 16 March that it was adding nine populations of Pacific Northwest salmon to the endangered species list, the agency had barely begun to consider the question of how, exactly, to save this regional icon. The fish face threats from many quarters, including water pollution, dam spillways, and logging practices that harm the river ecosystems. Which threats should government officials spend precious dollars trying to address? No single field study can provide the data needed to answer this question. Instead, NMFS scientists must rely, at least in part, on a technique called population viability analysis (PVA).

First developed more than 20 years ago, PVA has become "conservation biology's greatest scientific contribution," according to Steven R. Beissinger, an ecologist at the University of California (UC), Berkeley. The technique focuses on the likely fate of a population and what factors can determine or alter that fate. In its most common form, PVA combines stochastic models of population dynamics with field data on a species and its habitat—everything from birth and death rates to the frequency of natural disasters to predict how long a given population will persist under given circumstances. PVA has had some notable achievements, such as helping to identify measures for boosting grizzly bear populations in Yellowstone National Park. And as one of the few predictive tools ecologists can call on, PVA has become "practically mandatory in planning for endangered species," says Michael Gilpin of UC San Diego.

But increasingly, PVAs are being attacked as too simplistic, overly demanding of data, error-prone, and hard to verify. "Even good PVAs are almost always fraught with very serious statistical problems," says Mark S. Boyce, an ecologist at the University of Wisconsin, Stevens Point. "The confidence intervals are enormous, the error bars explode into the future, and they're very rarely field-tested." Still, Boyce says, ecologists must make do with this imperfect approach to predicting species survival, because it's the best they've got. "At the moment," he says, "there's no other choice."

To assess the state of the art of what PVA pioneer Michael Soulé of Tile Wildlands Project in Hotchkiss, Colorado, calls conservation biology's "flagship industry," 330 scientists gathered here last month for the first ever major conference on the technique.* They discussed hurdles facing attempts to extend PVA to cover a wider range of species, and how to factor in the behavior of our own species. And in an important development, one scientist described how he crash-tested PVA models in the lab, a practice that could help ecologists refine the technique.

**Growing Pains**

For decades empirical studies of wildlife populations resembled stock market analyses, with graphs projecting future trends based solely on historical upturns and downturns. That began to change in 1978 when Mark Shaffer, then a Ph.D. student at Duke University, examined the fate of Yellowstone's grizzly bears, which had been on the endangered list since 1967 and were under increasing stress from tourists. His analysis for the first time incorporated randomly occurring demographic and environmental events, such as unusually low birth rates or sparse food supplies, into a computer model of population growth. From this hybrid, Shaffer, now at the nonprofit Defenders of Wildlife, estimated the likelihood that bear populations of a given size would survive over given periods of time. From this he derived the "minimum viable population," which he defined as the smallest bear population with a 95% probability of surviving 100 years.

Shaffer's analysis, coupled with field data, revealed that the foremost factor determining how long a bear population would survive is the death rate of adult females.

Spurred by this finding, federal officials adopted new rules in 1983 that, among other things, blocked tourists from areas in Yellowstone frequented by mothers with cubs. Since then, the grizzly population has increased 5% per year, and some experts say the bear could be removed from the endangered list. "By all accounts," says Boyce, "the success has a lot to do with the PVAs."

Shaffer's approach appealed to the U.S. Forest Service, which by law must maintain "able populations" of vertebrates in the national forests (*Science*, 26 March, p. 1996). The agency held PVA workshops in the early 1980s, kindling broader interest in the technique. Subsequently, the appearance of software—freeware, at first, then commercial programs—for running PVAs on a personal computer got everybody



*Population Viability Analysis: Assessing Models for Recovering Endagered Species, 15–16 March.

doing it. Some ecologists, however, started using the models willy-nilly, even when data were lacking. Barnstorming consultants "would fly in, get together local experts in a species, try to take the information and plug it into the software, and get a result," Beissinger complains. These often slapdash efforts, he says, triggered a backlash from wildlife managers and "did not do PVA any good."

Beissinger and fellow Berkeley ecologist Dale R. McCullough set up last month's conference in part to address these concerns. A main thrust was stretching PVAs to cover a broader range of life-forms—a daunting task, argued Daniel F. Doak of UC Santa Cruz. Endangered plants, he noted, might seem to be obvious subjects for PVA, but their life cycles can defy analysis. For one, legions of individuals can lie dormant for years in natural seed "banks." Impossible to measure and therefore invisible to a routine PVA, these banks can germinate en masse after unpredictable events like floods or fires, confounding PVA predictions. Doak was "not optimistic" that the problem could be solved.

Scientists have also struggled with what some regard as a glaring oversimplification in most PVAs: The models ignore the genetic problems of small populations, such as the accumulation of harmful genes or the loss of beneficial gene variants. Zoos and captive-breeding programs, which work with very small populations, would like to include these factors in PVAs. Russell Lande of Oregon State University in Corvallis argued that ecologists should seek ways to modify PVA equations to reflect the loss of genetic fitness. Indeed, he suggested, negative "genetic factors may be operating at larger population sizes than we thought." But Beissinger and Berkeley colleague M. Ian Westphal argued last year in the *Journal of Wildlife Management* that eroding genetic fitness is unlikely to be the deciding factor in a population's fate—thus modelers should continue to ignore it. The importance of genetics in PVA, says Boyce, is the "hottest debate" in the field.

In perhaps the most ambitious effort to broaden PVA, ecologists are trying to factor humankind into their models. Because hunting, habitat loss from development, and other human acts often are the biggest threats to a species, these researchers believe that neglecting social forces in modeling a species' fate is nothing less than foolish. In one ongoing effort, Philip S. Miller and his colleagues at the World Conservation Union are augmenting PVAs with estimates of human population growth and land use to model how the habitat destroyed in Rwanda's civil war may affect mountain gorilla populations.

**Test to Extinction**

Even as modelers seek ways to make PVAs more complex and realistic, critics decry a lack of empirical verification. A key problem is that no ecologist wants to experiment with natural populations in danger of extinction. If PVAs predict a rapid demise, says Gilpin, "we immediately take steps to make sure those predictions don't come true."

One way to sidestep this constraint is to arrange lab extinctions. At the meeting, ecologist Gary E. Belovsky of Utah State University in Logan unveiled results from the first long term extinction experiment. For 4 years, Belovsky's team monitored more than 600 containers of brine shrimp, keeping track of how long each population would last before going extinct. Inventorying each population every 2 to 3 days, they eliminated the sampling errors that plague ecologists in the field. The experiment hinged on the initial complement of adults in each population, as well as a container's carrying capacity, set by the food supply. Belovsky spiced some containers with environmental stochasticity by randomly varying the food supply, creating more true-to-life conditions.

After the final population winked out last December, the Utah State team compared the results to predictions from five PVA models including two software packages. Under stochastic conditions, for example, most models underestimated the average time brine shrimp population lasted—that is, the models predicted that the populations were less resilient than they actually were. In addition, the initial population size, a focus of many PVAs, turned out to be less important than the container's carrying capacity in determining persistence. Belovsky found that simpler models performed better, which some ecologists took as a hint that making PVAs more complicated may backfire.

Despite PVAs spotty showing, Belovsky says he's not ready to abandon the technique "We might like to think these results would carry over to other species," he says, "but we just don't know." He plans to take his extinction experiments a step further by adding corridors between the containers, creating a set of distinct brine shrimp populations in linked habitat patches—metapopulation—that will allow migration and will resemble real life more closely.

Scientists agree that they must improve PVAs quickly, as the technique is likely to gain wider use in federal policy making on the fates of species. In deciding to add the nine salmon populations to the endangered list, says Robin Waples, director of NMFS's Conservation Biology Division in Seattle, the agency used the "smattering" of existing PVA but relied on other information including expert opinion. Restoring the populations, an undertaking that could take years and cost billions of dollars, will require examining all the factors that limit recovery. Waples hopes PVAs will play a bigger role as the agency ferrets out the "the baddest factors on the block."

With the future of Pacific salmon and other important species hanging in the balance, Katherine Ralls of the Smithsonian Institution, Beissinger, and two other ecologists issued a plea to the community to help draft guidelines for conducting PVAs and quality standards for evaluating them. Ralls suggests for instance, that ecologists make it a common practice to ignore all models lacking error bars or discussions of the limits of the data and conclusions. It may take some time to forge a consensus, but Beissinger says he's confident that guidelines and standards will emerge. No doubt, he concedes, it's time to refurbish the 20-year-old technique. But, he adds dryly, "it is also notable that none of us are ready to get rid of PVA."

# A Symphony of Bacterial Voices

*by Evelyn Strauss*

*Bacteria can communicate with members of both their species and others, thereby allowing them to coordinate their activities.*

Hum by yourself in an auditorium and the room will swallow the sound. But wait until the rest of the chorus shows up, and together you can fill the hall with music. Some activities require a group to make an impact. Indeed, even microbes behave as if they understand that there's power in numbers.

Over the past 5 years, scientists have found that a wide variety of bacteria gather a crowd before acting in unison—a phenomenon that's been dubbed quorum sensing. "Bacteria want to do some things when they're alone and others when they're in a community," says Bonnie Bassler, a microbial geneticist at Princeton University. "Sensing and responding to high population density allows them to switch to a new set of tasks."

In some cases, doing so benefits not just the microbes but also the creatures they inhabit. The classic example is light emission by the luminescent bacterium *Vibrio fischeri*, which lives in a specialized organ in host animals such as the Hawaiian bobtail squid. The light that is turned on when the organ is sufficiently filled with bacteria eliminates the squid's shadow as it swims in the moonlight making it hard for its predators to see it. Thus, the animals live longer, and the bacteria get a home.

In many other cases, though, the results of quorum sensing render bacteria harmful to their hosts. Some bacteria, when their population is large enough, congregate in slimy mats called biofilms, which have been implicated as a cause of many human infections (see Review by Costerton, Stewart, and Greenberg on p. 1318); others produce virulence factors, proteins that allow pathogenic bacteria to exploit their host (see Review by Galan and Collmer on p. 1322).

As researchers tease out the molecular details of how bacteria sense and respond to their neighbors, they may learn how to foul up the counting systems of specific strains. This could lead to ways to thwart the ability of pathogenic bacteria to cause illness without generating side effects by wiping out indigenous flora, as current broad-spectrum antibiotics usually do. "Instead of sterilizing the human host, you're trying to make it unfit for the particular bacteria to survive," says microbiologist Peter Greenberg of the University of Iowa, Iowa City. Conversely, research on quorum sensing could spawn methods to enhance desired activities, such as the production by plant bacteria of antibiotics that protect their host plants from disease.

Because recent research suggests that bacteria can take a census of other species as well as their own, understanding quorum sensing will likely also reveal insights into microbial ecology. "In most natural environments, bacteria live in complex mixed communities," says Greenberg. "Understanding how these develop and organize is going to be critical in attempts to control and manipulate them."

### The Light at the End of the Tunnel

Microbiologists got their first inklings that bacteria might be counting their neighbors in the late 1960s when J. Woodland Hastings, currently at Harvard University, and his colleagues noticed a perplexing pattern of bioluminescence in *V. fischeri* and various other marine bacteria. During early growth, these bacteria remained dim. But when their cells reached a certain density, their luminescence shot up.

Kenneth Nealson and Terry Platt, working with Hastings at Harvard, soon traced the effect to some unknown signaling substance that the bacteria secrete into their growth medium. But that substance remained elusive until the early 1980s. Then, using as an indicator a natural *V. fischeri* strain that can't make the signal but does respond to it, Anatol Eberhard of Ithaca College in New York and his colleagues purified the bacterium's quorum-sensing signal, a small molecule called an acylated homoserine lactone (acyl-HSL).

Over the next several years, researchers learned that *V. fischeri's* quorum-sensing system has two key components: an enzyme that makes the acyl-HSL plus a protein that detects the chemical and responds to it by activating specific genes. As the bacteria multiply within a confined space, the enzyme, called LuxI, produces ever-increasing amounts of acyl-HSL. Eventually the chemical, which diffuses freely into and out of the bacterial cells, reaches a critical concentration at which it activates the second component, a protein called LuxR. The activated LuxR in turn binds to specific stretches of regulatory DNA adjacent to the *V. fischeri* genes needed for bioluminescence and turns them on. This system ensures that the bacteria produce light only when large numbers are present, keeping them from wasting energy when their population is too small to emit a visible glow.

By the mid-1990s, microbiologists had found that many other bacteria that are, like *V. fischeri*, members of the gram-negative group use the same general acyl-HSL scheme for quorum sensing. (Gram-negative bacteria are so called because they don't retain the purple Gram stain.) And bacteria use other signals as well. "It was originally thought that this weird communication system found in fish light organs was just some kind of fluky thing," says Gary Dunny, a microbiologist at the University of Minnesota, Minneapolis. "But it turns out that it's probably quite the norm for bacteria."

About a dozen human pathogens are among the bacteria so far known to possess an acyl-HSL quorum-sensing system. Take, for example, *Pseudomonas aeruginosa*, a common cause of hospital-acquired infections and of lung infections in people with cystic fibrosis. *P. aeruginosa* is hard to eradicate because it forms biofilms, which shelter the bacteria from antibiotics, detergents, and the host's immune system. Greenberg's team has shown that acyl-HSL production by the bacteria is what triggers their congregation in biofilms.

In addition to infecting humans, many gram-negative bacteria live on plants and in the soil, where they have such beneficial effects as protecting plants from infections or helping fix nitrogen, as well as harmful ones such as causing wilting diseases or crown gall tumors. Some of these bacteria, too, use acyl-HSLs for quorum sensing. In work described in the November 1998 issue of *Molecular Plant-Microbe Interactions*, for example, microbiologist Stephen Farrand of the University of Illinois, Urbana-Champaign, and his colleagues surveyed a total of 106 bacterial strains, representing seven different genera that live on various plants.

The researchers found that about half, representing all seven genera, produce acyl-HSLs and have the biochemical machinery to activate gene expression in response to acyl-HSL signals. They haven't yet demonstrated directly that the acyl-HSLs activate the gene expression or that the bacteria use them for quorum sensing, but it is likely. Farrand says: "If the bacterium goes to the trouble of making an acyl-HSL, my guess is it has—or had [sometime in its past] —a quorum-sensing system that uses it as a signaling molecule."

The discovery that multiple strains of plant bacteria make similar signaling molecules raised the possibility that the bacteria can hear members of other species as well as their own. To find out if such eavesdropping occurs, Leland S. Pierson III, a microbiologist at the University of Arizona, Tucson, and his colleagues turned to a plant associated bacterium, *P. aureofaciens*. This microbe uses its own acyl-HSL quorum-sensing signal to determine when to make an antibiotic that helps it suppress its microbial competitors and also provides a service for its wheat plant hosts—protecting them against the fungal "take-all" disease.

In work reported in last November's *Molecular Plant-Microbe Interactions*, Pierson and his colleagues showed that production of the antibiotic, called phenazine, can also be turned on by signals from dozens of other bacterial strains. They isolated 700 strains of bacteria from wheat roots, let them grow in culture, and exposed a *P. aureofaciens* strain that doesn't itself make an acyl-HSL to the culture fluid in the lab. In 8% of the cases, signals in the fluid—which the group has since shown to be acyl-HSLs—stimulated the *P. aureofaciens* to make phenazine.

Pierson then went on to show that the same kind of cross talk can take place in the wild. His group prepared five sets of seeds, each coated with one of five different bacterial strains that had activated the phenazine biosynthetic gene in the lab plus a strain of *P. aureofaciens* equipped with a "reporter" gene that would produce an easily detectable product—it raises the freezing temperature of water—in response to acyl-HSLs. The researchers then planted the seeds and 10 days later looked to see whether the reporter bacteria on the seedling roots had reacted to signals from the five test strains. They found that all had produced positive effects. "There's a potential for communication between unrelated populations of bacteria in the real world," says Pierson.

One reason bacteria might want to intercept their neighbors' messages, he says, is that "a large amount of signal suggests that other bacteria are growing and happy. That tells the bacteria that this is a great place to be." Another is competition. As Pierson puts it, "a plant root is not Club Med. There are limited nutrients, and a bacterium needs to know who else is there so it can make decisions about how to expend energy and succeed in that environment." In *P. aureofaciens's* case, this means making antibiotics to inhibit the growth of competing organisms.

With all this chatter, it might be hard for one species to hear its own members count. Work by Princeton's Bassler and Michael Silverman of the now defunct Agouron Institute in La Jolla, California, suggests that one way bacteria solve this problem is by using multiple quorum-sensing systems. The researchers have shown that the luminescent genes of the marine bacterium Vibrio *harveyi* are regulated by an elaborate mechanism that involves two discrete quorum-sensing systems. One of these uses an acyl-HSL signal, although both its synthesis and the responding molecule differ from those of more conventional acyl-HSL quorum-sensing systems. But the second uses a different, as yet unidentified signal.

Activation of either system is enough to turn on the light-producing genes, so Bassler wondered why there should be such redundancy. Experiments she has since performed with Iowa's Greenberg provided an answer. They found that a strain of *V. harveyi* that can sense the acyl-HSL but not the second signaling substance turns on light production only in response to signals from *V. harveyi* and a few other bacteria. Conversely, a strain of *V. harveyi* that can't sense the acyl-HSL of quorum system one but can pick up the second type of signal responds to a broad range of bacteria. "I think *V. harveyi* has system one for intraspecies communication and system two for interspecies communication," says Bassler. "Maybe the combination provides it with a way to know its proportion of the total bacterial population. It could then adapt appropriately by turning on different sets of genes, depending on whether it's alone or in a consortium.

### *E. coli* and *S. typhimurium* Catch Up

The discovery that *V. harveyi* employs unconventional molecules —non-acyl-HSL signalers and non-LuxI/LuxR proteins for making and responding to the signaling compounds—may also help resolve a paradox: *Escherichia coli*, a gram-negative bacterium, has served as a prototype for understanding much about bacterial life, but even though quorum-sensing systems have been turning up all over the place, this organism didn't appear to have one —at least not one involving acyl-HSLs. New findings from Bassler's group now suggest that *E. coli* and its close relative *Salmonella typhimurium* may have quorum-sensing systems similar to *V. harveyi's* second system.

Last year, the researchers showed that *E. coli* and *S. typhimurium* make a substance that activates luminescence gene expression in a *V. harveyi* mutant that can sense signals only through its second system. Since then, Bassler's team identified a gene involved in producing the substance in *V. harveyi*, *E. coli*, and *S. typhimurium*. [The results are in the 16 February issue of *Proceedings of the National Academy of Sciences (PNAS)*.] Similar genes—and presumably similar signaling systems—seem to be widespread. By surveying the databases, Bassler and her colleagues have discovered related genes in over a dozen other species of bacteria.

The signaling substance this gene helps produce may tell *E. coli* to go on the attack once its population reaches a critical level. Working with the O157:H-7 strain of *E. coli*, which has gained a justifiably bad reputation as a dangerous food-borne pathogen, James Kaper's team at the University of Maryland School of Medicine in Baltimore has shown that the gene Bassler identified produces a substance that induces the expression of this organism's type III secretion system. This system produces a "molecular syringe," which injects bacterial factors critical for successful infection into animal host cells.

And *E. coli's* linguistic abilities may not stop there. In work reported in the 13 April issue of *PNAS*, microbiologist Philip Rather of Case Western Reserve University in Cleveland and his colleagues have uncovered what appears to be another quorum-sensing system in *E. coli*.

Like other quorum-sensing bacteria, *E coli* may use its multiple systems to sense both its own numbers and those of other species. These pathogens might use other microbial "noise" to figure out when they've reached their destination. "Maybe the're picking up signals from normal bacteria in the intestine, says Kaper. "They then recognize it as an environment in which they want to make this type III secretion system."

### The Gift of Tongues

Still more quorum-sensing languages have turned up in the gram-positive bacteria, which do take up the Gram stain. "We're beginning to discover that bacteria have a whole language, and acylhomoserine lactone is just one word. *E coli* has other words it can use. And gram-positives use another dialect," Greenberg says. One case in point is *Staphylococcus aureus*, which causes infections ranging from small skin abscesses to life-threatening conditions such as endocarditis and toxic shock.

When this pathogen infects its hosts, it activates a set of genes that encode a variety of protein-degrading enzymes and toxins that help it disseminate in the host and wreak havoc on the immune system. There has been some disagreement about the identity of the signaling molecule that activates these genes, but the latest evidence, published in the 15 February issue of *PNAS* by microbiologist Richard Novick's group at New York University Medical Center in Manhattan, shows that a short peptide can do the job.

In that work, the Novick team synthesized peptides they had previously implicated in quorum sensing by *S. aureus* and showed that they activate the microbe's virulence genes. The fact that the researchers could duplicate the effect with completely synthetic peptides confirms that the molecules are the quorum-sensing signals, Novick says.

The Novick team has also found that although the peptides from different *S. aureius* strains share basic chemical features, they divide into four groups based on variations in amino acid sequences and length. And because peptides from one group can inhibit virulence gene expression in *S. aureus* strains of another, the finding opens the door to developing new drugs for treating *S. aureus* infections.

Indeed, exploiting such dialects may turn out to be a broader strategy for turning bacteria's quorum-sensing systems against them. Researchers have found that the acyl-HSL signals of gram-negative bacteria have side chains that often vary from one bacterium to another, and some acyl-HSLs can inhibit the activity of the LuxR-like proteins of other species.

New results from Greenberg's team suggest another potential way of inhibiting quorum-sensing signals in these organisms. The enzymes that synthesize the acyl-HSL signals use a substrate that organisms from bacteria to humans exploit to construct a variety of molecules. As a result, trying to inhibit those enzymes might seem an unpromising strategy for shutting down a quorum-sensing system. But Greenberg and his colleagues found a specific molecular feature that distinguishes the acyl-HSL-synthesizing enzyme in *P. aeruginosa*, the biofilm-forming pathogen, as the group reports in the 13 April issue of *PNAS*. "This leads to the idea that you could find analogs that don't bind to proteins in humans but do bind specifically to this family of [microbial] proteins," says Greenberg. Such compounds might thwart the production of the signaling molecule and thus of *P. aeruginosa* biofilms.

Techniques for scrambling or shutting down quorum-sensing systems could be valuable against some nonbacterial pathogens, too. Quorum-sensing systems are now turning up in more complex, nucleated cells. "Often ideas are discovered in bacteria and if they're great, eukaryotes use them too," says Greenberg. Next week at the American Society for Microbiology meeting, for example, William Goldman, a microbiologist at Washington University School of Medicine in St. Louis, will present evidence that the fungus *Histoplasina capsulatum*, which causes histoplasmosis, a flulike respiratory disease in humans, can sense its own numbers.

Goldman and his colleagues have found that the fungal cells in dense cultures are adorned with a particular sugar whose presence correlates with the virulence of the fungus. But when they diluted those cells in fresh culture medium, newly budding cells did not display the sugar. "It continues to stay off until they reach a certain cell density and then —wham—everyone turns it on," says Goldman. If, however, the researchers add fluid from dense cultures to the diluted fungi, the buds express the sugar right away. "Here's a trait that's clearly related to virulence, and it's turning on and off in response to cell density," he adds.

At the same time that researchers are nailing down the biochemical details of the quorum-sensing systems, the many new findings are opening unexplored areas. Given bacteria's knack for languages, Greenberg wants to know, for example, whether they speak to animals or plants, and if so, whether the microbe or the host benefits. If the host can understand or even hear the bacteria, it might "get a leg up in responding to the impending infection," he says. Alternatively, the bacterium may trick the host into responding inappropriately.

And now that it seems clear that different bacterial species can hear each other, "we need to figure out the role of this cross communication in the normal ecology of the strains," says Arizona's Pierson. For example, he says, it might help the organisms coordinate tasks, with one strain producing an enzyme that inactivates an antibiotic while another secretes a substance that prevents the community from drying out. "Different species of bacteria have different capabilities, but they work together," says Bassler. "They're behaving like multicellular organisms. That's really smart"—but perhaps no smarter than might be expected of organisms that orchestrate their activities by performing in concert.

# Coming to Grips with the World's Greenhouse Gases

*by Karen Schmidt*

*Karen Schmidt is a writer in Washington, D.C.*

*Scientists are mounting an ambitious effort to trace the flux of carbon dioxide between land and air; their success will help decide the fate of the Kyoto climate change treaty.*

Forester Matt Delaney always dreamed of working in the tropics, so when he was asked to help measure carbon stored in 634,000 hectares of humid forest in Bolivia, he jumped at the chance. During 6 months last year, he and 10 Bolivian foresters fought their way through an area twice the size of Rhode Island to sample trees, soil, and ground cover at 625 sites. Along the way, they had to contend with relentless bees and the odd piranha attack.

"There was never anything easy about it," says Delaney, who works for Winrock International, a nonprofit in Morrilton, Arkansas. But the survival school-cum-carbon project is paying off. The team, he says, now has a good sense of how much carbon dioxide has been sucked out of the atmosphere by this primeval forest—necessary baseline for tracking changes in carbon storage if the world were to warm in the coming decades.

Such information is not meant to fill almanacs: Scientists and policy-makers of all political persuasions perceive the world's forests as a vast sponge for $CO_2$ belched out by factories, cars, and other societal sources. Canny business leaders view carbon storage as a way to offset industrial $CO_2$ emissions, thus balancing their carbon books and meeting reduction goals set for each country under the climate change treaty that 174 nations in Kyoto, Japan, agreed to last December. Environmental groups, meanwhile, see forest preservation as a savior for staving off global warming. Forests are one of the few ways to take $CO_2$ out of the atmosphere—there's no technology on the horizon to do that," says Tia Nelson, senior policy adviser at the Nature Conservancy in Washington, D.C. Researchers are only now grasping the monumental task of understanding the capacity of Earth's surface—primarily its oceans and forests—to absorb greenhouse gases. To track $CO_2$ flux between land and air, scientists are ramping up a global network of monitoring stations: their data could help in assessing whether nations are adhering to the Kyoto treaty. And in two years engineers will undertake a novel pilot project to pump $CO_2$ into the Pacific Ocean as a possible way to reduce atmospheric gas concentrations. But no scientific issue is thornier than gauging how much $CO_2$ can be socked away in forests.

Negotiators hammering out the details for implementing the Kyoto accord have agreed that changes in forest cover since 1990 can be counted for—and against—a nation trying to meet its $CO_2$ emissions obligations. But the treaty is hazy about how to calculate forest carbon stocks and whether nations can use forestry projects in developing countries to claim carbon credits. Efforts to clarify the $CO_2$ forest connection have just begun in earnest, with a U.N. workshop scheduled for September, and the results will be critical to the treaty's success. "If policy-makers are not well informed about the carbon cycle," says Michael Apps of the Canadian Forest Service in Edmonton, Alberta, "they will fail to meet the [treaty's] objective of stabilizing greenhouse gas concentrations in the atmosphere."

### Seeing the Forest for the Trees

The Koyto agreement requires 38 nations to come up with a baseline assessment for 1990 of their greenhouse gas emissions from all sources—including changes in land use—and of their carbon stocks. Then, they must reduce emissions by an average amount (7% for the United States) during the years 2008 to 2012. These targets will be adjusted for "verifiable changes in carbon stocks" because of "direct human-induced activities of afforestation, reforestation and deforestation since January 1, 1990." In broad terms, researchers estimate that deforestation currently results in about 20% of $CO_2$ emissions worldwide, and that afforestation—planting new forests—and reforestation draw $CO_2$ from the atmosphere. But those generalizations provide a shaky basis for binding international commitments that have huge economic consequences.

One big problem is that scientists are now 8 years late—and counting—in setting the 1990 baseline. "I'm not aware of any country that has a scientifically rigorous inventory for forests in 1990," says Darren Goetze, a biophysicist at the Union of Concerned Scientists. Even for the United States, he says, a national picture of forest density was compiled from satellite data only 5 years ago—a picture that's still being filled out by ground measurements. This uncertainty, says Dan Lashof, a senior scientist at the Natural

Resources Defense Council, could result in countries being rewarded for savings that simply result from more accurate field data. "This could undermine any significance of the Kyoto protocol," he says. What's needed to plug the gaps in knowledge about carbon sequestration in forests, Lashof says, is a comprehensive system for monitoring carbon stocks that combines remote sensing with ground-based sampling—including a global effort to track $CO_2$—and which provides consistent data from country to country.

The $10 million project in Bolivia is one of the largest attempts to replace the generalizations with real numbers for a specific tropical forest. The Noel Kempff Climate Action Project, as it is called, came about in 1997 when an unlikely coalition of The Nature Conservancy, the Bolivian government, a Bolivian foundation called Friends of Nature, and three U.S. companies (American Electric Power, British Petroleum America, and PacifiCorp) bought logging rights and began training local people on how to protect and use the forest—for example, by harvesting hearts of palm and breeding orchids instead of clearing the land to grow corn. The project is guided by mounting evidence that carefully managed forests can soak up $CO_2$.

Data gathered last year suggest that this forest swath sequesters about 15 million metric tons of carbon that would have been released over the next 3 decades if timber companies, farmers, and other developers were allowed to deforest land. To arrive at that figure, the team first had to measure carbon in the forest's trees, leaf litter, ground cover, and soil. Next, they estimated how much carbon would have been lost to development. To do that, they sampled nearby farm fields—carbon paupers compared to mature forest—to project how much $CO_2$ would be released if 13,000 hectares of forest were razed for cropland, and they estimated how much carbon would have been lost to proposed logging. By both stemming carbon loss and perhaps allowing the sink to continue accumulating, the coalition hopes to use the predicted carbon savings as "credits" to meet targets for reducing their $CO_2$ emissions.

But it is far from certain the strategy will work. Any benefits could be erased if conserving forest carbon in one place simply shifts deforestation elsewhere. Such concerns were on the table in Kyoto, where negotiators hotly debated whether to give credits and debits for forest alterations at all. Many parties believed that "by allowing forest sinks to be counted, it's almost like giving permission to extract more fossil fuels and do more oil exploration," says Jennifer Morgan, climate policy officer at the World Wildlife Fund. Because forest carbon stocks could be a deciding factor in whether a country meets its emissions target, she says, "we ought to be very cautious" about how this part of the treaty is implemented.

### A Matter of Definition

World bank environmental chief Robert Watson, chair of the United Nations' Intergovernmental Panel on Climate Change, agrees on the need for caution. To make the treaty work, he says, nations must adopt environmentally sound definitions of forest alteration—an unclear definition could backfire. For example, if "deforestation" leaves out destruction from fire, and "reforestation" is defined as planting trees where forest used to grow naturally, then some countries might encourage burning down forests and replanting them later for credit. This would result in a net carbon loss to the atmosphere because the new forest could take decades to store as much carbon as the original forest. Scientists will discuss possible definitions at the September workshop.

Experts agree that those definitions and the calculations for determining whether countries meet the Kyoto goals must take into account different forest types. "If you have a forest like part of the Brazilian Amazon that regrows slowly . . . and is inefficient to harvest and use, then it may be best to leave it protected," says Gregg Marland of the Oak Ridge National Laboratory in Tennessee. But for fast-growing forests that can be harvested easily, such as in the southeastern United States, he says, "then it's best to make forest products" that sequester carbon or displace other carbon-intensive materials such as cement.

The global picture may seem overwhelming, but sequestration efforts such as the Noel Kempff project in Bolivia suggest that it is possible to take steps now. Whether such schemes will count under the treaty is expected to be a hot topic at a treaty meeting in Buenos Aires in November, in part because most Latin American countries—which already have forest conservation programs they want to bolster—are lobbying for it. But Nelson of the Nature Conservancy predicts the issue will not resolve soon: "Working out how to do these projects is very complicated, because there's so much room for interpretation, so many agendas, and so much at stake."

# Research Problems in Biology

# RESEARCH PROBLEMS IN BIOLOGY



 

# Introduction to the Teacher

This section of the *Teacher's Resource Book* provides suggestions for a variety of problems that students can investigate on their own or in small groups. It is meant to serve as a springboard for open-ended laboratory work, giving students experience in designing, planning, and carrying out their own experiments. In implementing this program, you should keep in mind two important considerations: safety and materials. Before proceeding with this part of the laboratory program, review the Guidelines for Laboratory Safety in the Annotated Teacher's Edition of the textbook.

**Safety**

One of the primary objectives of the program is for students to develop their own procedures. Therefore, few specific procedural instructions have been included. Neither safety notes nor precautions appear in the discussion of the Research Problems except for the general safety admonishment at the end of each discussion (in contrast to the specific directions provided for the regular laboratory exercises). For this reason, it is essential for students to submit a written experimental design that shows comprehension of the concepts involved and contains a complete procedural plan. The plans should demonstrate students' awareness of the hazards inherent in the procedures and the precautions necessary to prevent harm. It is your responsibility to review students' plans carefully, to discuss with each student any necessary changes, and to approve the revised plan before a student begins work. Be sure students show in their written plans that they are familiar with the hazards and that they understand the necessary precautions for each chemical with which they plan to work. Similarly, be sure students show that they are familiar with any biohazards inherent in the materials they are using and that they take all necessary precautions. Some of the equipment that students may plan to use also presents hazards. These hazards include, but are not limited to electrical shocks, sharp edges, recently bent or fire-polished hot glass tubing, other hot surfaces (such as hot plates), blending equipment, bright lights, and loud noises. Be sure students have identified any such hazards and have planned for the appropriate precautions. Have students review the information in the textbook about laboratory safety before they begin any independent work.

**Materials**

The hazards and precautions associated with using chemicals that may be used in each research project are summarized on the labels of the chemical containers and are described in more detail in the Material Safety Data Sheet for each chemical. Do not allow students to use any chemical unless they understand the hazard and precautionary information. If you do not have a Material Safety Data Sheet for a chemical, do not use the chemical or mixture of chemicals until you obtain the information. (Obtain Material Safety and Data Sheets from your supplier. All suppliers of hazardous chemicals are required by law to provide this information to their customers.)

# Designing an Experiment

Science often is described as both a body of knowledge and the process of gathering that knowledge. The methods and the number of steps used in gathering scientific information may vary from one investigator to the next, but scientific methods usually involve the alternation of two types of activities: the observational and the explanatory. This process is generally cyclical because new observations usually result in new or refined explanations. These explanations, in turn, suggest further new observations or ideas.

In pursuing science-related activities, keep in mind the limitations and assumptions of the field, as well as the attitudes that underlie the spirit of honest inquiry. Science deals only with objects or events that can be observed objectively, and it is limited to the natural world or universe.

Science depends on several assumptions about the realities of our universe and the human ability to comprehend these realities. The pursuit of science-related activities also requires the researcher to develop certain attitudes. An effective researcher asks questions, has a disciplined curiosity, is open-minded, withholds judgment, respects evidence, displays intellectual honesty, behaves responsibly, and understands his or her own competence and limitations.

Before planning a biological investigation, the researcher first recognizes a question or a problem to be studied. Research problems often are identified as a result of observing puzzling situations. For example, one may observe that in a lodgepole pine forest, little if any undergrowth exists. Why? Do the trees block available sunlight? Do the trees affect the pH of the soil? Such inquiry may lead to an investigation. A great deal of preliminary planning and investigation may occur before the problem is stated in a formal manner.

**Considerations for the Student**

Before you contemplate a problem to investigate in the school laboratory, you should be aware of the factors that may influence your selection of a research topic. Consider the following:

1. Your study may need to conform to a time schedule specified by your teacher and/or building schedule. Also, your particular choice of topic may already have defined parameters, such as a particular season or month.
2. Your teacher may provide a list of topics from which you must choose. This will set limits on the content areas that can be investigated.
3. Give careful consideration to the type of equipment you will need. Some items may not be available in your school laboratory. Check on equipment and supplies before you engage in extensive planning. Keep in mind that you may have to share certain items, such as microscopes and collecting equipment.
4. Be sure that the organisms or specimens you need are available. Living material may be available only during certain times of the year. Plan accordingly, and whenever possible, use specimens that can be collected or purchased locally. Be aware of regulations regarding the use of live animals.
5. Your problem must be specific enough that you can collect adequate data. Do not try to tackle too large a problem at once. For example, instead of trying to determine all the ecological relationships in a local pond, concentrate on the effect of varying light quality on freshwater plankton populations.
6. Be sure you understand all the safety considerations involved in your research proposal. There are no Caution reminders in Research Problem discussions. Instead, you must develop your own *written* safety plan. Find out about chemical safety by studying labels and Material Safety Data Sheets of the chemicals you plan to use. You must be aware of hazards associated with the biological materials and the equipment you plan to use, and you must know the precautions you need to take. Your experimental design must include descriptions of the precautions you will use, and you must demonstrate to your teacher that you understand

the potential harm and that you know the precautions to take to prevent that harm. Without such precautions, the design of an experiment that may use dangerous chemicals or microorganisms or may involve dangerous laboratory procedures could cause serious harm to you or your classmates. Review the safety guidelines in your textbook.

**Preparation**

To avoid designing laboratory or field experiments around uncertain or hazy hypotheses, gather as much information as possible about the problem or the issue. Usually, you can accomplish your preparation through library research. Consult the card catalog (or computer catalog) and any available journals or abstracts in your field of interest. Anyone who does not bother to review previous literature runs the risk of duplicating someone's effort or mistakes. Whenever you make notes about an article, record title, author, date, publisher, journal, page numbers, and all other pertinent information. Index cards work very well for abstracting information from articles and journals. If you are using sources from the Internet, also include the relevant URLs.

**Formulate a Hypothesis**

After you have gathered background information and made preliminary observations, formulate a hypothesis. A hypothesis is merely a tentative explanation that you propose to account for or to explain observed phenomena. Any hypothesis selected or formulated must be testable. Remember, the primary purpose of designing scientific experiments is to test proposed hypotheses. Such experiments generally test hypotheses by testing the validity of the predictions or conclusions derived from the hypothesis. Also keep in mind that hypotheses are never proven. They are either supported or not supported by the data or experimental results.

The methods for gathering data to test hypotheses are quite varied. Numerical (quantitative) data may be based on observation and/or experiment. Observation is a passive process by which the observer attempts to record data without interfering with the process under study. Experimentation, on the other hand, is active. The experimenter attempts to control the experimental situation. A great advantage of experimentation is that the researcher can study the effects of specific treatments while controlling other variables. Experiments also readily reveal cause-effect relationships.

**Experimental Design**

The ideal design is one that yields the most relevant and reliable information with the least expense and effort. A good experiment hinges on careful planning. Fortunately, an experimental design tests itself: The process often goes wrong when the design is faulty. Before an attempt is made to collect data for interpretation, the experimenter must test the proposed procedures and techniques. Initial adjustments and refinements are common components of designing an experiment. Keep Murphy's Law in mind as you design your experiment: If anything can go wrong, it will.

A common design pattern is comparative experimentation. This involves setting up at least two equivalent groups of units. Specific treatments may be given to one or more of the groups while the control group is subjected to identical conditions except that it does not receive the treatment(s). Any extraneous factors will affect all groups, and the observed differences among groups at the end of the experiment may be attributed to the effect of the treatment(s).

Regardless of the type of biological study or design pattern, you must (1) determine what variables are present and how they will be controlled, (2) determine what measurements will be made and the precision needed in such measurements, (3) decide what sampling methods are to be used and (4) consider the problem of sample size. In general, large samples are more accurate than small ones.

The particular method(s) of testing a hypothesis will certainly be influenced by the nature of the problem being investigated. Here are some examples of experimental design that you may use.

1. Simple experiment (before and after) with one group and no control.

   observation 1 → treatment → observation 2

   An observation is made, treatment is applied, and another observation is made. The data collected from such an experiment may be unreliable because variables are not controlled and the cause-effect relationship cannot be clearly established. For example, in studying the effect of physical exercise on $CO_2$ production in one person, you might measure the content of a breath sample immediately before and shortly after exercise. However, simple designs such as this one may yield inconclusive data. You cannot be certain whether the treatment or some other extraneous factor has an effect during the experiment.

2. Comparative experiment with one control and one experimental group.

   The members of each group are determined at random. Observations are made, and treatment is given only to the experimental group. After treatment, observations are made, and data are collected from both the control and experimental groups.

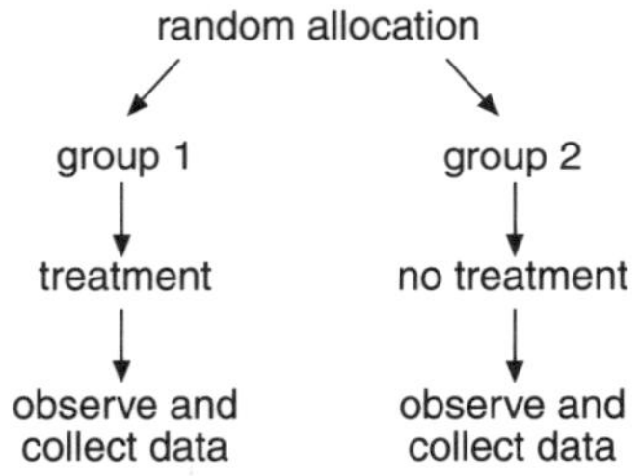


   For example, does plant X root extract affect germination of species Y seeds? In designing an experiment to investigate this problem, two equal and randomly selected groups of seeds could be germinated. One group, the control, would be given distilled water; while the other experimental group would receive distilled water plus the root extract. Data could be collected on the number of seeds that germinated and their growth (height, weight).

3. Comparative experiment with one control and two or more experimental groups.

   Again, members of each group are determined at random. Each experimental group receives a different treatment after the initial observations are made, and the control group receives no treatment. After treatment,

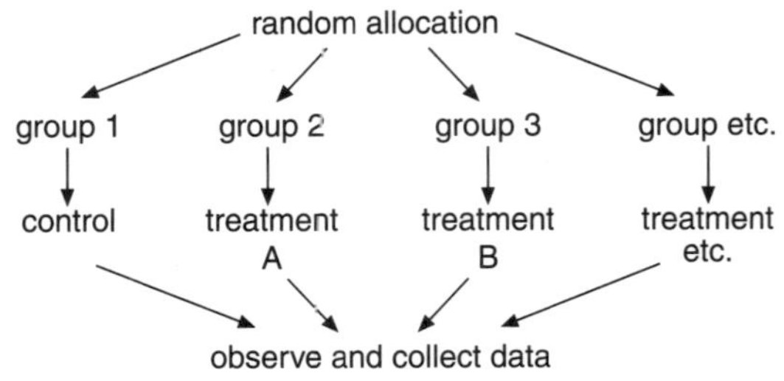


additional observations are made, and data are collected from all groups.

   For example, what is the effect of varying concentrations of algaecide X on planktonic crustacean populations? A water sample containing crustaceans is distributed among four containers. One container is designated as the control, and the other three are experimental groups. The number of crustaceans is roughly equal among all four tanks. One experimental group is given a commercially recommended algaecide dose, a second group receives a dose 5 times that of the first, and the third group receives roughly 10 times the original dose. Population counts are performed periodically.

4. Nonrandom comparative experiment with the nonrandom assignment of members into two or more homogeneous groups.

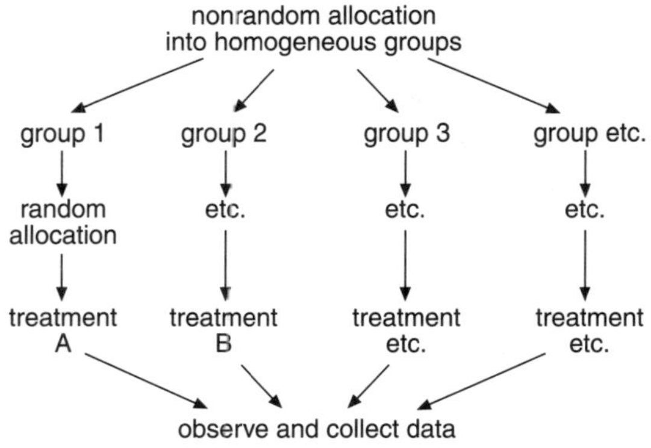


   This pattern may be employed when a variable within the experimental population, such as initial weight or height, may influence results. For example, what are the effects of diets A, B, and C on weight gain in newly weaned male rats? Because initial weight may influence weight gain, the young rats are first nonrandomly placed into groups based on their weight (heavy, medium, or light). The rats in each group are then randomly assigned one of three test diets. Weight gains are recorded periodically over a 28-day period.

**Data Collection**

As part of the design process, you should carefully consider the results or data to be collected from your experiment. It would be a serious error and possibly a waste of valuable time to collect data from an experiment before planning how it is to be organized or analyzed. You may need some background in statistical analysis when planning your research. Use the following checklist as you plan your experiment:

1. Is the problem specific and clearly stated?
2. Is the problem stated in the form of a question?
3. Is the hypothesis testable as it is stated?
4. Is the hypothesis logically consistent with existing data and knowledge?
5. Is the design clearly written?
6. Have you identified all the hazards and determined the precautions you will take to prevent an accident?
7. What procedures will you follow in case there is an accident, no matter how minor?
8. Are the variables in the experiment identified?
9. Is there an attempt to control the variables?
10. Are controls used?
11. Are the sampling methods reliable? Be sure the sample size is adequate. Take care that the samples are as unbiased as possible.
12. Is there an indication of the data to be measured?
13. Is the experiment practical from the standpoints of class time, availability of materials, required skills, and background knowledge?

▷ Chapter 4 Autotrophy: Collecting Energy from the Nonliving Environment

# Factors That Affect Leaf Stomate Behavior

## Background

*Gas exchange in plants occurs through tiny pores in the underside of the leaf, called stomates. Usually, stomates open during the day and close (or partially close) at night. Although a great deal of research has been conducted to discover what controls stomate behavior, as yet no single scheme explains all experimental data. Upon exposure to light, potassium ions seem to accumulate quickly within the guard cells, increasing the osmotic pressure and opening the stomates. The assumption is that active transport mechanisms are involved, but the precise mechanism is still uncertain.*

## Suggested Problems

Using a minimum of laboratory equipment, you can research various questions involving stomates. For example, do numbers of stomates (per square centimeter) vary within a species? You could make various comparisons such as sun leaves versus shade leaves, leaves from plants in polluted areas versus leaves of plants from unpolluted areas, or leaves developing in nutrient-stressed environments versus nonstressed environments. You could examine whether stomate behavior is affected by environmental stress factors such as acid rain; a mineral or nutrient deficiency (potassium, nitrogen, and so forth); disease; or the feeding of animals.

## Suggested Approach

Determine your study topic, and develop a written experimental design that includes safety measures. Do not begin your experiment until your teacher has approved the plan.

You can study stomates of living leaves by making stomate impressions. This simple technique involves spreading a thin film of Duco™ cement over the lower leaf surface (where stomates are generally concentrated). After a few minutes, this substance dries, forming a thin, transparent film that may be carefully peeled off. Impressions of stomates on the film then can be observed under a dissecting microscope. To clarify the imprints, the film could be mounted in a drop of stain such as Lugol's iodine solution or methylene blue.

## Safety

Make sure you understand the hazards and precautions described on the labels and the Material Safety Data Sheets for all the chemicals you plan to use in your investigation. First, review your procedure to be sure you have included all the necessary precautions you will follow in your use of these chemicals. Second, review your procedure to be sure you plan to use the necessary precautions to prevent accidents that may be caused by the equipment you will use. Third, review your procedure to be sure your plans include the precautions you will take to control any biological hazards involved in your work. Always ask your teacher to explain any hazards that you do not understand and to help you plan the precautions you will take to prevent harm from an accident. Report all accidents, no matter how seemingly insignificant, to your teacher.

## Special Equipment/Techniques

1. Use of microscope
2. Preparing stomate impressions

## References

Arms, K. and P. Camp. *Biology*, 3rd ed. Chicago, IL: Saunders College Publishing, 1987.

Elias, T. S. and H. S. Irwin. "Urban Trees." *Scientific American*, November 1976.

Kimball, J. *Biology*, 5th ed. Menlo Park, CA: Addison Wesley, 1983.

Morholt, E., P. Brandwein, and J. Alexander. *A Sourcebook for Biological Sciences*, 2nd ed. New York: Harcourt, Brace and World, Inc., 1966.

Sunquist, F. "The Strange, Dangerous World of Folivory." *International Wildlife*, Vol. 21, No. 1, p. 4, January/ February 1991.

# Measuring Dissolved Oxygen

## Background

*Dissolved oxygen (DO) is probably the most important factor in maintaining healthy ecosystems in lakes and rivers. Most aquatic organisms require certain levels of dissolved oxygen for their survival. Some aquatic animals, such as trout, need medium to high levels of dissolved oxygen; whereas others, such as carp and catfish, survive well in waters having relatively low oxygen levels. Depletion of the DO content can cause major changes in the diversity of species. Many types of pollution-intolerant aquatic animals may disappear, leaving a few more tolerant species, such as aquatic worms and certain fly larvae. Algae and anaerobic organisms also may become abundant in waters having low DO levels.*

*The accumulation of organic wastes is a primary factor in the depletion of dissolved oxygen. Such wastes may be in the form of sewage, agricultural runoff, discharges from food-processing plants, or natural sources such as falling leaves or decaying aquatic vegetation. Aerobic microorganisms consume large amounts of oxygen in decomposing these wastes. In nutrient-rich lakes and rivers that have an abundance of algae and plant life, large fluctuations in the levels of dissolved oxygen may occur during a 24-hour period. From morning through afternoon, DO levels may rise because of photosynthesis by algae and plants. At night, however, after photosynthesis stops, plants, animals, and many microorganisms continue to consume oxygen in respiration. Just before dawn, the DO may reach critically low levels. Four milligrams per liter is the minimum amount of DO needed for warm-water fish such as bluegill, bass, and pike.*

*One measure of the amount of organic matter in water that aerobic microorganisms break down is the biochemical oxygen demand (BOD). This involves calculating the quantity of oxygen used by all microorganisms that are directly or indirectly involved in the aerobic oxidation of organic matter. This test is run over a 5-day period at 20°C. The results are expressed in parts per million (ppm) of dissolved oxygen consumed. A BOD of 100 means that 100 milligrams of oxygen were consumed by 1 L of a sample over a 5-day period at a temperature of 20°C. BOD ranges of 1–3 ppm for natural surface waters are considered very clean to moderately clean. By comparison, untreated municipal sewage may have a BOD of about 600 ppm. Procedures for running a BOD test are given in references by Mitchell and Stapp (1988) and by the American Public Health Association (1985).*

## Suggested Problems

How do DO levels compare in a local lake or river during different periods of the day? How do DO levels compare in a stream or river above and below a suspected source of organic waste pollution? (Collect samples at the selected sites for a BOD determination.) What are the DO levels in fast-moving rapids and slow-moving areas of a stream? (Survey each area for kinds of aquatic organisms.) Associate seasonal variations (plant life, temperature, weather) with DO levels.

## Suggested Approach

Determine your study topic, and develop a written experimental design that includes safety measures. Do not begin your experiment until your teacher has approved the plan. Have your teacher explain and demonstrate the techniques used in testing for dissolved oxygen.

## Safety

Make sure you understand the hazards and precautions described on the labels and the Material Safety Data Sheets for all the chemicals you plan to use in your investigation. First, review your procedure to be sure you have included all the necessary precautions you will follow in your use of these chemicals. Second, review your procedure to be sure you plan to use the necessary precautions to prevent accidents that may be caused by the equipment you will use. Third, review your procedure to be sure your plans include the precautions you will take to control any biological hazards involved in your work. Always ask your teacher to explain any hazards that you do not understand and to help you plan the precautions you will take to prevent harm from an accident. Report all accidents, no matter how seemingly insignificant, to your teacher.

**Special Equipment/Techniques**

1. Use of a dissolved-oxygen test kit
2. Water-sampling techniques

**References**

Andrews, W. A., ed. *Environmental Pollution.* Englewood Cliffs, NJ: Prentice-Hall, Inc., 1972.

Andrews, W. A., ed. *Freshwater Ecology.* Englewood Cliffs, NJ: Prentice-Hall, Inc., 1972.

Caduto, M. J. *Pond and Brook: A Guide to Nature Study in Freshwater Environments.* Englewood Cliffs, NJ: Prentice-Hall, Inc., 1985.

McKay, C.P.A., Jr. "Oxygen Supersaturation in Ice-covered Antarctica." *Science*, Vol. 225, No. 5042, January 17, 1992.

Mitchell, M. K. and W. B. Stapp. *Field Manual for Water Quality Monitoring.* Dexter, ME: Thomson-Shore, Inc., 1988.

Needham, J. G. and P. R. Needham. *A Guide to the Study of Freshwater Biology.* San Francisco: Holden-Day, 1962.

Viola, F. J. "Science in the Dead Zone." *Sea Frontiers*, Vol. 40, No. 4, August 1994.

▷ Chapter 5 Cell Respiration: Releasing Chemical Energy

# Measuring Respiration Rates in Aquatic Organisms: Effects of Selected Variables

## Background

*Cellular respiration is the primary energy-releasing process in living cells. During the process, molecular oxygen is used and $CO_2$ and water are produced. In aquatic organisms, specialized respiratory structures, or gills, are involved in the exchange of $CO_2$ and $O_2$. In some aquatic insects, tracheal gill surfaces are ventilated by beating, or pulsating, of the gills. In fishes such as the goldfish, ventilation occurs by movement of the gill cover (operculum). One method of estimating the rate of respiration in aquatic animals involves counting the number of tracheal gill beats (in an aquatic insect) or the number of gill-cover movements (in a fish) over a standard unit of time.*

*Consistent results in counting the gill beats of some aquatic insect larvae have been reported. Hellgrammites, which are larvae of the dobsonfly, are comparatively large and may measure up to 90 mm (see Figure 3.1). They are fierce predators and live in shallow, swift-moving streams. They may be collected in local streams or purchased at some bait shops.*

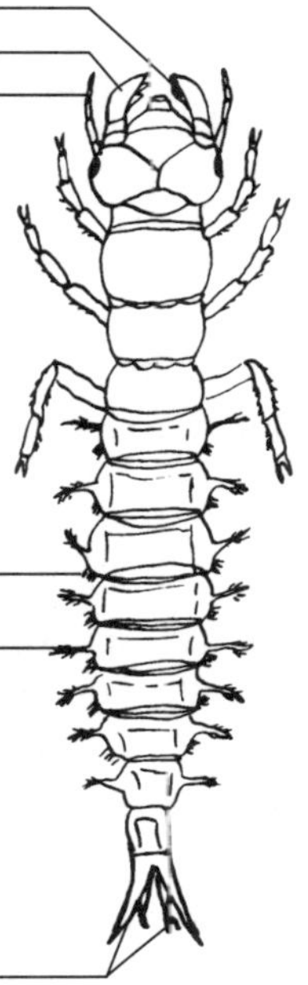


**FIGURE 3.1** A hellgrammite is the aquatic larval form of the dobsonfly.

### Suggested Problems

What is the effect of varying temperature on rate of respiration? Does the visual presence of a predator affect respiration rate? Does the respiration rate vary under dark and light conditions? What is the effect of varying salt concentration on rate of respiration?

### Suggested Approach

Determine your study topic, and develop a written experimental design that includes safety measures. Do not begin your experiment until your teacher has approved the plan.

To test the temperature variable, measure the rate of respiration at 5° increments within a 5–35°C range. An acclimation period (about 5 minutes) should precede each measurement of respiration. Try to obtain at least three readings per animal at each temperature. Place closeable plastic bags of ice or hot water in the aquarium to keep the temperature within a 1–2° range of what is desired.

The effect of the visual presence of a predator may be tested by placing the test animal in a transparent plastic bag or glass jar filled with water. After measuring the rate of respiration, submerge the container and animal in a larger aquarium unit that contains one or more natural predators. Observe and record behavioral responses, and again check the rate of respiration. Make sure that water temperatures in the two containers are the same.

To compare the light variable, place the animal in a water solution that contains a pH indicator such as phenolphthalein or bromothymol blue. As $CO_2$ is released into the solution, the lowered pH will cause the indicator to change color. After a specified period, the animal may be removed and the solution titrated with 0.04% NaOH. Add this base drop by drop until the original color returns. Compare the final color with a control container that has the same volume of indicator as that of the original test solution. The number of micromoles of $CO_2$ produced may be estimated by multiplying the milliliters of base (NaOH) needed to reach the end point by 10. (One milliliter is approximately equal to 25 drops.)

The salt variable may be studied by counting gill ventilation movements or by measuring the $CO_2$ production by the titration method described for the light variable. Prepare salt (NaCl) concentrations that range from 0% to 8% as follows: 0.0%, 1.0%, 2.0%, 4.0%, and 8.0%.

**Safety**

Make sure that you understand the hazards and precautions described on the labels and the Material Safety Data Sheets for all the chemicals you plan to use in your investigation. First, review your procedure to be sure you have included all the necessary precautions you will follow in your use of these chemicals. Second, review your procedure to be sure you plan to use the necessary precautions to prevent accidents that may be caused by the equipment you will use. Third, review your procedure to be sure your plans include the precautions you will take to control any biological hazards involved in your work. Always ask your teacher to explain any hazards that you do not understand and to help you plan the precautions you will take to prevent harm from an accident. Report all accidents, no matter how seemingly insignificant, to your teacher.

**Special Equipment/Techniques**

1. Culture methods for hellgrammites, daphnia, or brine shrimp
2. Preparation of solutions of specified percentage concentrations
3. Titration techniques

**References**

Hoover J. J., K. L. Gage, and M. S. Paulissen. "Hellgrammite Respiration—Temperature's Role in Ectotherm Physiology." *American Biology Teacher*, Vol. 22, No. 7, pp. 39–42, January 1988.

Lawson, C. A. and R. E. Paulson, Eds. *Laboratory and Field Studies in Biology.* New York: Holt, Rinehart and Winston, Inc., 1960.

Morholt, E., P. Brandwein, and J. Alexander. *A Sourcebook for Biological Sciences*, 2nd ed. New York: Harcourt, Brace and World, Inc., 1966.

Taylor, M. F. "Simple Tools for Measuring Anaerobic and Aerobic Respiration." *American Biology Teacher*, Vol. 49, No. 7, pp. 439–441, October 1987.

 

▷ Chapter 6 Cell Structures and Their Functions

# Microbiology: Sampling the Environment for Microorganism Abundance

## Background

*Microorganisms are everywhere in the environment. In the air, for example, microbes that are dormant spores may be carried on dust and various other particles. Fortunately, most of these dormant forms of bacteria and fungi do not cause disease in humans.*

*In water supplies, a vast array of harmful materials (including microorganisms such as bacteria, viruses, and protozoans) may be transported in the water. Organisms transmitted through water usually grow in the intestinal tract and then leave the body in the feces. If fecal pollution of water supplies occurs, pathogens may be transmitted when the contaminated water is consumed. Indicator organisms are used to check water supplies. The most widely used indicator of organisms associated with the intestinal tract is the coliform group of bacteria.*

## Suggested Problems

Bacteriological techniques may be used to investigate a number of questions dealing with microbes in the local environment. Which areas of the school or home have the greatest number of airborne microorganisms? Are coliform bacteria present in local water supplies?

## Suggested Approach

Determine your study topic, and develop a written experimental design that includes safety measures. Do not begin your experiment until your teacher has approved the plan.

A simple method for comparing airborne microorganisms uses an exposed petri dish. Sterile dishes containing nutrient agar (for bacterial forms) and/or sabouraud dextrose agar (for microfungi) are exposed for a brief period (10 minutes) in specified locations. This step simply involves removing the lid and allowing airborne spores to collect on the surface of the agar. After an incubation period of 24–48 hours, visible bacteria and/or mold colonies will appear where particles landed on the agar.

Mold spores may be captured by attaching a loop of sticky tape to a clean microscope slide. Hold the slide horizontally to capture the first few drops of a gentle rain. Later examine the surface of the tape under both low- and high-dry magnifications of a microscope.

Aquatic microlife may be captured by attaching clean microscope slides to a spring-loaded clothespin. Tie a length of fishing line to each clothespin, and suspend a set of slides in various aquatic habitats such as classroom aquariums, streams, lakes, marshes, swamps, ditches, and gutters. In some cases, you may wish to use an empty 1-gallon plastic milk jug as a float and weight the line with fishing-line weights. If outside sampling cannot be performed, use mayonnaise jars to hold field-collected water samples. Keep each set of slides immersed in the water samples for 2–3 weeks (or longer). Remove and examine one slide from the set as a wet mount. Stain the others with Gram's stain.

## Safety

Make sure you understand the hazards and precautions described on the labels and the Material Safety Data Sheets for all the chemicals you plan to use in your investigation. First, review your procedure to be sure you have included all the necessary precautions you will follow in your use of these chemicals. Second, review your procedure to be sure you plan to use the necessary precautions to prevent accidents that may be caused by the equipment you will use. Third, review your procedure to be sure your plans include the precautions you will take to control any biological hazards involved in your work. Always ask your teacher to explain any hazards that you do not understand and to help you plan the precautions you will take to prevent harm from an accident. Report all accidents, no matter how seemingly insignificant, to your teacher.

**Special Equipment/Techniques**

1. Methods of preparing sterile medium in petri dishes
2. Sampling methods for microorganisms
3. Use of incubator
4. Microlife identification
5. Diversity index data collection

**References**

Brock, T. and K. Brock, *Basic Microbiology with Applications*, 2nd ed. Englewood Cliffs, NJ: Prentice Hall, Inc., 1978.

BSCS. *Investigating Your Environment.* Menlo Park, CA: Addison-Wesley, 1975.

Creager, J. C., J. G. Black, and V. E. Davison. *Microbiology Principles and Applications.* Englewood Cliffs, NJ: Prentice Hall, 1990.

Jahn, T., E. Bovee, and F. Jahn. *How To Know the Protozoa*, 2nd ed. Dubuque, IA: Wm. C. Brown Publishers, 1979.

Millipore. *Experiments in Microbiology.* Bedford, MA: Millipore Corp., 1969.

Mitchell, M. and W. Stap. *Field Manual for Water Quality Monitoring.* Dexter, ME: Thomson-Shore, 1988.

Morholt, E., P. Brandwein, and J. Alexander. *A Sourcebook for Biological Sciences*, 2nd ed. New York: Harcourt, Brace and World, Inc., 1966.

Prescott, G. W. *How To Know the Freshwater Algae*, 3rd ed. Dubuque, IA: Wm. C. Brown Publishers, 1978.

Rainis, K. *Nature Projects For Young Scientists.* New York: Franklin Watts, 1989.

▷ Chapter 6 Cell Structures and Their Functions

# Microbiology: Testing for Possible Antimicrobial Substances

## Background

*An antimicrobial agent is a chemical that destroys or inhibits growth of microbes. Antimicrobial agents include a wide variety of synthetic chemicals as well as natural products. Agents that kill organisms are often called cidal. Cidal agents that kill a broad spectrum of target organisms are called germicides. Germicides, in turn, are often divided into two broad groups: antiseptics and disinfectants. Antiseptics kill most microorganisms but not necessarily their spores; they may be applied to skin or mucous membranes (not necessarily internally). A disinfectant is an agent that kills microorganisms but not necessarily their spores; it is not safe for application to living tissues. Disinfectants and germicides commonly act by denaturing the proteins of microbial cells so that they are no longer functional. Disinfectants also cause cell membrane damage, or lysis.*

*Agents that inhibit growth (reproduction) of only some types of microorganisms are termed static. Usually, dilution of cidal agents decreases their effectiveness, rendering them static, not cidal, in activity.*

*An antibiotic is a chemical substance produced by a microorganism that can destroy or inhibit the growth of other microorganisms. Antibiosis is the antagonism of one organism toward another. An example of this is the production of antibiotics.* Bacillus subtilis *produces the antibiotic bacitracin, which will inhibit the growth of other bacteria, such as* Micrococcus luteus. *Although most natural antimicrobial agents are extracted from fungi* (Penicillium) *and certain bacteria* (Bacillus, Streptomyces), *plants also are known to produce such substances. Tomatine from tomato leaves, for example, shows fungicidal activity. A variety of plant extracts (from leaves, stems, roots, flowers, and/or seeds) may be tested for antimicrobial activity. Varieties of seeds that appear resistant to molds in germination tests may be good prospects.*

*For centuries, people have believed that garlic has medicinal properties. Grave diggers who buried victims of the plague in France in 1721 supposedly escaped infection by drinking a concoction of garlic and wine. This same concoction can still be purchased in French apothecary shops.*

*Although trace elements of certain heavy metal ions are essential for the growth of most microorganisms, higher concentrations may exhibit antimicrobial activity. The element copper serves as an excellent example. Copper is incorporated in the plasma membranes of some microorganisms. However, excess copper or its inorganic salts can inhibit the growth of bacteria and fungi. Many over-the-counter salves and ointments contain heavy metal compounds that are used as topical bacterial growth inhibitors.*

*Many household products use phenolic compounds as antiseptics, disinfectants, and preservatives. Creosols (methyl-phenols) are used in soaps and disinfecting products. Amyl-m-creosol and thymol (5-methyl-2-isopropyl-phenol) are used in some mouthwashes. Latex paints commonly employ p-chloro-m-creosol as a preservative. Read product labels to identify the use of these compounds in as many product types as possible and to conduct tests assessing their bacterial growth-inhibiting effects.*

## Suggested Problems

What common products contain compounds that inhibit the growth of bacteria and fungi? What are these compounds, and how do they work? What concentrations of these compounds are required to produce the desired effect? How do excess amounts of certain trace minerals affect bacterial growth?

## Suggested Approach

Determine your study topic, and develop a written experimental design that includes safety measures. Do not begin your experiment until your teacher has approved the plan.

Antibiosis: Inoculate a petri dish containing a mixture of tryptic soy agar and 1 mL of an active *M. luteus* culture broth with a small amount of *B. subtilis.* Incubate the plate upside down at 25°C.

Paper discs may be used for sensitivity testing. Place discs that have been soaked in the product being tested on nutrient agar surfaces that have been inoculated with the test microorganism. If antimicrobial activity exists, a clear zone of inhibition will appear around the disc. Control discs should be soaked in sterile distilled water.

**Safety**

Make sure you understand the hazards and precautions described on the labels and the Material Safety Data Sheets for all the chemicals you plan to use in your investigation. First, review your procedure to be sure you have included all the necessary precautions you will follow in your use of these chemicals. Second, review your procedure to be sure you plan to use the necessary precautions to prevent accidents that may be caused by the equipment you will use. Third, review your procedure to be sure your plans include the precautions you will take to control any biological hazards involved in your work. Always ask your teacher to explain any hazards that you do not understand and to help you plan the precautions you will take to prevent harm from an accident. Report all accidents, no matter how seemingly insignificant, to your teacher.

**Special Equipment/Techniques**

1. Use of items such as beakers, funnels, filters, pipets, mortar and pestle, petri dishes, bacteriological tubes, and sterilizer (pressure cooker)
2. Use of basic bacteriological techniques:
   (a) Preparation of culture media
   (b) Sterilization
   (c) Plating
   (d) Inoculation

**References**

Brock, T. and K. Brock. *Basic Microbiology with Applications*, 2nd ed. Englewood Cliffs, NJ: Prentice Hall, 1978.

BSCS. *Investigating Your Environment.* Menlo Park, CA: Addison-Wesley, 1975.

Creager, J. C., J. G. Black, and V. E. Davison. *Microbiology Principles and Applications.* Englewood Cliffs, NJ: Prentice Hall, 1990.

Gause, G. F. *Search For New Antibiotics.* New Haven, CT: Yale University Press, 1960.

Morholt, E., P. Brandwein, and J. Alexander. *A Sourcebook for Biological Sciences*, 2nd ed. New York: Harcourt, Brace and World, Inc., 1966.

Rainis, K. *Nature Projects for Young Scientists.* New York: Franklin Watts, 1989.

Singleton, P. and D. Sainsbury. *Dictionary of Microbiology.* New York: John Wiley & Sons, 1978.

▷ Chapter 7 Transport Systems

# Factors That Affect Transpiration

**Background**

*Transpiration is the evaporation of water from plant surfaces, particularly from the leaves. In complex plants, water absorbed from the soil is carried upward in xylem tissue through the stem to the leaves. The process of transpiration is partly responsible for this movement of water to the leaves. Water molecules exist as continuous columns in xylem vessels that extend from the roots to the leaves. As water evaporates from leaf surfaces, a pull is created at the ends of the water columns. This transpiration pull is transmitted throughout the length of the water columns because of the cohesive forces between the water molecules. Although the water absorbed by a plant is chiefly used in photosynthesis, most (up to 98%) is lost to the atmosphere through transpiration. About 90% of the water lost by transpiration escapes through the stomates. The remaining 10% is lost through the cuticle.*

**Suggested Problems**

What are the effects of wind, light intensity, and temperature on the rate of transpiration? What is the effect of applying mineral oil or treating leaves with commercial preparations designed to clean and/or polish leaves of ornamental plants?

**Suggested Approach**

Determine your study topic, and develop a written experimental design that includes safety measures. Do not begin your experiment until your teacher has approved the plan.

Construct a potometer such as the one shown in Figure 6.1. Use a scalpel to remove a leafy branch from a potted plant such as *Coleus*, *Geranium*, *Tradescantia*, or *Zebrina*. CAUTION: Scalpels are sharp. Use care in handling. Place the cut end of the shoot in a large pan of water, and cut off about 3 cm of the stem (while holding it under the water) to remove any air that may have been trapped in the xylem elements. Use a pipet to fill the potometer with water until it begins to overflow. Quickly insert the leafy shoot into the rubber tubing. Avoid trapping air bubbles in the system. Fill the pipet end of the potometer to the zero mark. Readings may be taken at specified intervals for 10–15 minutes.

The wind variable may be tested by placing the apparatus at varying distances from an electric fan. Check with your teacher about the proper use of electrical equipment in the lab. To determine the effects of varying light and temperature, place the potometer at varying distances from light and heat sources. The light source selected should not emit excessive heat. Fluorescent light units generate less heat than do incandescent sources.

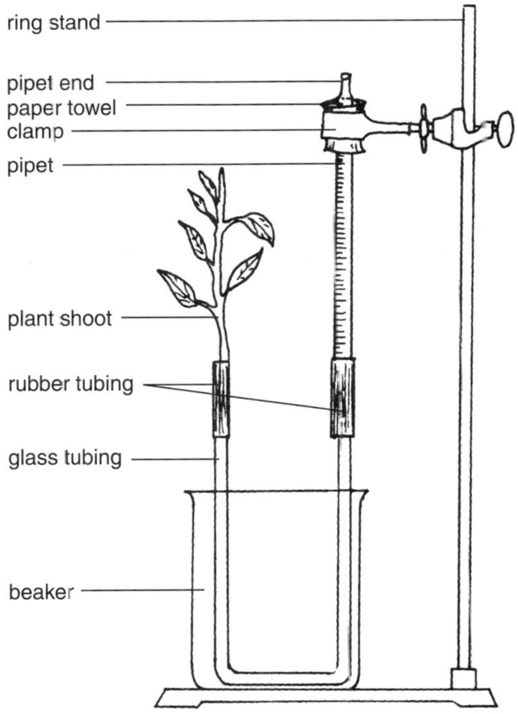


**FIGURE 6.1** Potometer setup

**Safety**

Make sure you understand the hazards and precautions described on the labels and the Material Safety Data Sheets for all the chemicals you plan to use in your investigation. First, review your procedure to be sure you have included all the necessary precautions you will follow in your use of these chemicals. Second, review your procedure to be sure you plan to use the necessary precautions to prevent accidents that may be caused by the equipment you will use. Third, review your procedure to be sure your plans include the precautions you will take to control any biological hazards involved in your work. Always ask your teacher to explain any hazards that you do not understand and to help you plan the precautions you will take to prevent harm from an accident. Report all accidents, no matter how seemingly insignificant, to your teacher.

**Special Equipment/Techniques**

1. Construction of a potometer
2. Reading a pipet

**References**

Abramoff, P. and R. Thomson. *Experimental Approach to Biology.* San Francisco: W. H. Freeman and Co., 1966.

Hummer P. J., Jr., A. Kaskel, J. Kennedy, and R. Oram. *Probing Levels of Life: A Laboratory Manual.* Columbus, OH: Charles E. Merrill Publishing Co., 1986.

Morholt, E., P. Brandwein, and J. Alexander. *A Sourcebook for Biological Sciences*, 2nd ed. New York: Harcourt, Brace and World, Inc., 1966.

▷ Chapter 10 Animal Growth and Development

# Factors That Affect Brine-Shrimp Development

## Background

*Brine shrimp,* Artemia salina, *are basic food sources for a large number of saltwater organisms. Brine shrimp produce large numbers of eggs, and some types of eggs are resistant to drying. These resistant eggs may remain viable for years.*

*Brine shrimp are able to withstand a wide range of salinity and are found either in the open ocean or in tide pools. Salt concentration of estuaries is about 2% compared to seawater, which is nearly 4%.*

*During its period of growth and development, a brine shrimp sheds its outer covering about 12 times. Juveniles have three pairs of appendages, and each pair is a different size (see Figure 7.1). Appendages are tipped with hairlike bristles. As development proceeds, the number of body segments and the number of hairs on the appendages increase.*

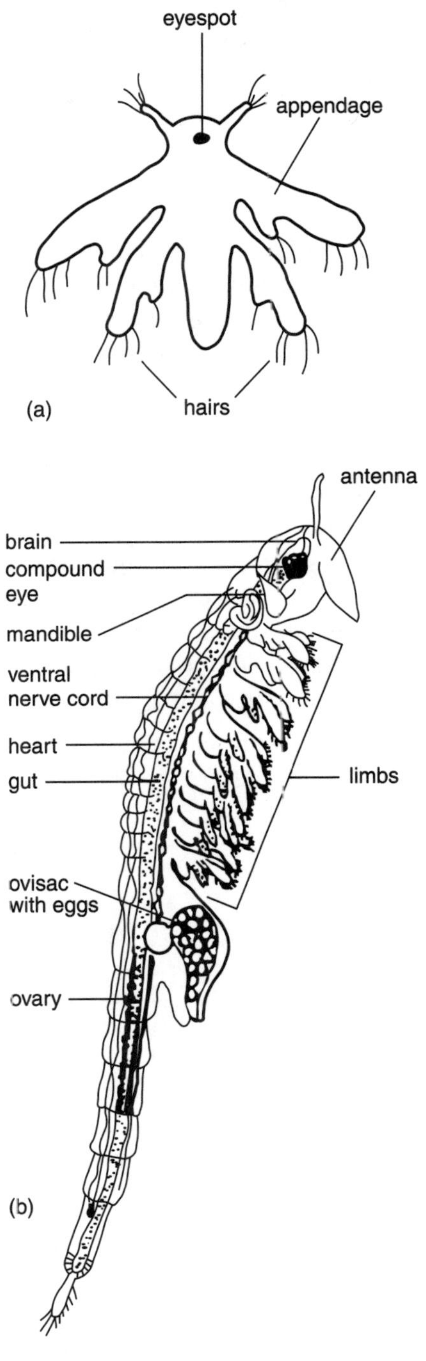


**FIGURE 7.1** (a) The larval form of the brine shrimp. (b) An adult female brine shrimp.

### Suggested Problems

What is the effect of varying salt concentration on brine-shrimp development? What is the effect of varying pH levels on brine-shrimp development? What is the effect of varying temperature on brine-shrimp development?

### Suggested Approach

Determine your study topic, and develop a written experimental design that includes safety measures. Check for regulations regarding use of live animals. Do not begin your experiment until your teacher has approved the plan.

Prepare salt (NaCl) concentrations ranging from 0.1% to 10.0% (for example, 0.1%, 0.5%, 1.0%, 2.0%, 4.0%, 8.0%, and 10.0%); and place them in small jars or dishes. Place a given quantity of brine-shrimp eggs (for example, 0.1 g/100 mL) in each dish. Using a compound microscope, observe the cultures every other day for at least 2 weeks. To observe organisms under the microscope, place a culture sample on a cavity slide and mix with a medium such as Detain II™ to slow the organisms' movements. You may want to record information on physical appearance and to estimate the average body length of the larvae using a stage or ocular micrometer.

Seawater typically ranges from pH 7.8 to pH 8.4. Prepare solutions of varying pH by adding 0.1 *M* HCl or 0.1 *M* NaOH to a given saltwater culture. Your teacher may demonstrate this procedure using a pH meter.

**Safety**

Make sure you understand the hazards and precautions described on the labels and the Material Safety Data Sheets for all the chemicals you plan to use in your investigation. First, review your procedure to be sure you have included all the necessary precautions you will follow in your use of these chemicals. Second, review your procedure to be sure you plan to use the necessary precautions to prevent accidents that may be caused by the equipment you will use. Third, review your procedure to be sure your plans include the precautions you will take to control any biological hazards involved in your work. Always ask your teacher to explain any hazards that you do not understand and to help you plan the precautions you will take to prevent harm from an accident. Report all accidents, no matter how seemingly unimportant, to your teacher.

**Special Equipment/Techniques**

1. Use of compound and dissecting microscopes
2. Preparation of solutions of specified percentage concentrations
3. Preparation of solutions of specified pH
4. Use of pH meter
5. Use of ocular or stage micrometer to measure length of specimens under the microscope

**References**

Browne, R. A. "Sex and the Single Brine Shrimp." *Natural History,* Vol. 102, No. 5, p. 34, May 1993.

Hummer, P., Jr., A. Kaskel, J. Kennedy, and R. Oram. *Probing Levels of Life: A Laboratory Manual.* Columbus, OH: Merrill Publishing Co., 1986.

Kaplan, E. H. *Problem Solving in Biology: A Laboratory Workbook.* London: Macmillan, 1970.

Morholt, E., P. Brandwein, and J. Alexander. *A Sourcebook for Biological Sciences,* 2nd ed. New York: Harcourt, Brace and World, Inc., 1966.

▷ Chapter 11 Plant Growth and Development

# Plant Mineral Requirements: The Effect of Foliar Feeding

**Background**

*The absorption of mineral nutrients is generally thought of as a function of the plant roots, but evidence indicates that plant leaves also may be involved. In fact, manufacturers of some plant fertilizers refer to foliar feeding—the application of nutrient solutions to plant leaves. Standard botany or plant physiology texts often describe symptoms of various mineral deficiencies. Table 8.1 provides a general summary of mineral-deficiency symptoms in plants.*

**Suggested Problems**

Can plant leaves absorb needed minerals in significant amounts? If so, are some types of leaves more efficient in absorption than others—those with thin cuticles, for example? Are factors such as temperature, humidity, concentration of the chemical in the spray, age of the plant, or age of the leaves involved?

**Suggested Approach**

Determine your study topic, and develop a written experimental design that includes safety measures. Do not begin your experiment until your teacher has approved the plan.

If plant leaves are able to absorb needed minerals, then those growing in a mineral-deficient soil medium should show no deficiency symptoms when they receive supplementary foliar feeding. Plants may be grown in a medium that is relatively free of minerals, such as a washed silica sand or a vermiculite, and watered with completer or mineral-deficient solutions. Plant groups may include the following:

1. Control 1: receiving complete nutrient solution applied to mineral-deficient soil medium
2. Control 2: grown in good soil for comparison
3. Mineral deficient: grown with a mineral-deficient solution (for example, without nitrogen, phosphorus, or potassium) applied to soil medium
4. Complete: receiving complete nutrient solution applied to only leaves and stem

Covers should be placed over the soil surface to prevent the solutions used in the feeding applications from draining into the root area. Special pot containers that have covers with a hole for the plant stem are available from some suppliers. The gap between the stem and lid opening may be sealed with modeling clay. Plastic coffee can lids may be similarly modified. Some suppliers offer mineral requirements kits containing premeasured packets of chemicals for preparing complete and mineral-deficient solutions.

**Safety**

Make sure you understand the hazards and precautions described on the labels and the Material Safety Data Sheets for all the chemicals you plan to use in your investigation. First, review your procedure to be sure you have included all the necessary precautions you will follow in your use of these chemicals. Second, review your procedure to be sure you plan to use the necessary precautions to prevent accidents that may be caused by the equipment you will use. Third, review your procedure to be sure your plans include the precautions you will take to control any biological hazards involved in your work. Always ask your teacher to explain any hazards that you do not understand and to help you plan the precautions you will take to prevent harm from an accident. Report all accidents, no matter how seemingly unimportant, to your teacher.

**Special Equipment/Techniques**

1. Planting pots with covers to prevent spray from contacting the soil
2. Preparation of complete and mineral-deficient solutions
3. Possible use of Spectronic 20 colorimeter for pigment analysis

**Table 8.1 Mineral Elements Required by Plants**

| Element | Function | Deficiency |
|---|---|---|
| **Macronutrients** | | |
| Calcium | Component of cell walls; enzyme cofactor; involved in cell membrane permeability; encourages root development | Characterized by death of growing points |
| Magnesium | Part of the chlorophyll molecule; activator of many enzymes | Development of pale, sickly foliage, an unhealthy condition known as chlorosis |
| Nitrogen | Component of amino acids, chlorophyll, coenzymes, and proteins; an excess causes vigorous vegetative growth and suppresses food storage and fruit and seed development | Early symptom is yellowing of leaves, followed by a stunting in growth of all parts of the plant |
| Phosphorus | Promotes root growth and hastens maturity, particularly of cereal grains | Underdeveloped root system; all parts of the plant stunted |
| Potassium | Enzyme activator; production of chlorophyll | Pale, sickly foliage |
| Sulfur | Component of some amino acids | Chlorosis; poor root system |
| **Micronutrients** | | |
| Boron | Pollen germination; regulation of carbohydrate metabolism | Darker color; abnormal growth; malformations |
| Chlorine | Evolution of oxygen during photosynthesis | Small leaves; slow growth |
| Copper | Component of some enzymes | Lowered protein synthesis |
| Iron | Needed for chlorophyll production | Chlorosis, appearing first in youngest leaves |
| Manganese | Activates Krebs cycle enzymes | Mottled chlorosis |
| Molybdenum | Part of enzymes for nitrate reduction | Severe stunting of older leaves |
| Zinc | Component of some enzymes; internode elongation | Small leaves; short internodes |

**References**

Arms, K. and P. Camp. *Biology*, 3rd ed. New York: Saunders, 1987.

Carolina Biological Supply Co. *Plant Mineral Requirements* (leaflet available with Mineral Requirements Kit). Burlington, NC: Carolina Biological Supply Co.

Crafts, A. S. *Translocation in Plants.* New York: Holt, Rinehart, and Winston, 1961.

Eberhard, C. *Biology Laboratory: A Manual to Accompany Biology*, 3rd ed. by K. Arms and P. Camp. New York: Saunders College Publishing Co., 1987.

Meyer, B. S., D. B. Anderson, and R. H. Bohning. *Introduction to Plant Physiology.* Princeton, NJ: D. Van Nostrand Co., 1960.

Morholt, E., P. Brandwein, and J. Alexander. *A Sourcebook for Biological Sciences*, 2nd ed. New York: Harcourt, 1966.

▷ Chapter 11 Plant Growth and Development

# Investigating Possible Chemical Inhibition by Plants

## Background

*Chemicals may play a significant role in competition among certain higher plants. Some plants can limit the distribution of other plants by producing toxic chemicals. One such instance is the effect of black walnut trees on nearby grass. Leaves, fruits, and roots of walnut trees produce a substance that prevents the growth of many nearby plants that may be competitors. Researchers have demonstrated that chemical inhibitors produced by certain shrubs in the California chaparral accumulate in the soil, thus inhibiting the growth of seedlings in a wide belt around each shrub. Such chemical interactions may affect species diversity. Chemical inhibition, which is called allelopathy, has been demonstrated in a wide variety of plant tissues.*

### Suggested Problems

Do plants in the local area provide evidence of allelopathy? Are some plant groups more sensitive than others to possible chemical inhibitions (for example, monocots versus dicots)?

### Suggested Approach

Determine your study topic, and develop a written experimental design that includes safety measures. Do not begin your experiment until your teacher has approved the plan.

To test for allelopathy, place germinating seeds in the presence of an extract from plant tissues that possess a suspected inhibitory effect. Control groups may include untreated seeds and seeds treated with a plant-tissue extract that is known to be nonallelopathic. To obtain an extract, use a blender to grind plant leaves in a small amount of distilled water. Filter the slurry through cheesecloth or coarse filter paper. The filtrate, or extract, that you obtain may be added to a group of germinating seeds. For further study, extracts that show positive effects may be diluted to determine their relative potency.

Seeds such as radish, which germinate rapidly, will produce results in a few days. To germinate seeds, place them in a petri dish on a circle of filter paper saturated with distilled water. Identify plants in the local area that have possible allelopathic interactions. Plants growing in areas with few or no nearby species may be considered as a possible allelopathic species. Collect leaf samples from any such plants for preparation of the extract and testing.

### Safety

Make sure you understand the hazards and precautions described on the labels and the Material Safety Data Sheets for all the chemicals you plan to use in your investigation. First, review your procedure to be sure you have included all the necessary precautions you will follow in your use of these chemicals. Second, review your procedure to be sure you plan to use the necessary precautions to prevent accidents that may be caused by the equipment you will use. Third, review your procedure to be sure your plans include the precautions you will take to control any biological hazards involved in your work. Always ask your teacher to explain any hazards that you do not understand and to help you plan the precautions you will take to prevent harm from an accident. Report all accidents, no matter how seemingly insignificant, to your teacher.

### Special Equipment/Techniques

1. Techniques for seed germination
2. Use of blender and filtering techniques
3. Dilution techniques

### References

Cox, G. W. *Laboratory Manual of General Ecology.* Dubuque, IA: Wm. C. Brown, 1972.

Morholt, E., P. Brandwein, and J. Alexander. *A Sourcebook for Biological Sciences*, 2nd ed. New York: Harcourt, Brace and World, Inc., 1966.

▷ Chapter 11 Plant Growth and Development

# Factors That Affect Leaf Response in the Sensitive Plant *Mimosa*

## Background

*The sensitive plant* Mimosa pudica *provides an excellent example of rapid plant response to environmental stimuli. Factors such as darkness, touch, heat, and certain chemicals cause the leaflets to fold together. When* Mimosa *leaves are pinched or tapped, they quickly fold up. This change is due to a sudden loss of turgor in special cells at the base of the leaflet. Active transport of potassium ions may be involved in the water-balance process. When potassium ions move into the cells, turgor increases. Certain stimuli apparently cause the rapid loss of these ions and a subsequent loss of turgor.*

### Suggested Problems

Would a potassium deficiency influence response and/or recovery time? If so, could the response be restored through foliar feeding of the deficient mineral? Do response and/or recovery times vary with the age of the leaf? Does pruning or leaf removal affect the response mechanism?

### Suggested Approach

Determine your study topic, and develop a written experimental design that includes safety measures. Do not begin your experiment until your teacher has approved the plan.

The methods and procedures described in the research problem topic "Plant Mineral Requirements: The Effect of Foliar Feeding" may be employed to research potassium deficiency. Seeds could be planted at different time intervals to study effects of age of the plant and/or leaf. *Mimosa* seeds may be available at local nurseries.

### Safety

Make sure you understand the hazards and precautions described on the labels and the Material Safety Data Sheets for all the chemicals you plan to use in your investigation. First, review your procedure to be sure you have included all the necessary precautions you will follow in your use of these chemicals. Second, review your procedure to be sure you plan to use the necessary precautions to prevent accidents that may be caused by the equipment you will use. Third, review your procedure to be sure your plans include the precautions you will take to control any biological hazards involved in your work. Always ask your teacher to explain any hazards that you do not understand and to help you plan the precautions you will take to prevent harm from an accident. Report all accidents, no matter how seemingly insignificant, to your teacher.

### Special Equipment/Techniques

1. Methods of germinating seeds and maintaining plants
2. Foliar feeding techniques

### References

Kimball, J. *Biology*, 5th ed. Menlo Park, CA: Addison-Wesley, 1983.

Morholt, E., P. Brandwein, and J. Alexander. *A Sourcebook for Biological Sciences*, 2nd ed. New York: Harcourt, Brace and World, Inc., 1966.

 

▷ Chapter 24 Ecosystem Structure and Function

# Microhabitat Preference Studies

## Background

*Preference studies generally deal with the behavioral responses of organisms to factors in their microhabitat. Factors may be conditions such as temperature, light intensity, humidity, and substrate (the surface on which the organism lives). The researcher provides choices of various environmental conditions. Experiments may be structured by setting up a continuous gradient of a factor such as light intensity from one extreme to another or by offering a choice of two or more distinct levels of the factor. A variety of animals, both aquatic and terrestrial, can be used. These may include earthworms, daphnia, brine shrimp, isopods (sowbugs, and so forth), snails, or insects (mealworms, and so forth).*

## Suggested Problems

Do different organisms exhibit preferences for different microhabitats? Does the presence of other organisms affect the preferences?

## Suggested Approach

Determine your study topic, and develop a written experimental design that includes safety measures. Check for regulations dealing with the use of live animals. Do not begin your experiment until your teacher has approved the plan.

A control run that is identical to the experimental run should be carried out, free from the variable under study. This procedure will reveal any unsuspected factors that may influence the response of the animals during the test.

Some suggestions for establishing particular environmental gradient variables are the following:

1. Temperature: Use a long trough or tube that has one end enclosed with an ice pack and the other with a heating device, such as heat tape or a hot-water bottle. Check with your teacher about the correct procedures for using electrical equipment in the lab.
2. Light intensity: Use an overhead light source suspended above a chamber equipped with a lid that is half transparent and half opaque. A light source also may be placed at one end of a long chamber (tube) with the other end of the chamber darkened.
3. Humidity: A drying agent, such as silica gel or $CaCl_2$, may be placed at one end of a long tube; and a humidifying agent, such as a water source with a paper or cloth wick, may be placed at the other.
4. Substrate: Use a container that has a partitioned bottom to prevent mixing of components. The partition should extend to the substrate surface. Studies may involve variables such as particle size, humidity, color, or presence of a chemical in the substrate.

An interesting variation of the single-species preference test involves comparing the preference of a solitary organism with the preference shown by the organism when a second species, such as a potential predator or competitor, is present. Generally, the results of simple preference tests may be statistically evaluated by using a chi-square test.

## Safety

Make sure you understand the hazards and precautions described on the labels and the Material Safety Data Sheets for all the chemicals you plan to use in your investigation. First, review your procedure to be sure you have included all the necessary precautions you will follow in your use of these chemicals. Second, review your procedure to be sure you plan to use the necessary precautions to prevent accidents that may be caused by the equipment you will use. Third, review your procedure to be sure your plans include the precautions you will take to control any biological hazards involved in your work. Always ask your teacher to explain any hazards that you do not understand and to help you plan the precautions you will take to prevent harm from an accident. Report all accidents, no matter how seemingly insignificant, to your teacher.

**Special Equipment/Techniques**

1. Culturing of organisms selected for study
2. Construction of microhabitat test apparatus

**References**

Biermann, C. "Snails, Pill Bugs, Mealworms, and Chisquare." *American Biology Teacher*, Vol. 50, No. 2, pp. 109–111, February 1988.

Cox, G. W. *Laboratory Manual of General Ecology.* Dubuque, IA: Wm. C. Brown, 1972.

Follansbee, H. *Animal Behavior*, (BSCS) Laboratory Block. Boston: D.C. Heath and Co., 1965.

Morholt, E., P. Brandwein, and J. Alexander. *A Sourcebook for Biological Sciences*, 2nd ed. New York: Harcourt, Brace and World, Inc., 1966.

Warburg, M. R. "The Response of Isopods Towards Temperature, Humidity and Light." *Animal Behavior*, Vol. 12, pp. 175–186, 1964.

▷ Chapter 24 Ecosystem Structure and Function

# Feeding Preferences in Herbivorous Larval Insects

**Background**

*Insect larvae that feed on leaf tissue may be used for various kinds of feeding preference studies, such as preferences for different species of leaves and for variables associated with leaves from the same plant species. Some preference choices may involve leaf tissue from stressed plants versus nonstressed plants, sun leaves versus shade leaves, leaves treated with different levels of a toxic substance versus leaves that are untreated, or leaves from different species. Studies at the Boyce Thompson Institute for Plant Research at Cornell revealed that some pollution-stressed plants produce a chemical, glutathione, that not only appears to protect their tissues from pollutants but also makes them more appetizing to insects.*

**Suggested Problem**

When offered a choice between glutathione-soaked leaves and untreated ones, will insects show a preference?

**Suggested Approach**

Determine your study topic, and develop a written experimental design that includes safety measures. Check for regulations regarding use of live animals. Do not begin your experiment until your teacher has approved the plan.

Prepare discs of leaf tissue by using a cork borer to punch out areas of the leaf that do not have large veins. Different types of leaf discs (for example, treated versus untreated) may be suspended on pins and arranged around the edge of a circular container or enclosure, such as a petri dish. The two types of leaf discs should alternate. Place an equal number and mass of leaf discs in another container for dry biomass calculations.

Place insect larvae in the center of the enclosure, and allow them to select leaf discs for food. After a specified time (determined by their rate of consumption), the remaining plant tissue should be removed to determine the dry biomass. The percentage of dry mass consumed may be determined as follows:

$$\text{\% of dry mass consumed} = \frac{\text{original dry mass} - \text{remaining dry mass}}{\text{original dry mass}} \times 100$$

**Safety**

Make sure you understand the hazards and precautions described on the labels and the Material Safety Data Sheets for all the chemicals you plan to use in your investigation. First, review your procedure to be sure you have included all the necessary precautions you will follow in your use of these chemicals. Second, review your procedure to be sure you plan to use the necessary precautions to prevent accidents that may be caused by the equipment you will use. Third, review your procedure to be sure your plans include the precautions you will take to control any biological hazards involved in your work. Always ask your teacher to explain any hazards that you do not understand and to help you plan the precautions you will take to prevent harm from an accident. Report all accidents, no matter how seemingly insignificant, to your teacher.

**Special Equipment/Techniques**

1. Construction of arena for preference testing
2. Use of drying oven and balance

**References**

"Adding Insect to Injury." *Discover*, Vol. 6, No. 6, p. 10, June 1985.

Cox, G. W. *Laboratory Manual of General Ecology.* Dubuque, IA: Wm. C. Brown, 1972.

Morholt, E., P. Brandwein, and J. Alexander. *A Sourcebook for Biological Sciences*, 2nd ed. New York: Harcourt, Brace and World, Inc., 1966.

▷ Chapter 25 Change in Ecosystems

# Succession in Microcosms: Influence of an Environmental Variable

## Background

*Ecological succession can be thought of as the changes that occur through time among biological populations in a given area. Sometimes such a sequence of events follows a predictable pattern and leads toward a relatively stable, or climax, community. Depending on the particular organisms and type of area under study, ecological successions may occur over a period lasting from a few days or weeks to perhaps thousands of years. Populations of certain freshwater microorganisms may change rapidly over short periods of time, thereby lending themselves to laboratory studies of succession as well as to the experimental manipulation of the process.*

### Suggested Problems

Variables for testing succession patterns in freshwater microcosms may include light intensity; temperature; pH; and pollutants such as salt, detergent, oil, or fertilizer. How do the results compare with the control groups?

### Suggested Approach

Determine your study topic, and develop a written experimental design that includes safety measures. Do not begin your experiment until your teacher has approved the plan.

Pasteurize a sample of pond or lake water by heating it to 70°C. After allowing it to cool, inoculate it with several drops of rich culture containing protozoa, algae, waterfleas, or other similar organisms. Add a small amount of nutrient such as skim milk or rice grains. Divide the sample among two or more containers for culturing, and place the cultures under a light source. Allow the organisms to propagate for at least 2 days. Estimate the initial population densities for each category of organisms. A species-diversity index may also be calculated. Introduce the variable(s) to one or more cultures. Estimate the population densities and/or species diversity periodically for 3–4 weeks.

### Safety

Make sure you understand the hazards and precautions described on the labels and the Material Safety Data Sheets for all the chemicals you plan to use in your investigation. First, review your procedure to be sure you have included all the necessary precautions you will follow in your use of these chemicals. Second, review your procedure to be sure you plan to use the necessary precautions to prevent accidents that may be caused by the equipment you will use. Third, review your procedure to be sure your plans include the precautions you will take to control any biological hazards involved in your work. Always ask your teacher to explain any hazards that you do not understand and to help you plan the precautions you will take to prevent harm from an accident. Report all accidents, no matter how seemingly insignificant, to your teacher.

### Special Equipment/Techniques

1. Methods of culturing freshwater microorganisms
2. Methods for estimating populations using random sampling
3. Use of microscope

### References

Benton, A. and W. Werner, Jr. *Field Biology and Ecology*, 2nd ed. New York: McGraw-Hill, 1966.

BSCS. *Investigating Your Environment.* Menlo Park, CA: Addison-Wesley, 1975. (Diversity index)

Hart, C. W., Jr., and S. Fuller, Eds. *Pollution Ecology of Freshwater Invertebrates.* New York: Academic Press, 1974.

Morgan, A. H. *Fieldbook of Ponds and Streams.* New York: Putnam, 1930.

Morholt, E., P. Brandwein, and J. Alexander. *A Sourcebook for Biological Sciences*, 2nd ed. New York: Harcourt, Brace and World, Inc. 1966.

 

▷ Chapter 25 Change in Ecosystems

# Acid Rain Research

## Background

*Numerous studies cite the wide-ranging effects of acid rain on terrestrial and aquatic ecosystems. Of the many problems and topics that may be investigated, this guide focuses on the following:*

1. *The effect of acid rain on plant susceptibility to frost damage*
2. *The toxic effect of acid-soluble aluminum compounds*
3. *The effect of acid rain on leaf-litter decay*

*Some research on plant damage from frost indicates that acid rain increases levels of nitrate (a plant nutrient) and, thus, slows or delays dormancy processes. This delay, in turn, may cause perennial plant tissues to be more susceptible to winter-frost damage. Other research indicates that the concentration of aluminum compounds in many lakes (such as those in the Swedish West Coast region) rapidly increases with increasing water acidity. Acid rain may also affect leaf-litter decay. Observations indicate that the rate of decomposition of organic matter (leaf litter, and so forth) is linked to pH values of the water.*

## Suggested Problems

Do increased nitrate levels affect plant tolerance to frost? What effects do soluble aluminum compounds have on aquatic plants and animals? What effects do aluminum compounds have on the roots of terrestrial plants, soil, decomposers, and soil microfauna? Are bacterial and fungal decomposition processes inhibited at low pH? What effects does low pH have on aquatic animals that process organic litter?

## Suggested Approach

Determine your study topic, and develop a written experimental design that includes safety measures. Do not begin your experiment until your teacher has approved the plan.

To investigate frost damage, place a group of perennial plants, such as pine seedlings, in a fume hood and spray them with an acid rain solution that contains a mixture of dilute sulfuric acid and nitric acid: pH 4–5 (dilute 1 mL of 0.1 *M* nitric or sulfuric acid to 1 L). The plants should be treated with this solution for several weeks and then subjected to cold stress. Comparisons could be made with controls that receive the same treatment but controls would be sprayed with distilled water.

The toxicity of soluble aluminum compounds may be tested on a variety of aquatic or terrestrial organisms. Aquatic animals may be placed in containers having acid-rain solutions with acid-soluble aluminum compounds and without acid-soluble aluminum compounds. These groups may then be compared to controls that received neither acid rain nor aluminum compounds. Plants and/or soil microfauna may be studied by applying acid-rain watering solutions both with a selected soluble aluminum compound and without that aluminum compound.

To investigate the effect of lowered pH on the rate of leaf-litter decay in an aquatic ecosystem, recently fallen leaves may be added to aquarium units having water of varying pH. To determine dry leaf weights before and after treatment, use a drying oven (60–70°C). The effect of lowered pH on the activity of aquatic leaf-shredding animals also may be investigated. It may be possible to cover some leaves with fine mesh nylon screening to compare the effects of grazing by animals.

## Safety

Make sure you understand the hazards and precautions described on the labels and the Material Safety Data Sheets for all the chemicals you plan to use in your investigation. First, review your procedure to be sure you have included all the necessary precautions you will follow in your use of these chemicals. Second, review your procedure to be sure you plan to use the necessary precautions to prevent accidents that may be caused by the equipment you will use. Third, review your procedure to be sure your plans include the precautions you will take to control any biological hazards involved in your work. Always ask your teacher to explain any hazards that you do not understand and to help you plan the precautions you will take to prevent harm from an accident. Report all accidents, no matter how seemingly insignificant, to your teacher.

**Special Equipment/Techniques**

1. Preparing solutions of specified pH
2. Use of pH meter (optional)
3. Use of balance
4. Use of drying oven

**References**

Burton, T. M., R. M. Stanford, and J. W. Allan. "The Effects of Acidification on Stream Biota and Organic Matter Processing." *Acid Precipitation: Effects on Ecological Systems*, F. M. D'Itri, Ed. Ann Arbor: Science Publishers, 1982.

Cowling, E. B., "Acid Precipitation and Its Effects on Terrestrial and Aquatic Ecosystems." *Annals of the New York Academy of Sciences*, Vol. 338, pp. 540–555, 1980.

Drablos, D. and A. Tollan, Eds. *Ecological Impact of Acid Precipitation: Proceedings of International Conference, Sandeflord, Norway.* March 11–14, 1980. Johs. Grefshe Trykkerl A/S Mysen (Norway SNSF Project).

Environmental Studies Board, Committees to Review Methods for Exotoxicology, Commission on Natural Resources, National Research Council. *Testing for Effects of Chemicals on Ecosystems.* Washington, DC: National Academy Press, 1981.

Hendry, G. *Early Biotic Responses to Advancing Lake Acidification.* Acid Precipitation Series, Vol. 6. Boston: Butterworth, 1984.

Montastersky, R. "Acid Precipitation Drops in United States," *Science News*, Vol. 144, No. 2, p. 22, 10 July 1993.

Morgan, A. *Fieldbook of Ponds and Streams.* New York: Putnam, 1930.

Needham, J. G. and P. R. Needham. *A Guide to the Study of Fresh-water Biology*, 5th ed. San Francisco: Holden-Day, 1962.

Rand, G. and S. Petrocelli. *Fundamentals of Aquatic Toxicology.* New York: Hemisphere Publishing, 1985.

 

# Blackline Masters

# BLACKLINE MASTERS



 

# BLACKLINE MASTERS, continued

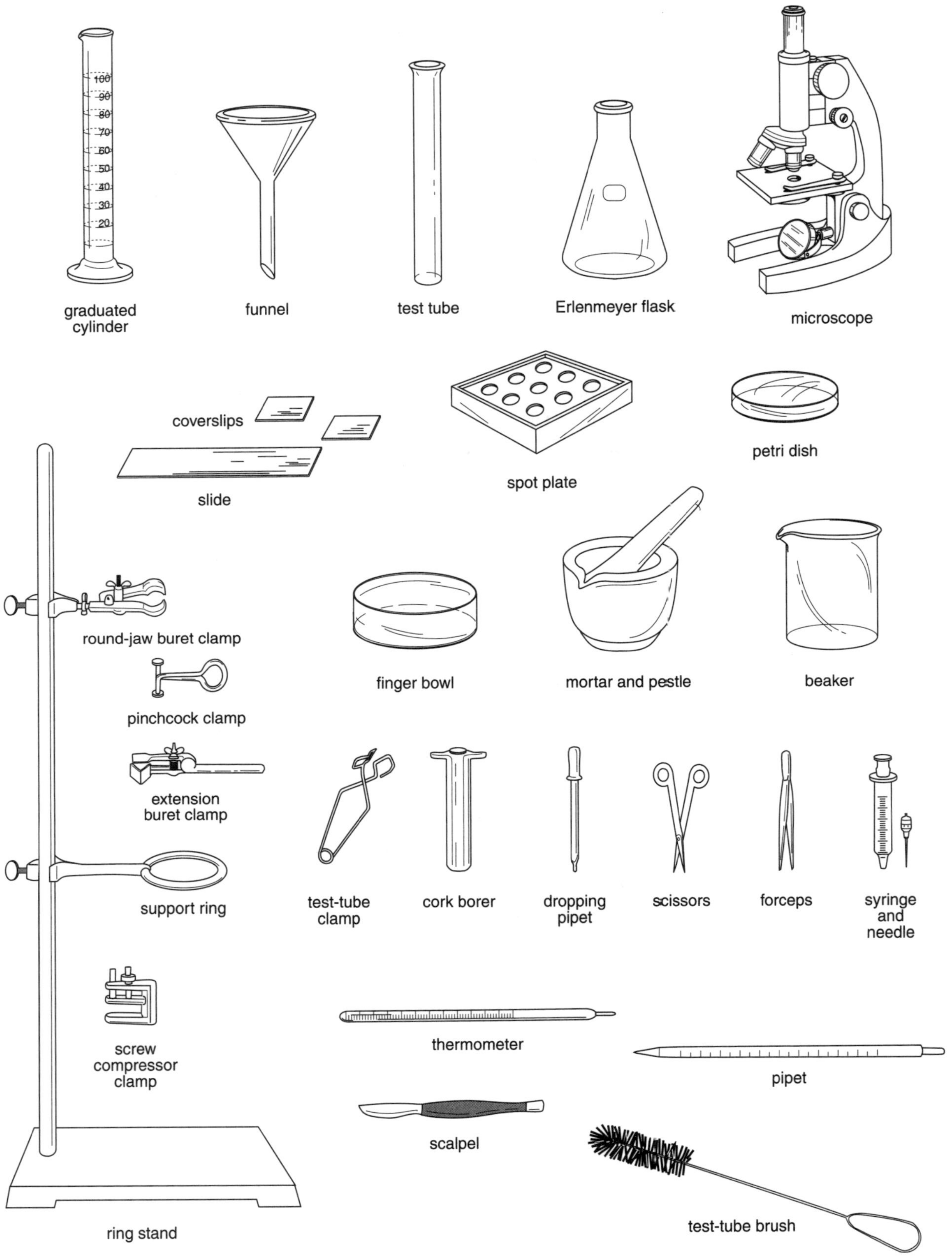
100
90
80
70
60
50
40
30
20
graduated cylinder
funnel
test tube
Erlenmeyer flask
microscope
coverslips
slide
spot plate
petri dish
round-jaw buret clamp
pinchcock clamp
extension buret clamp
support ring
screw compressor clamp
ring stand
finger bowl
mortar and pestle
beaker
test-tube clamp
cork borer
dropping pipet
scissors
forceps
syringe and needle
thermometer
pipet
scalpel
test-tube brush

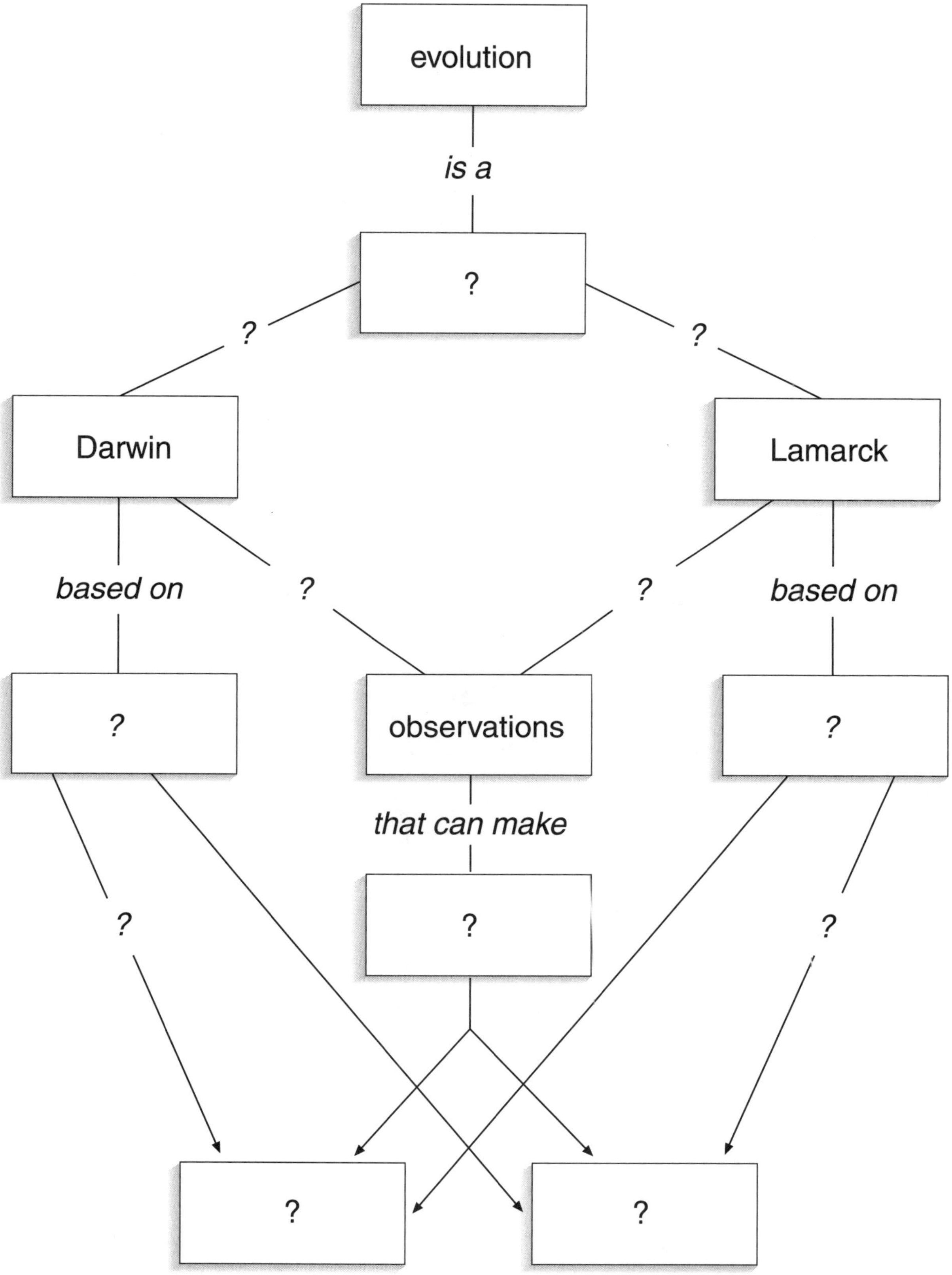
evolution
is a
?
?
?
Darwin
Lamarck
based on
?
?
based on
?
observations
?
that can make
?
?
?
?
?

Figure 1.8 *Covalent Bonds* ◆

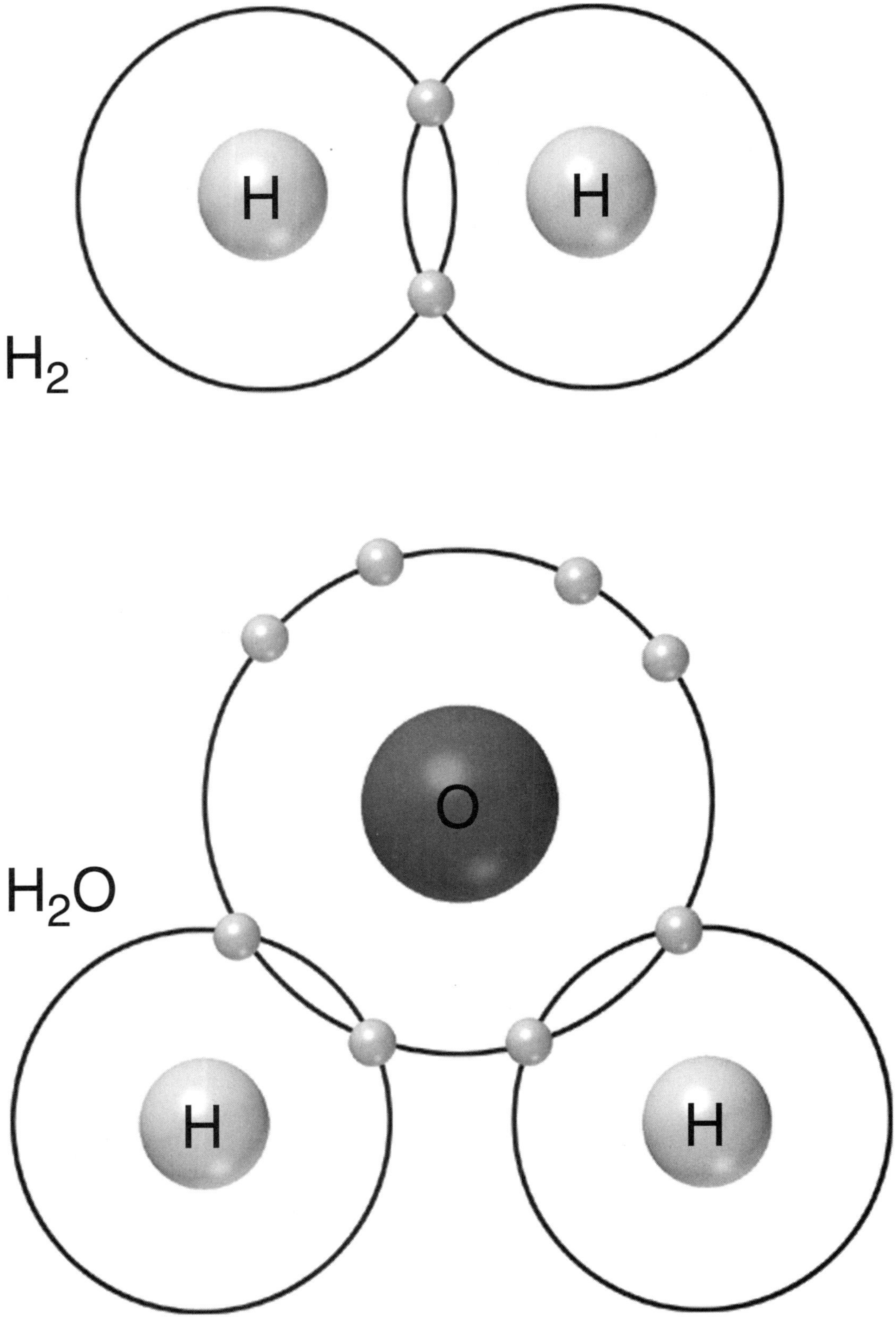


 

Figure 1.10 *Hydrogen Bonds* ◆

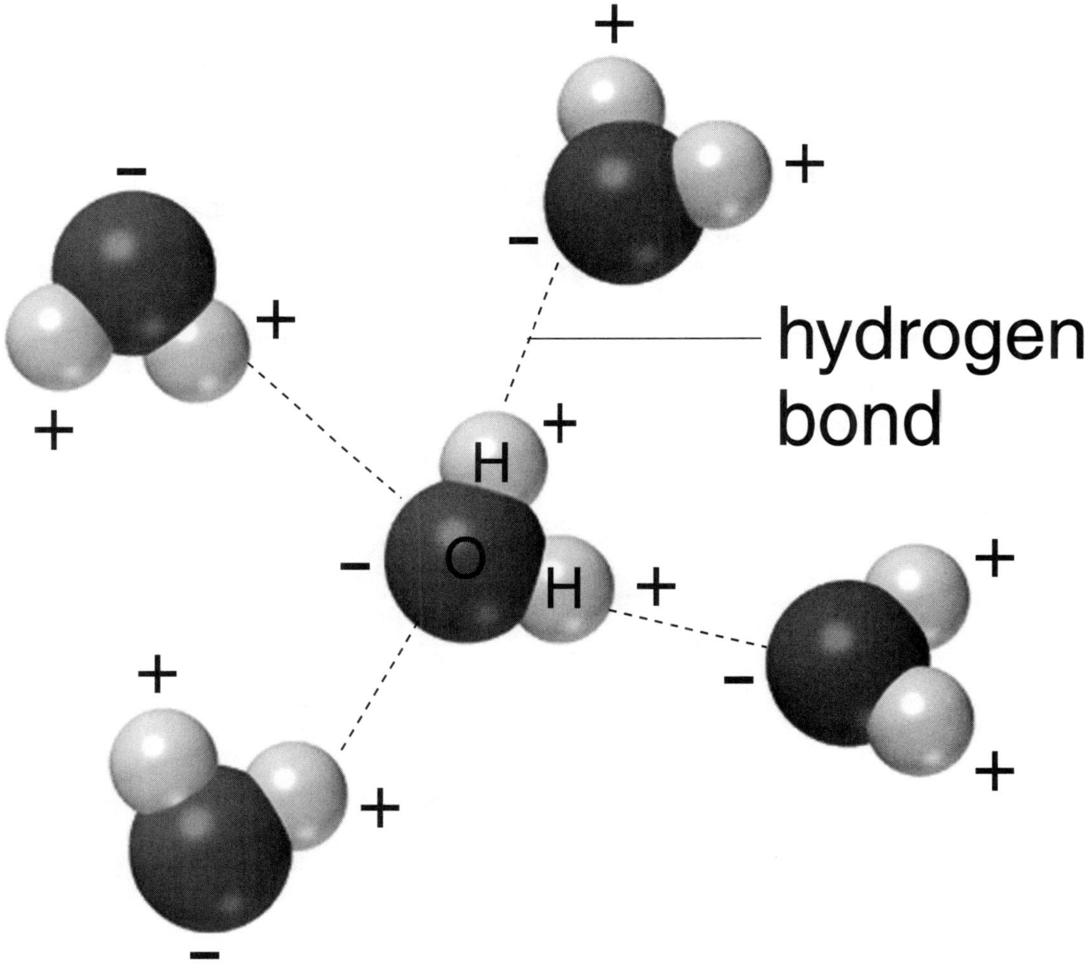

Figure 1.17 *Forms of Glucose* ◆

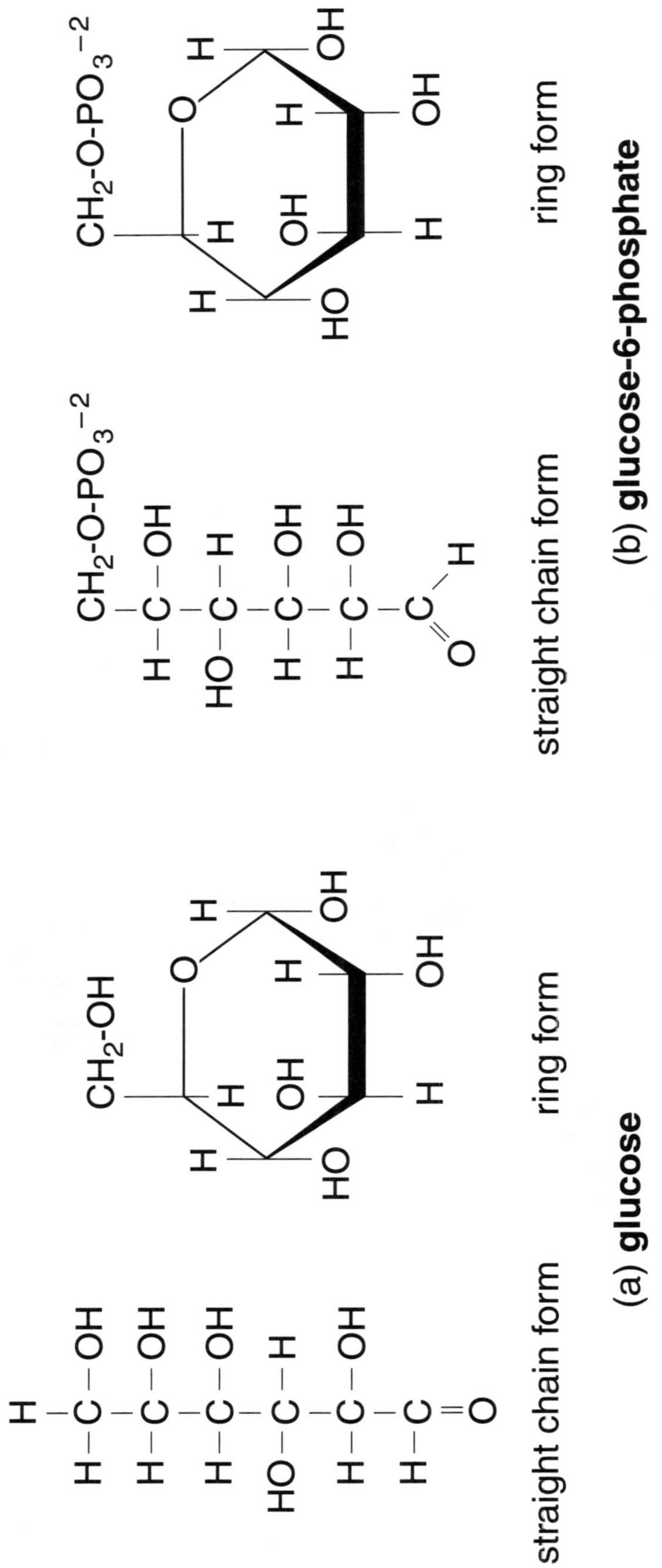

Figure 1.18 *Carbohydrate Molecules* ◆

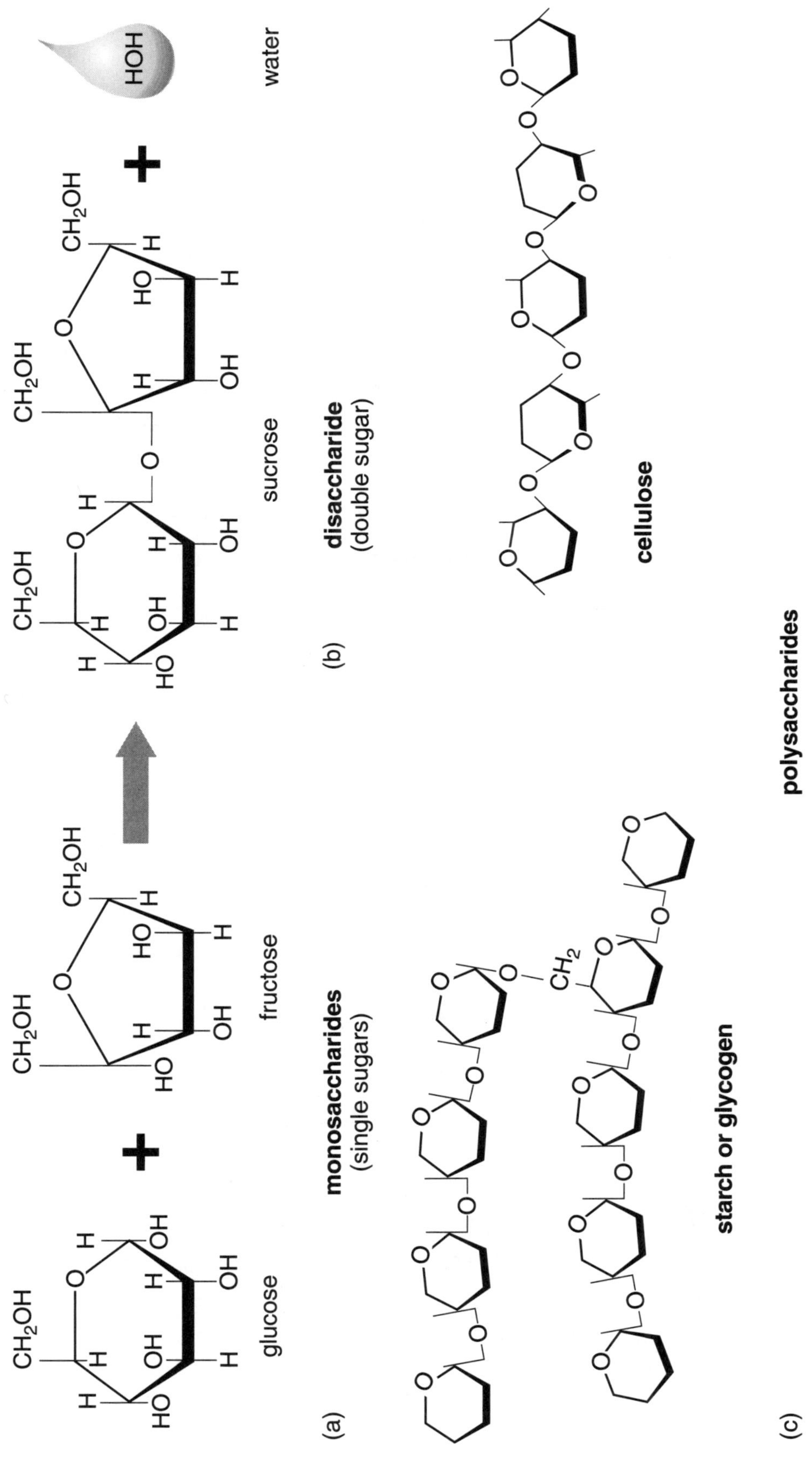

Figure 1.19 *Formation and Structure of Fats* ◆

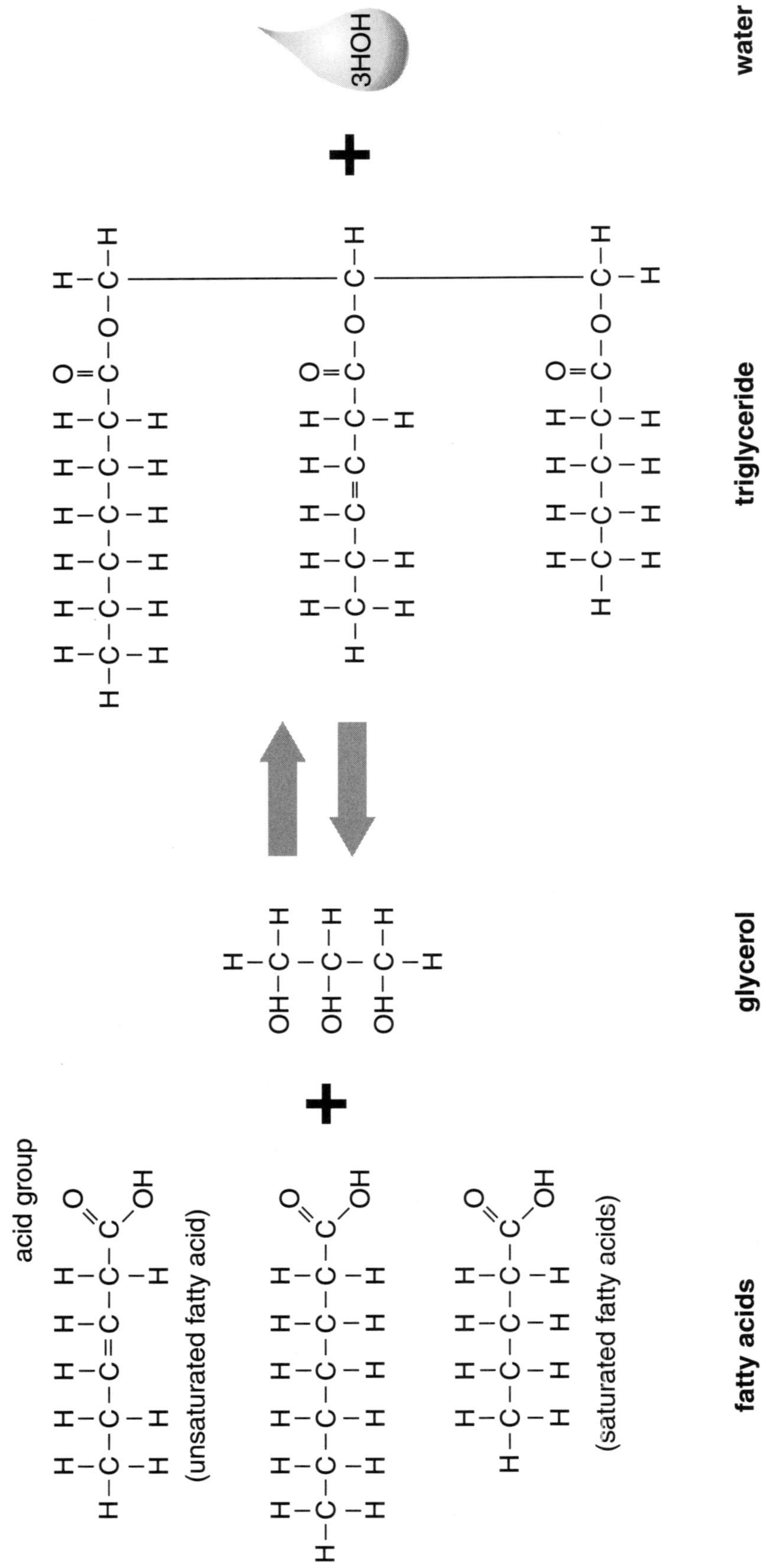

Figure 1.22 *Peptide Bonds and Polypeptides* ◆

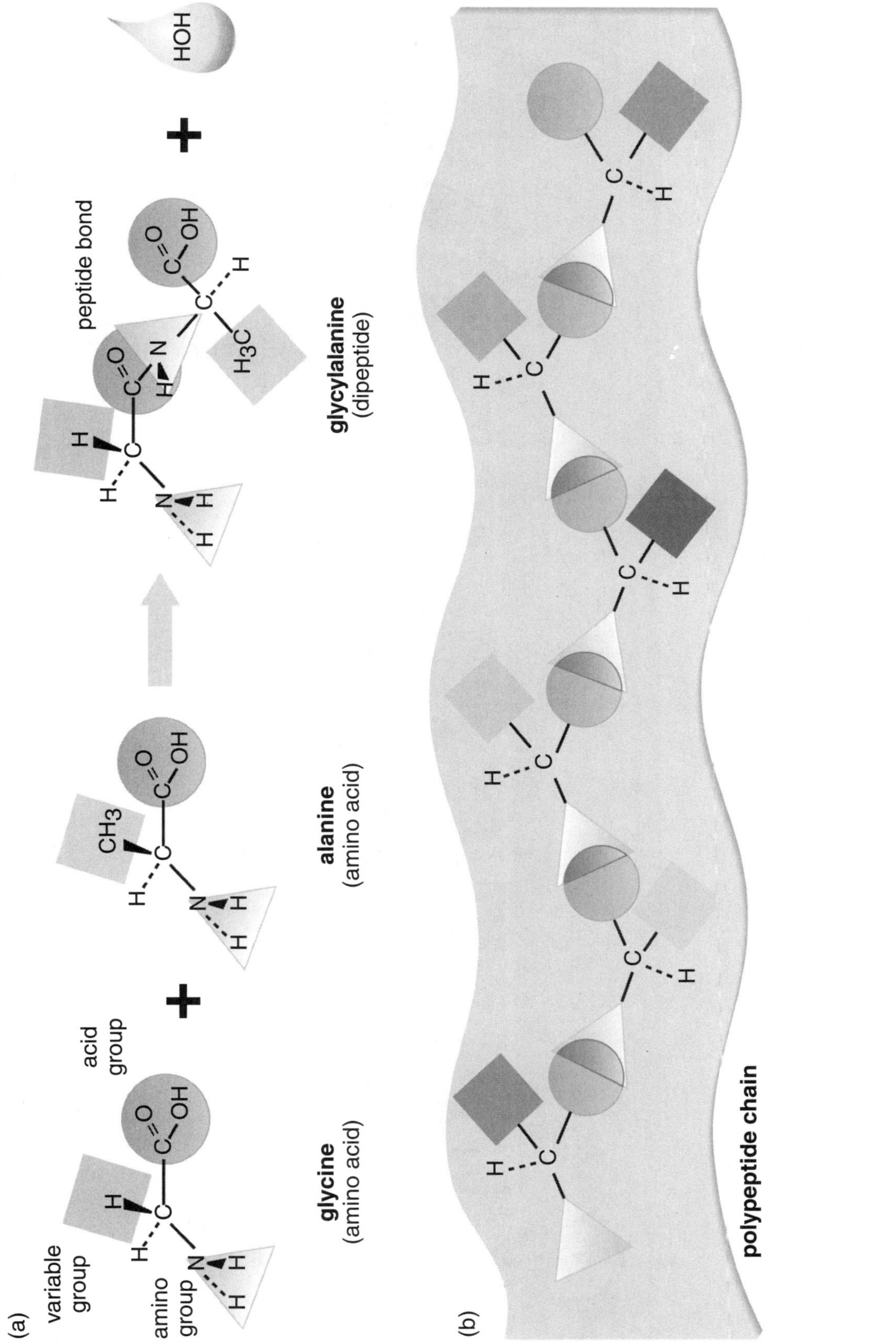

Figure 1.31 *The Structure of DNA* ◆

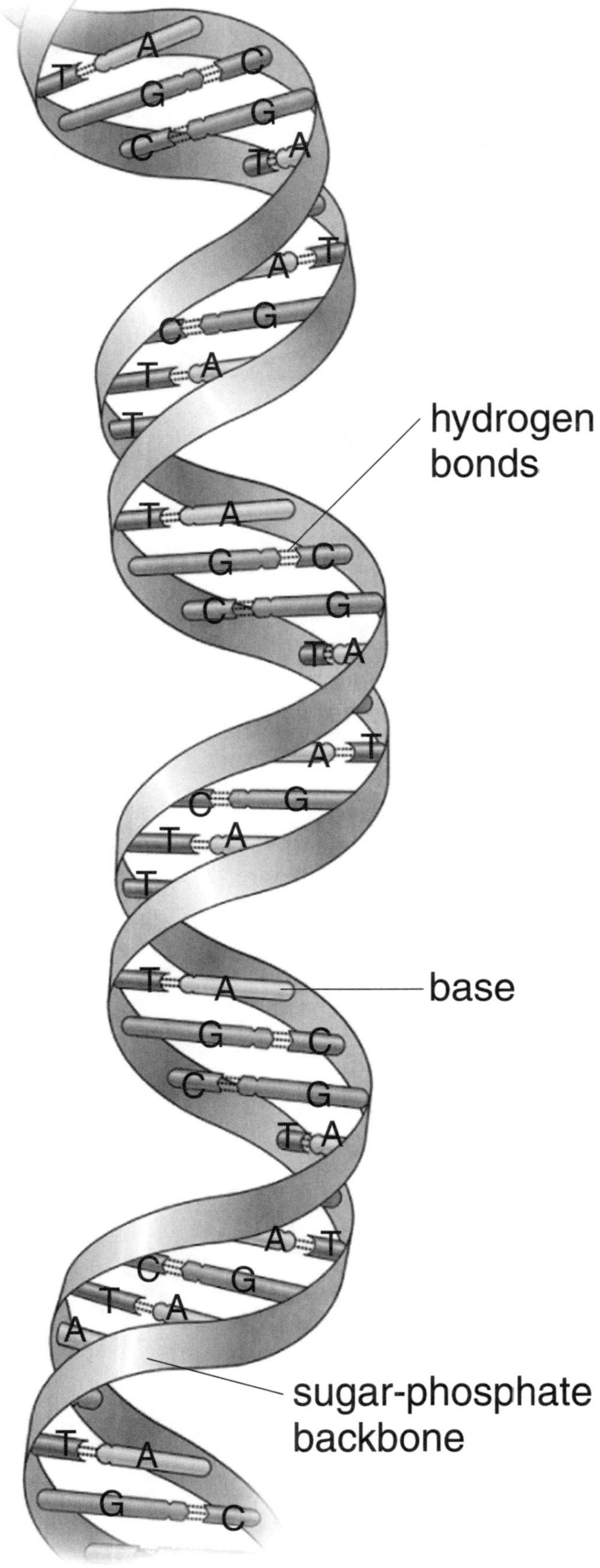

Figure 1.32 *The Structure of RNA* ◆

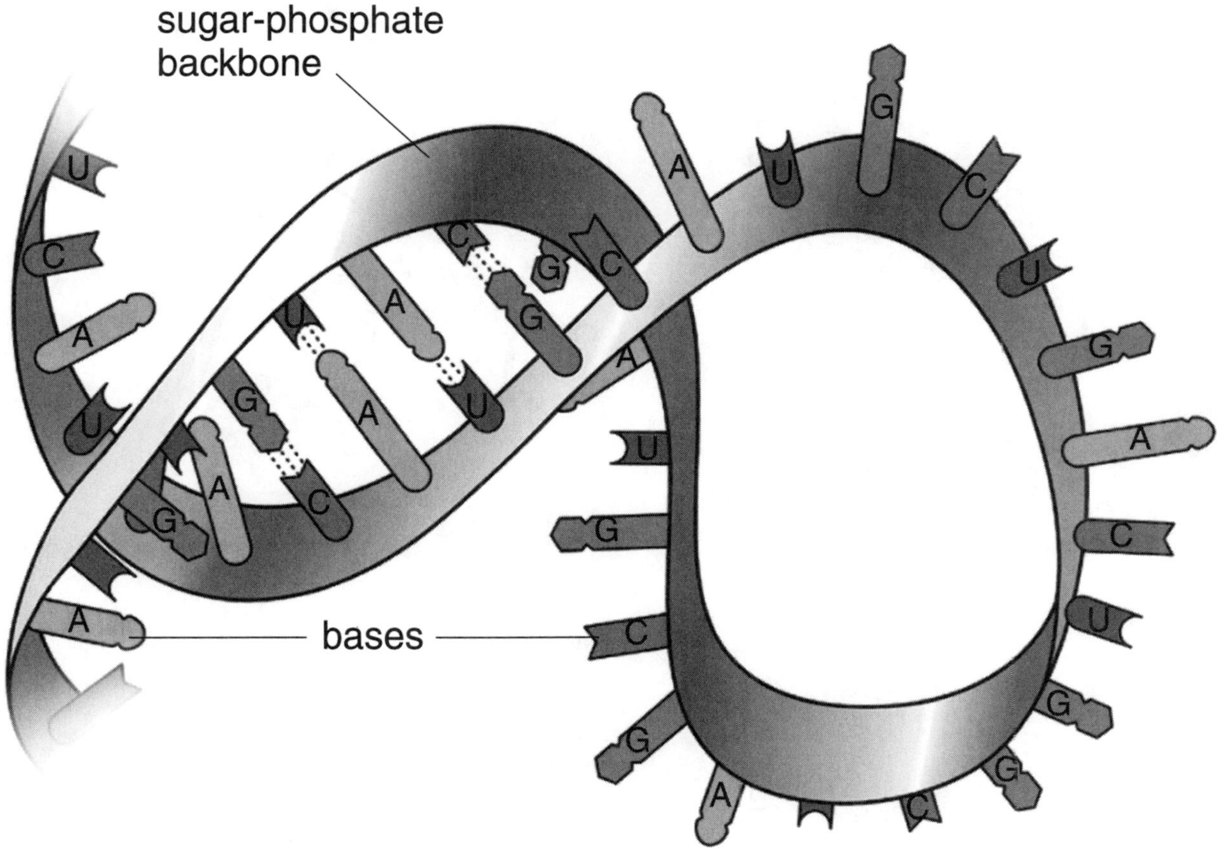

Figure 2.14 *Biosynthesis* ◆

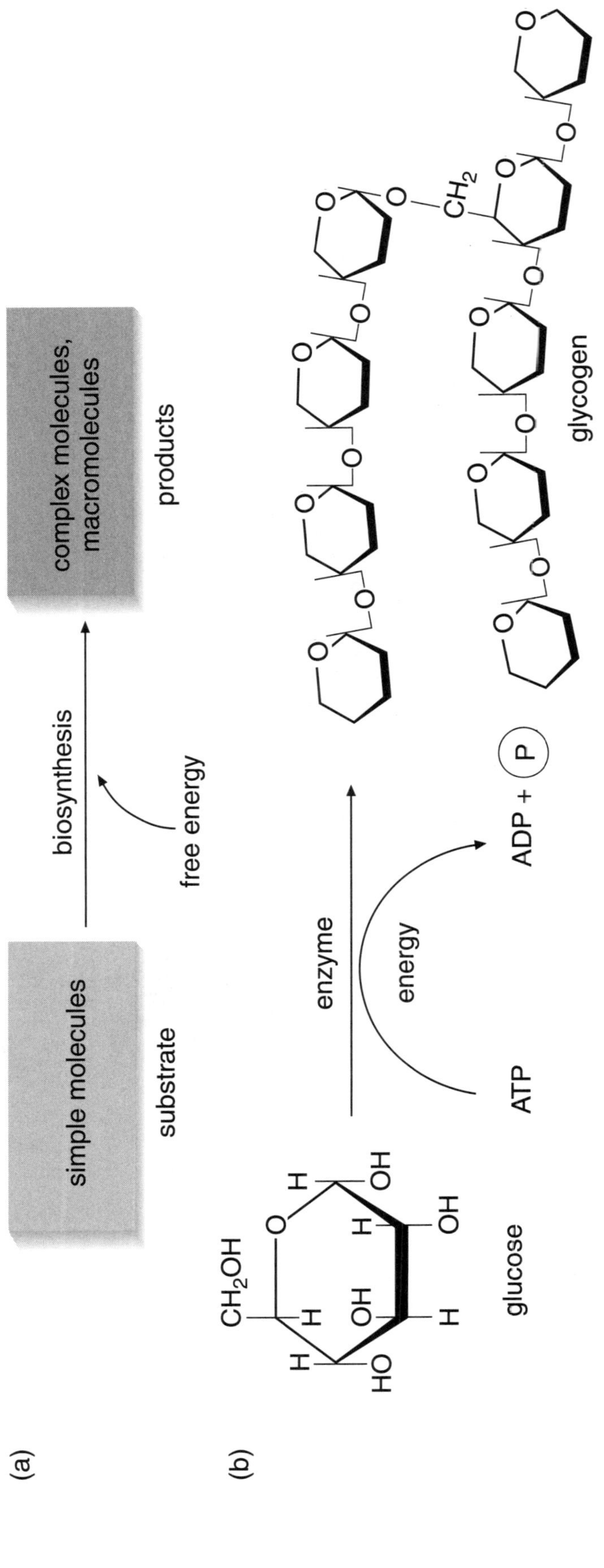

Figure 2.15 *Decomposition* ◆

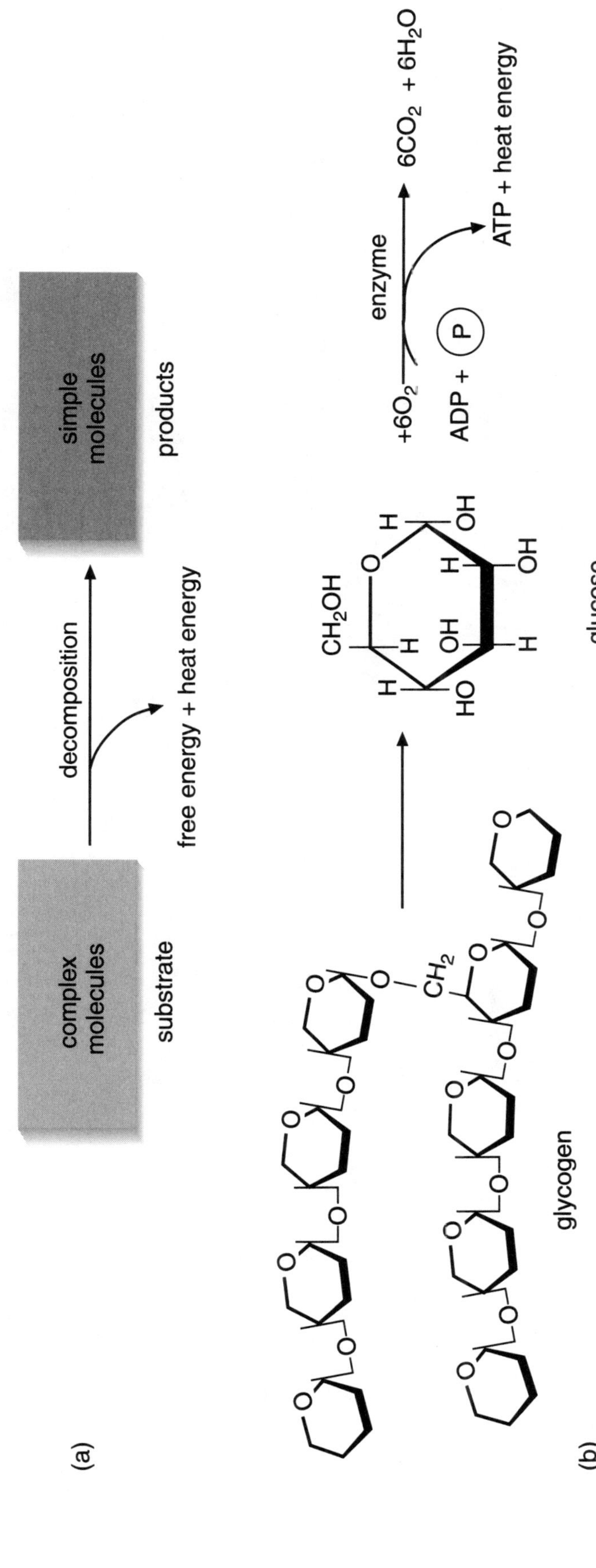

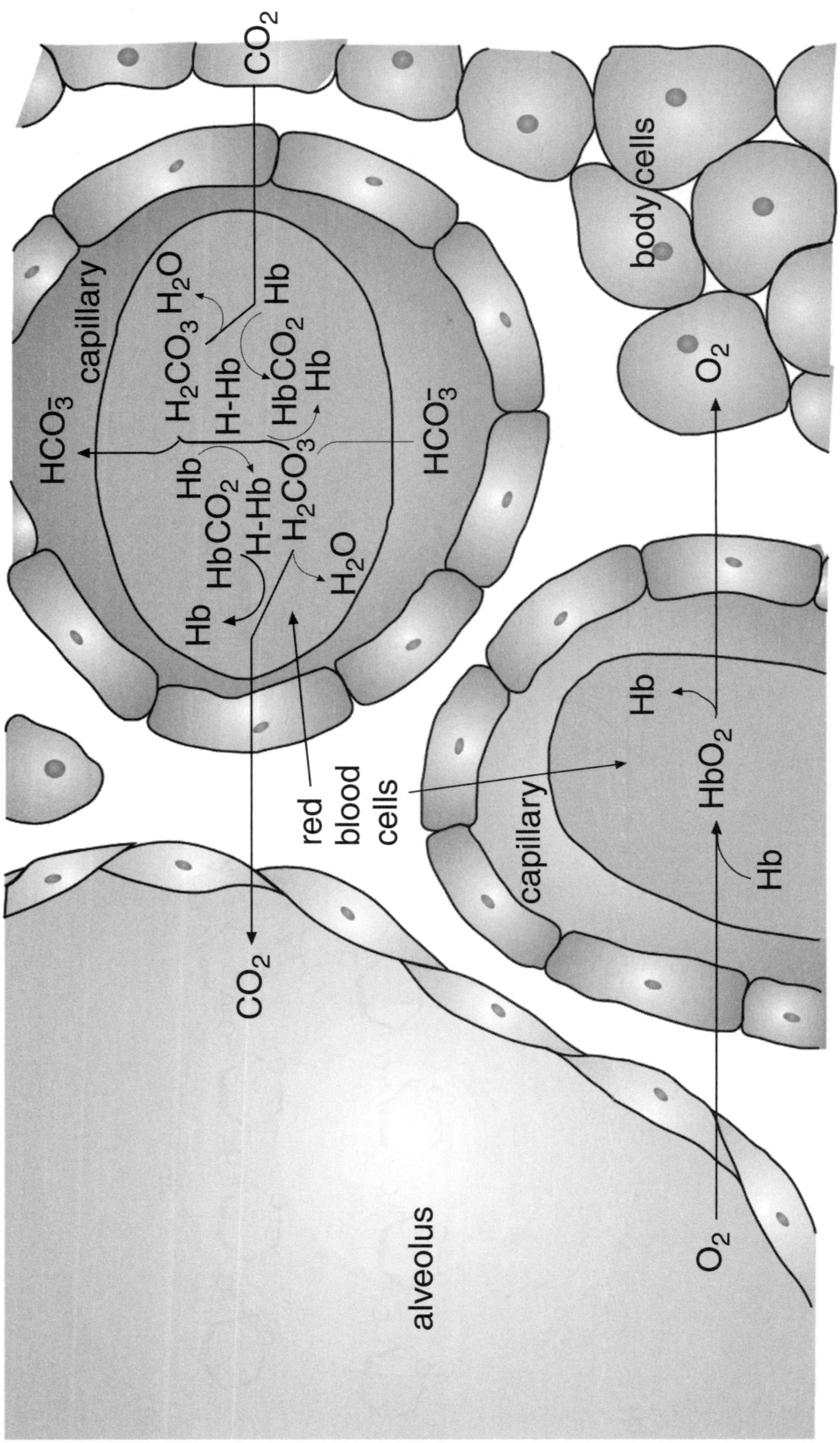

$CO_2$
capillary
$HCO_3^-$
$H_2CO_3$
$H_2O$
Hb
H-Hb
$HbCO_2$
$HCO_3^-$
body cells
$O_2$
Hb
$HbO_2$
red blood cells
capillary
$CO_2$
alveolus
$O_2$


 

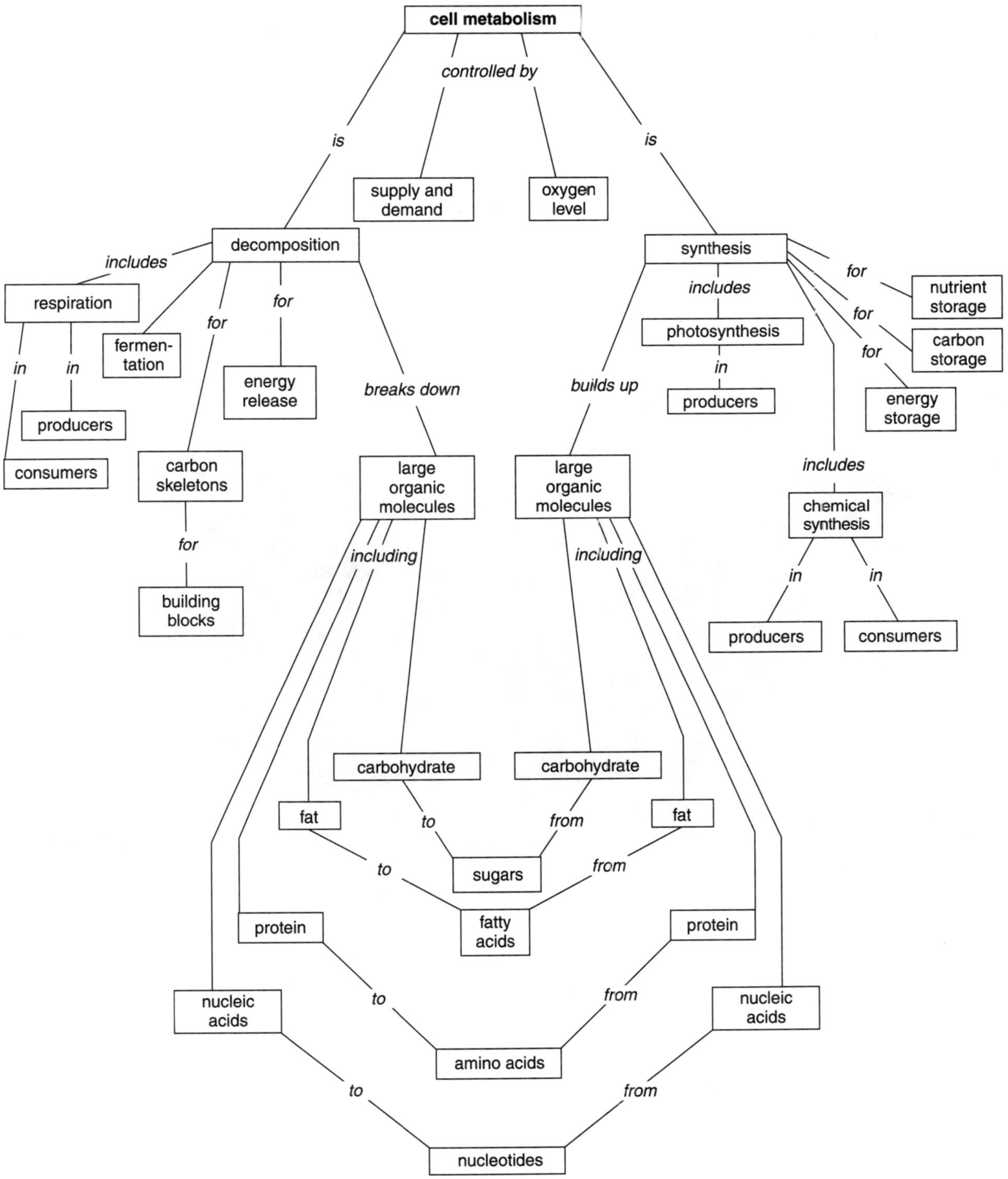

cell metabolism
controlled by
is
is
supply and demand
oxygen level
decomposition
synthesis
includes
respiration
fermen-tation
for
for
for
nutrient storage
for
carbon storage
for
energy storage
in
in
producers
consumers
energy release
breaks down
builds up
includes
photosynthesis
in
producers
carbon skeletons
for
building blocks
large organic molecules
large organic molecules
including
including
includes
chemical synthesis
in
in
producers
consumers
carbohydrate
carbohydrate
fat
to
from
fat
to
sugars
from
fatty acids
protein
protein
to
from
nucleic acids
nucleic acids
amino acids
to
from
nucleotides

Figure 3.3 *Cell Membrane* ◆

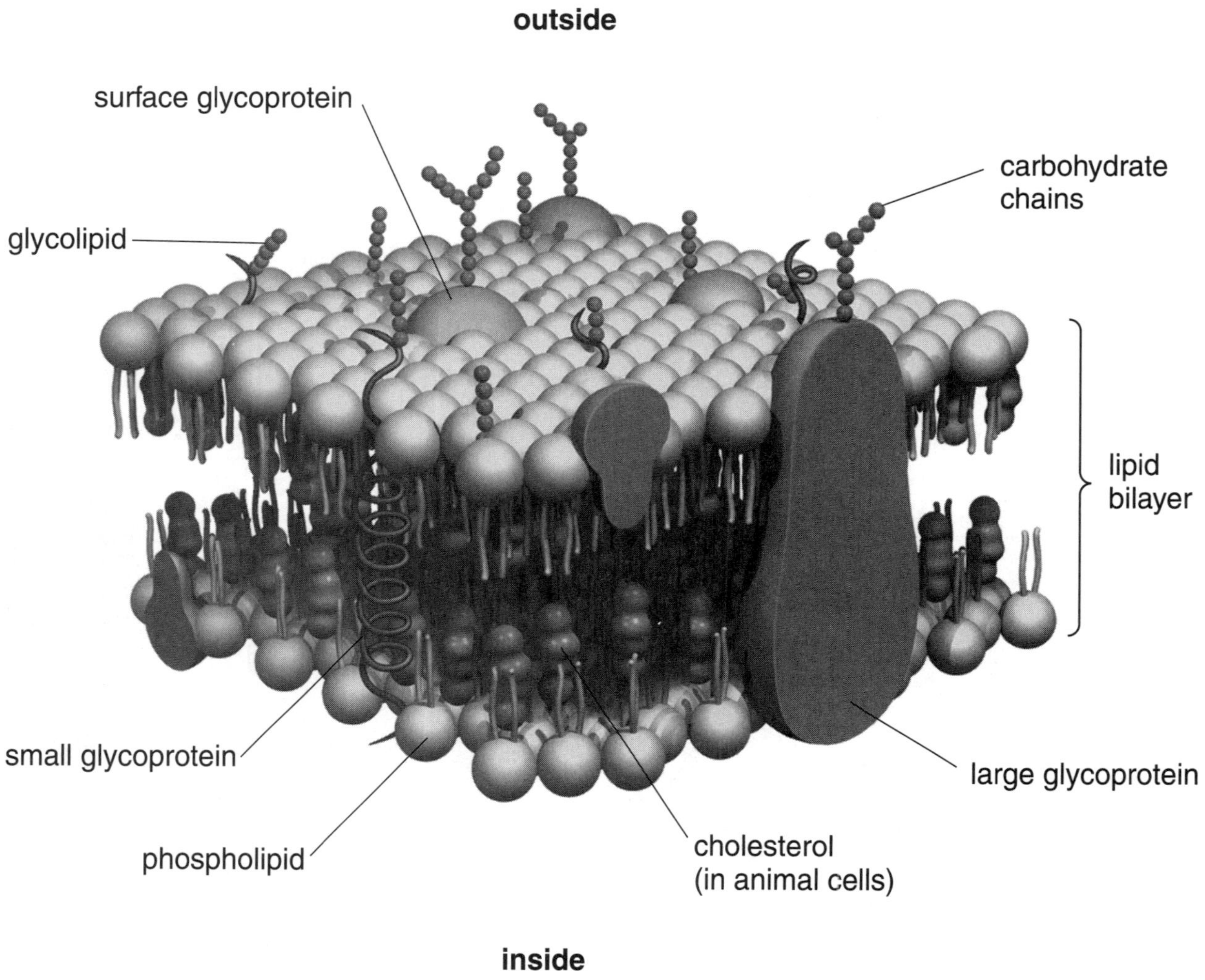

Figure 4.14 *Products of Photosynthesis*

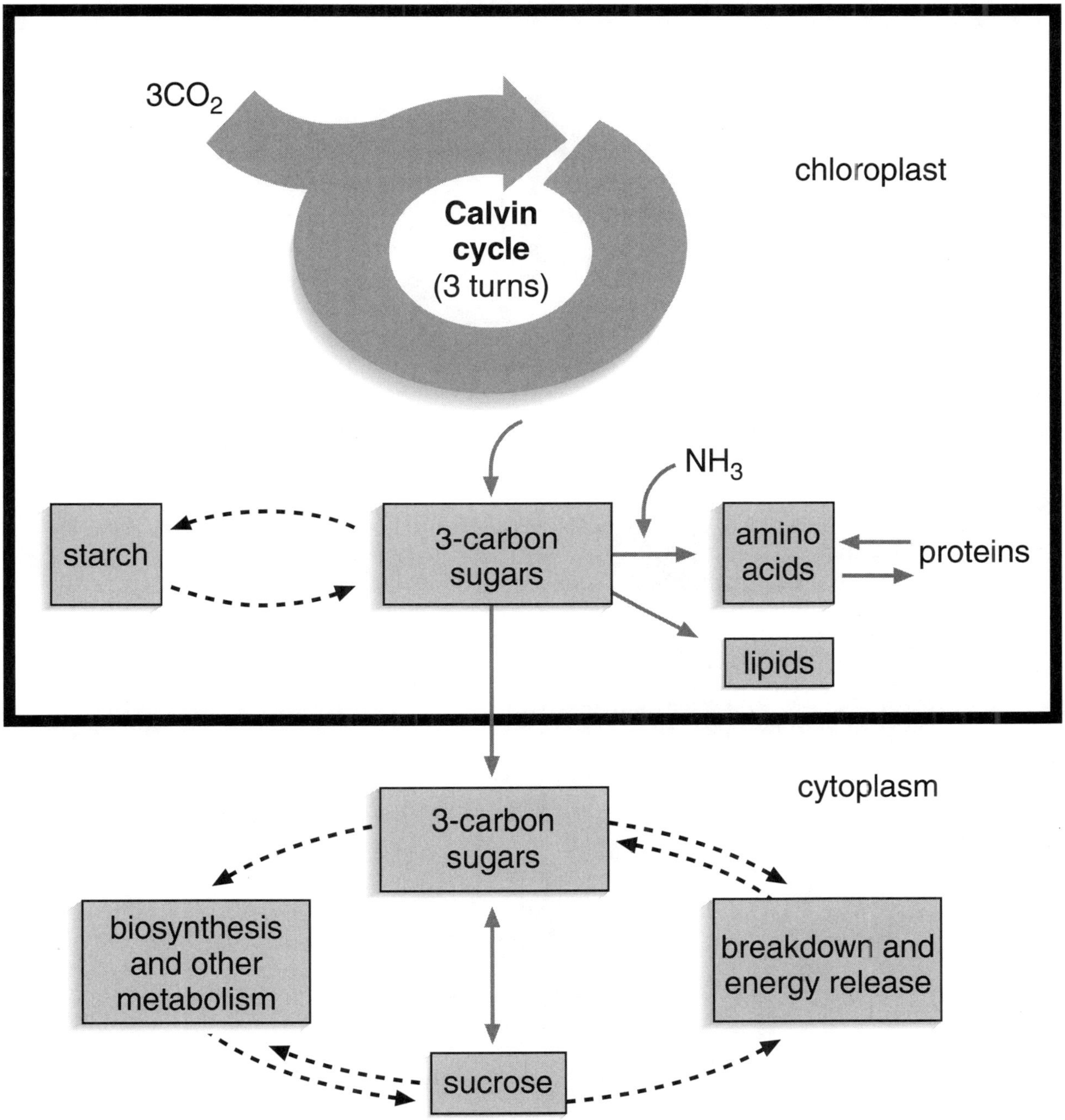

Figure 4.7 *Overview of Photosynthesis* ◆

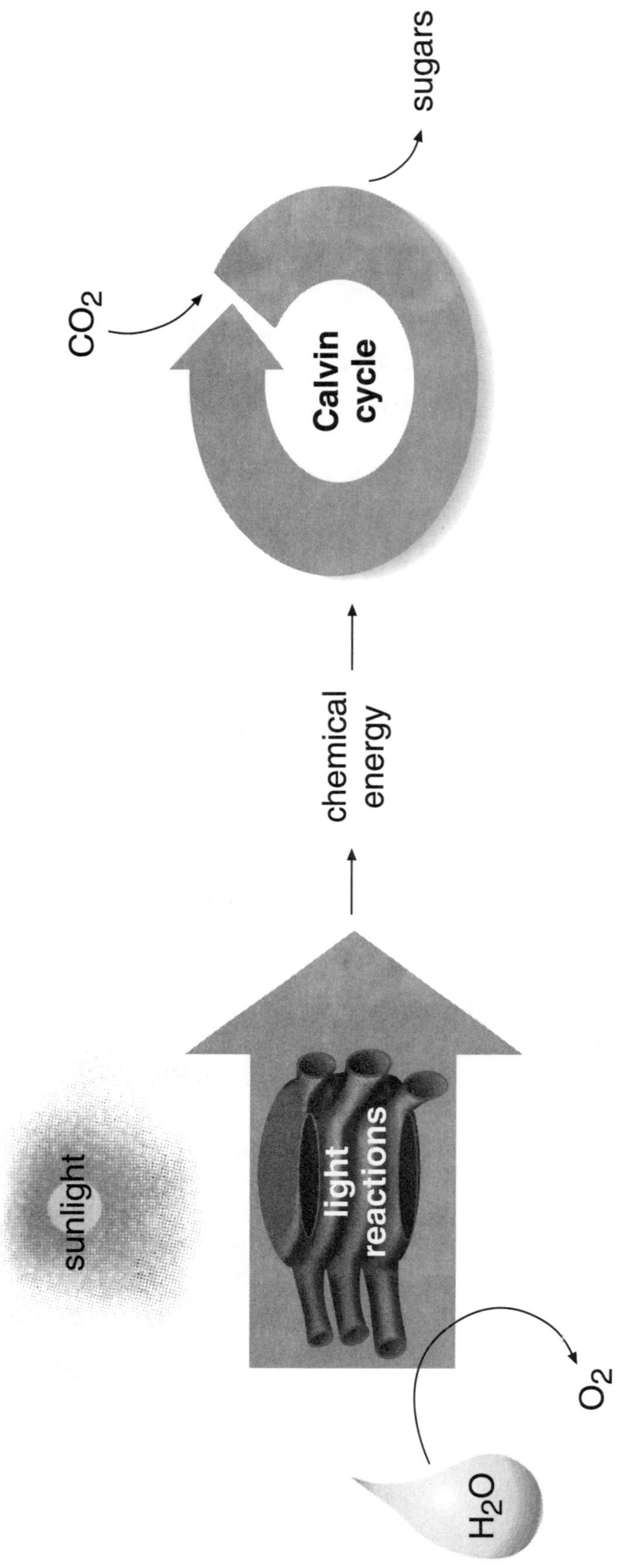

Figure 4.9 *Light Reactions* ◆

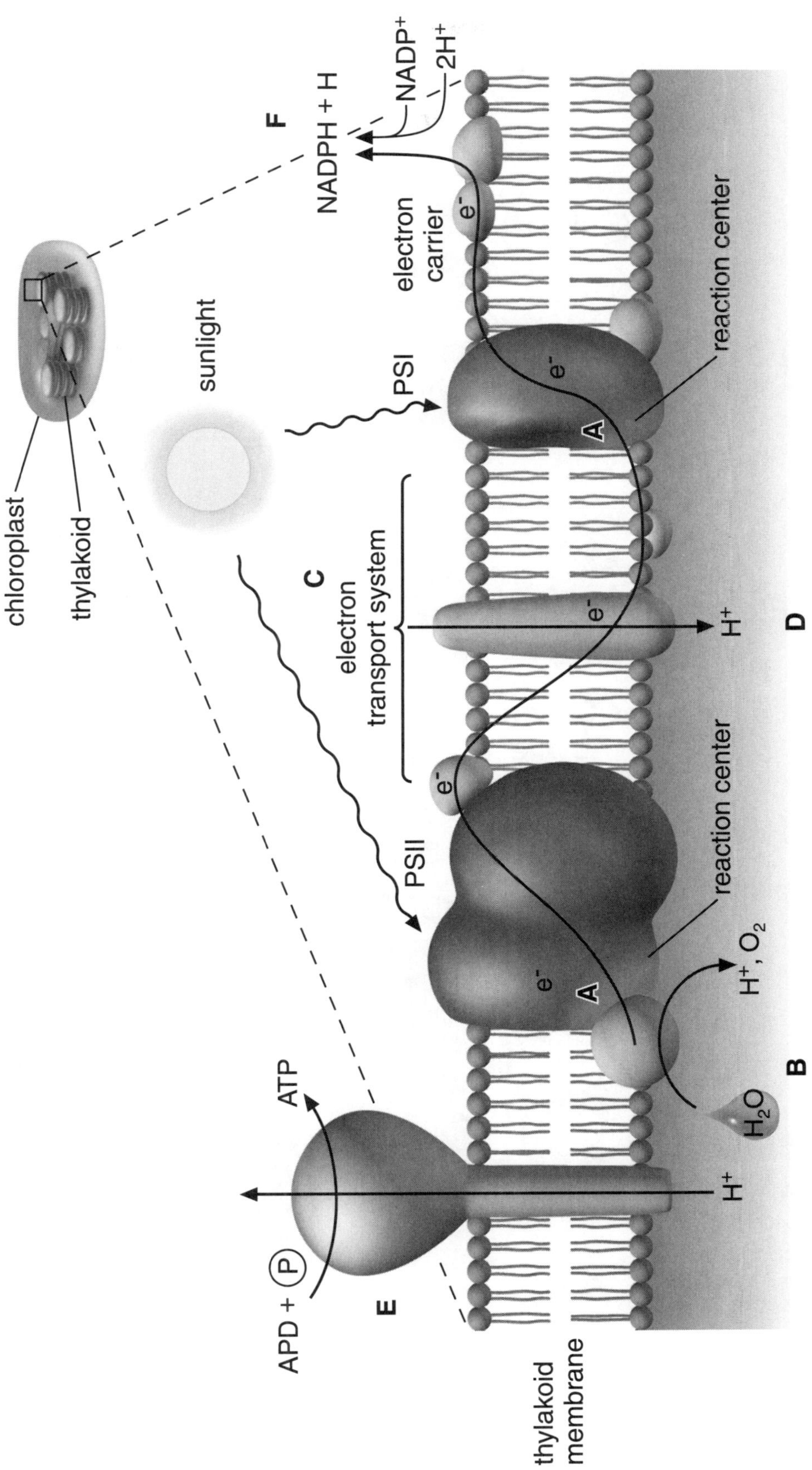

Figure 4.13 *Calvin Cycle*

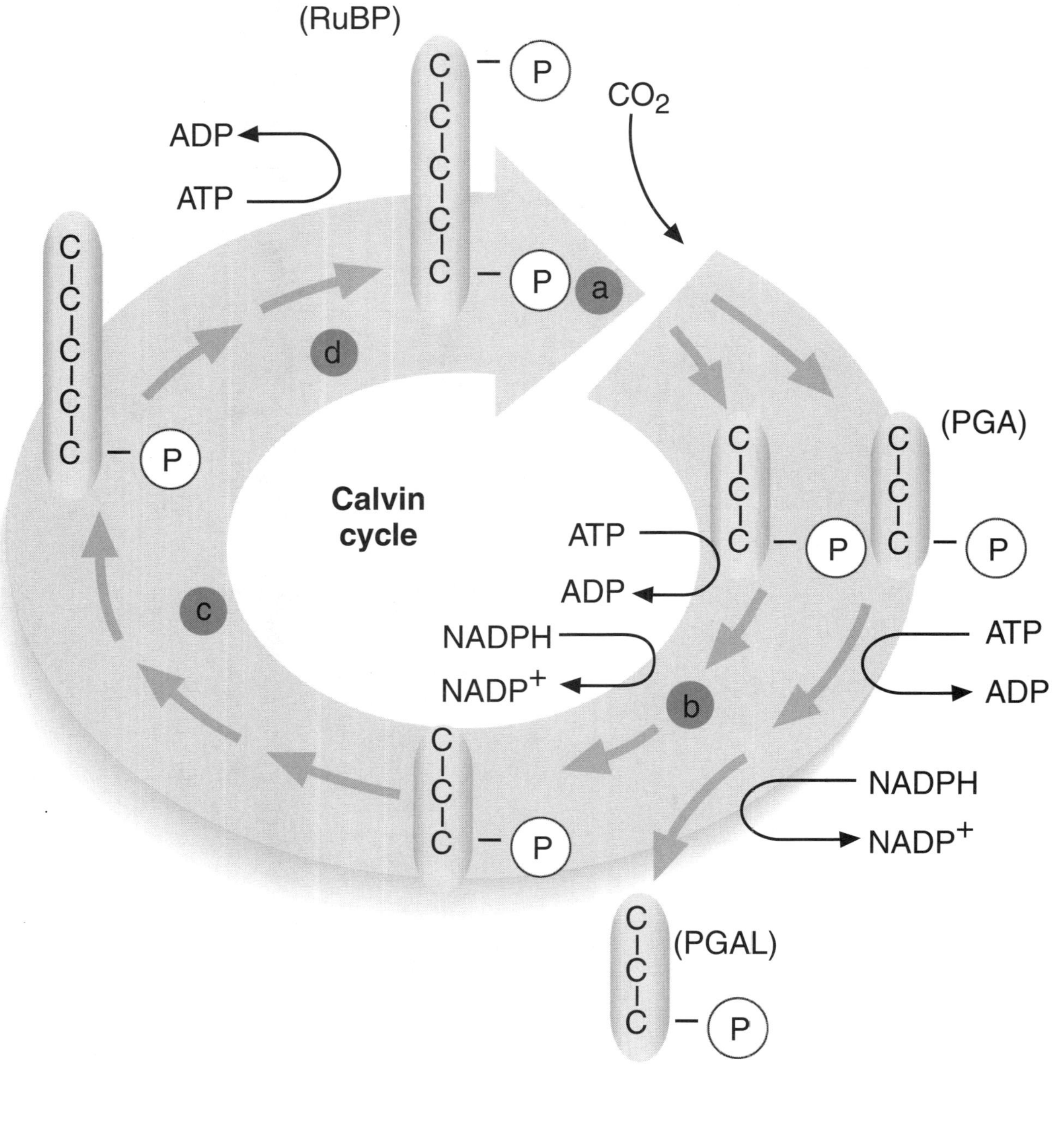

Figure 4.22 $CO_2$ *Fixation in* $C_4$ *Plants*

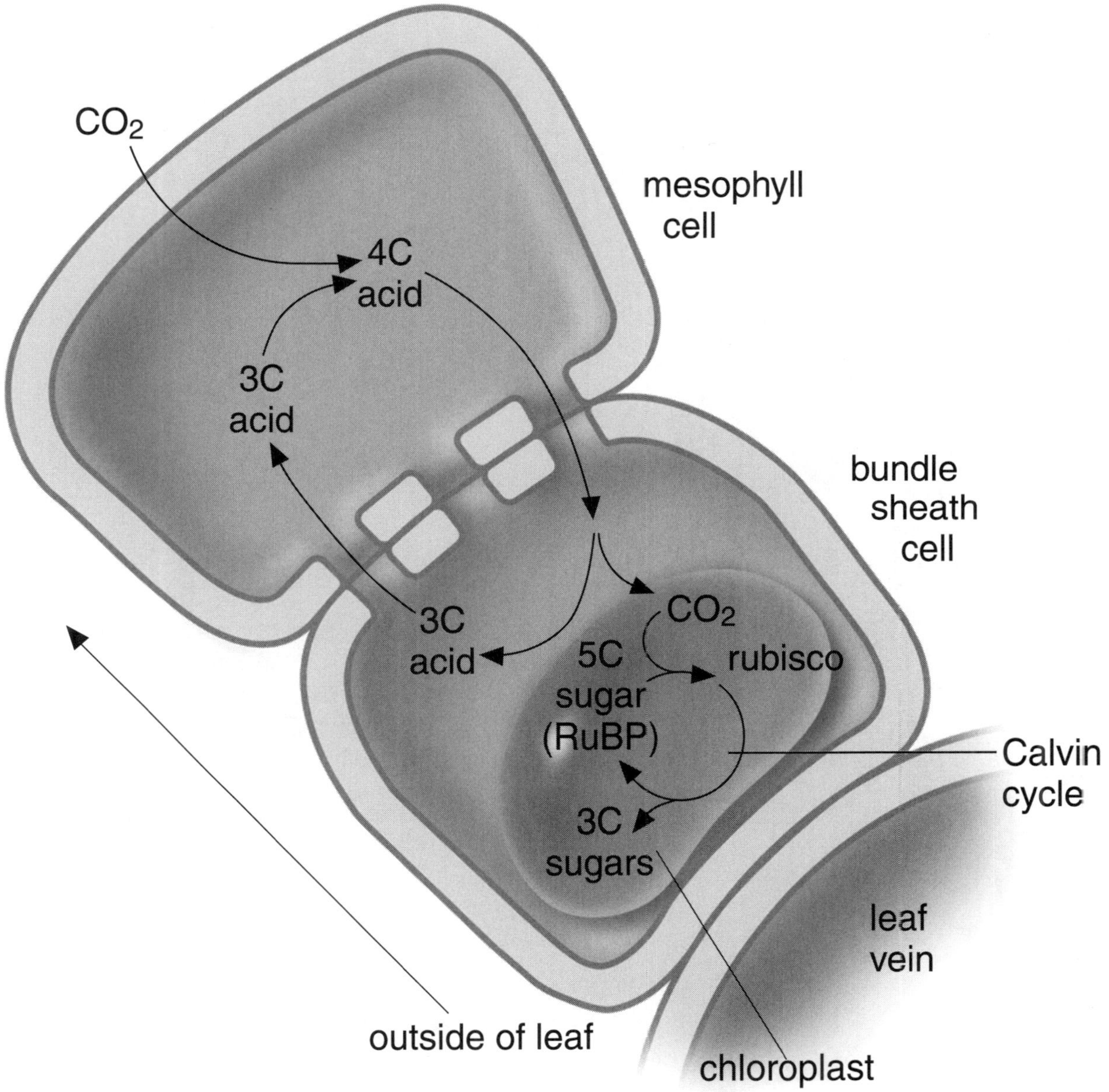

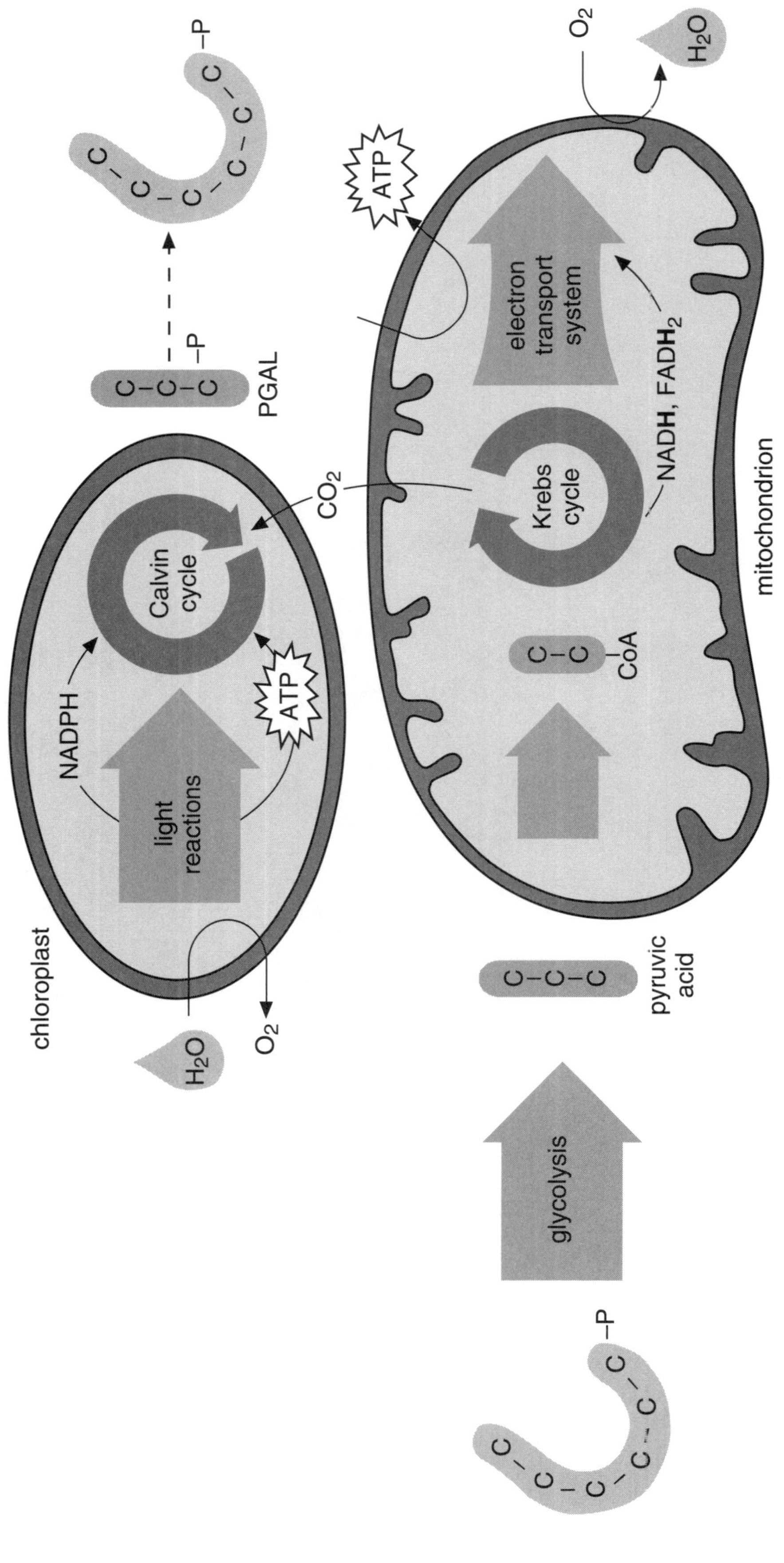

chloroplast
light reactions
NADPH
ATP
Calvin cycle
$CO_2$
$H_2O$
$O_2$
C–C–C–P
PGAL
C–C–C–C–C–C–P
glycolysis
C–C–C
pyruvic acid
C–C–CoA
Krebs cycle
NADH, $FADH_2$
electron transport system
ATP
$O_2$
$H_2O$
mitochondrion

Figure 5.4 *Stages in Respiration* ◆

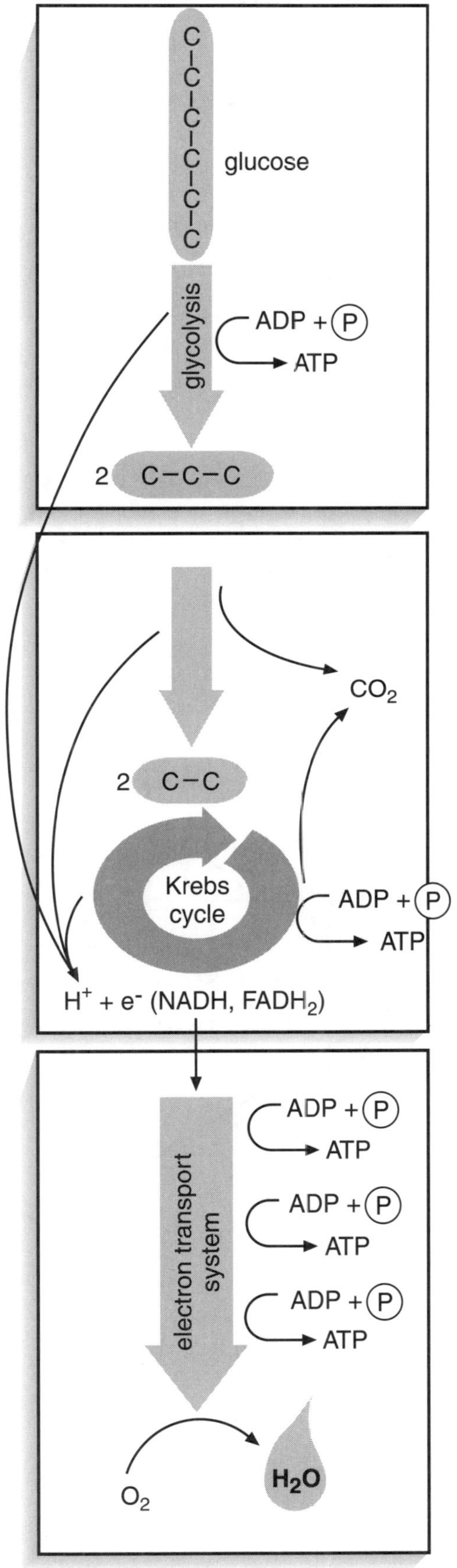

Figure 5.6 *Glycolysis* ◆

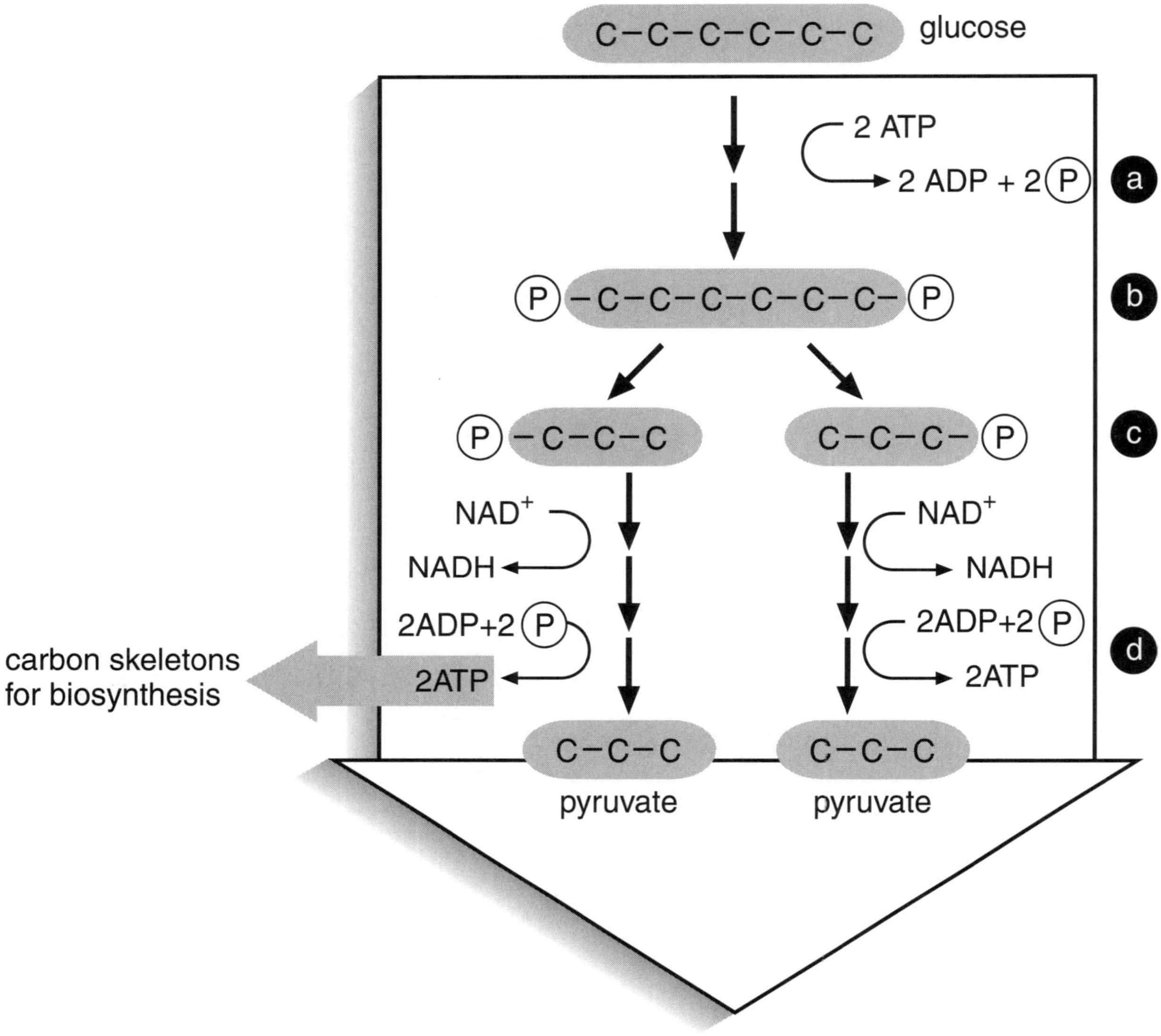


 

Figure 5.10 *Krebs Cycle*

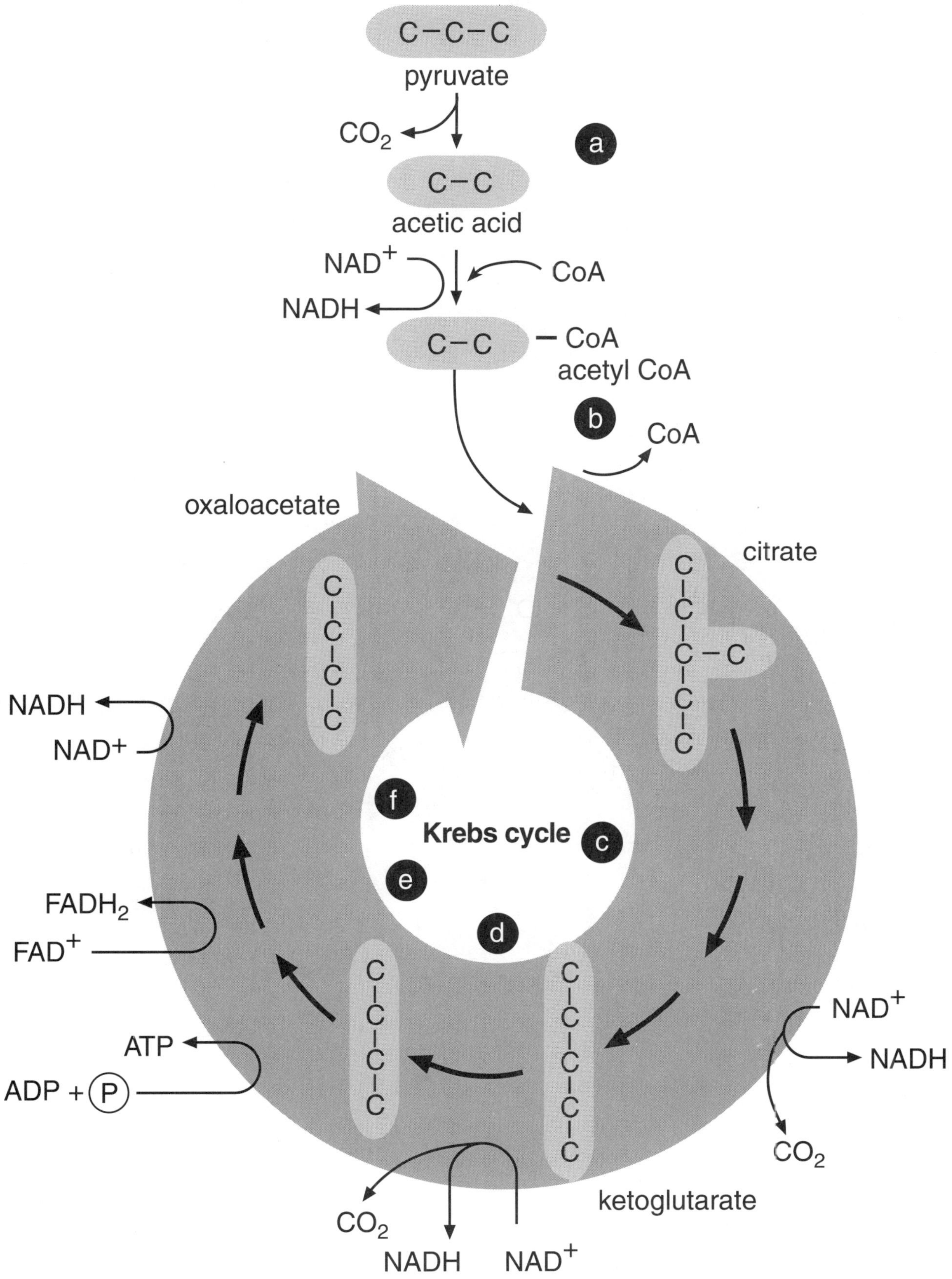

Figure 5.11 *Electron Transport System* ◆

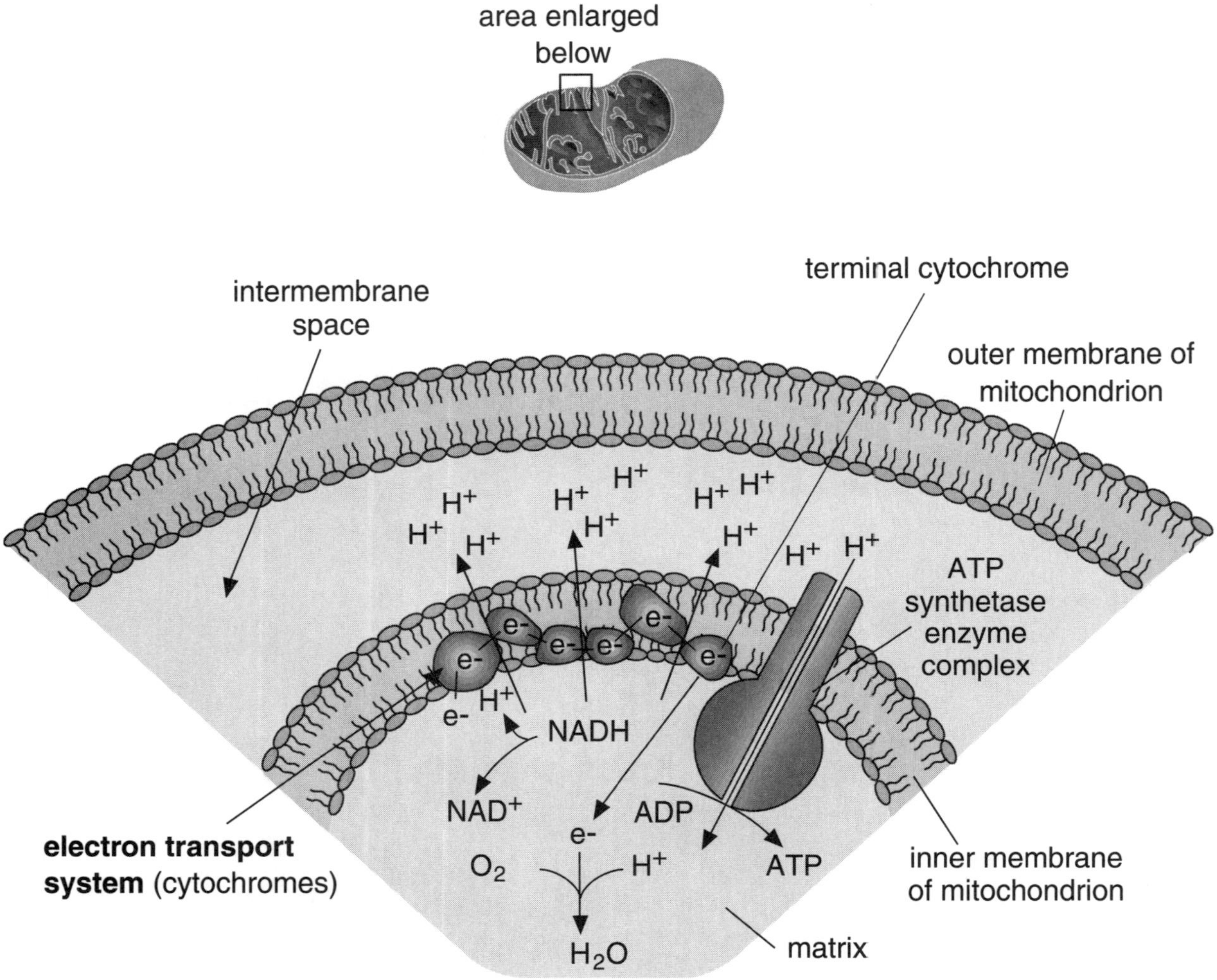

Figure 5.16 *The Krebs Cycle in Metabolism* ◆

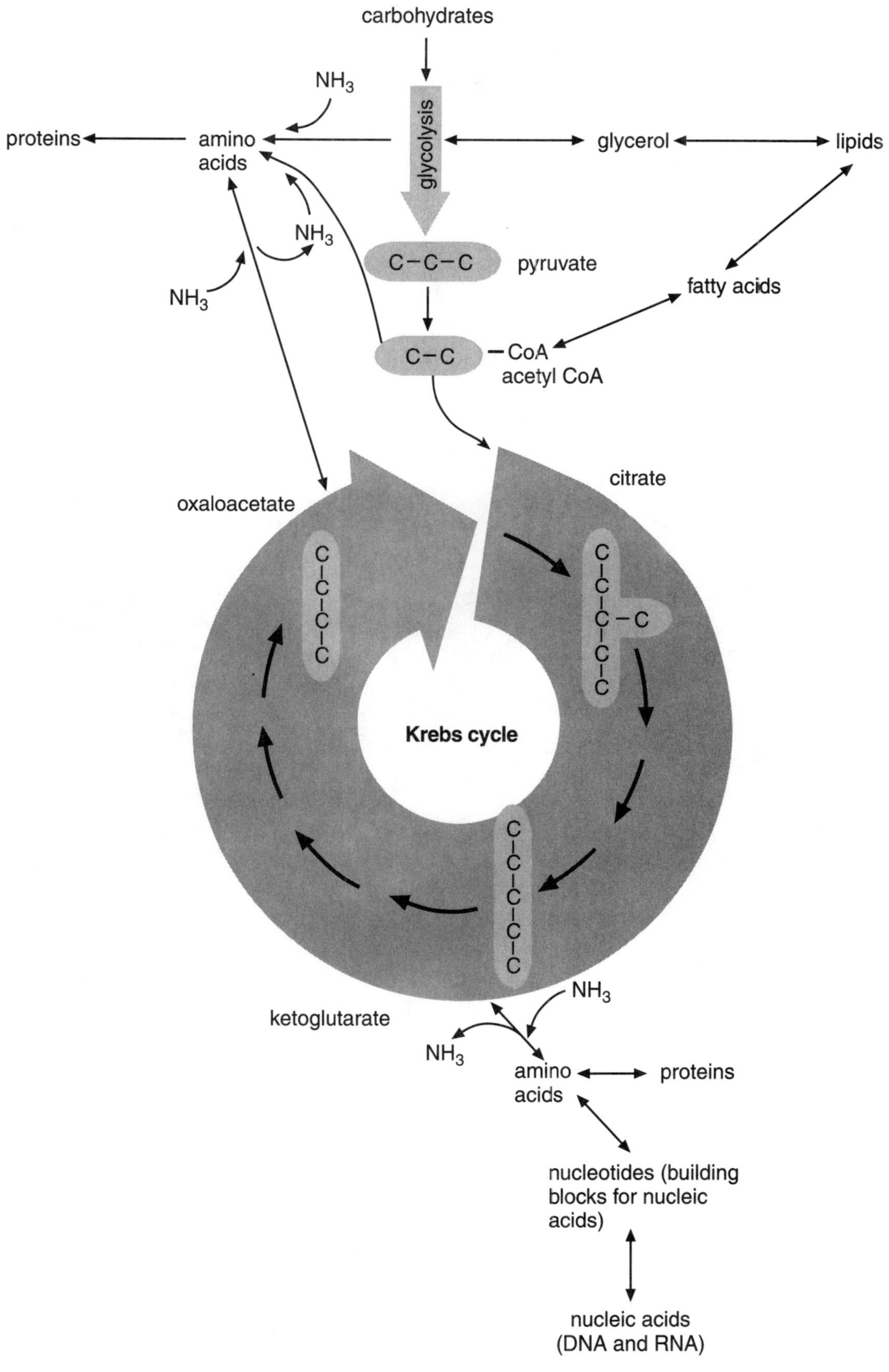

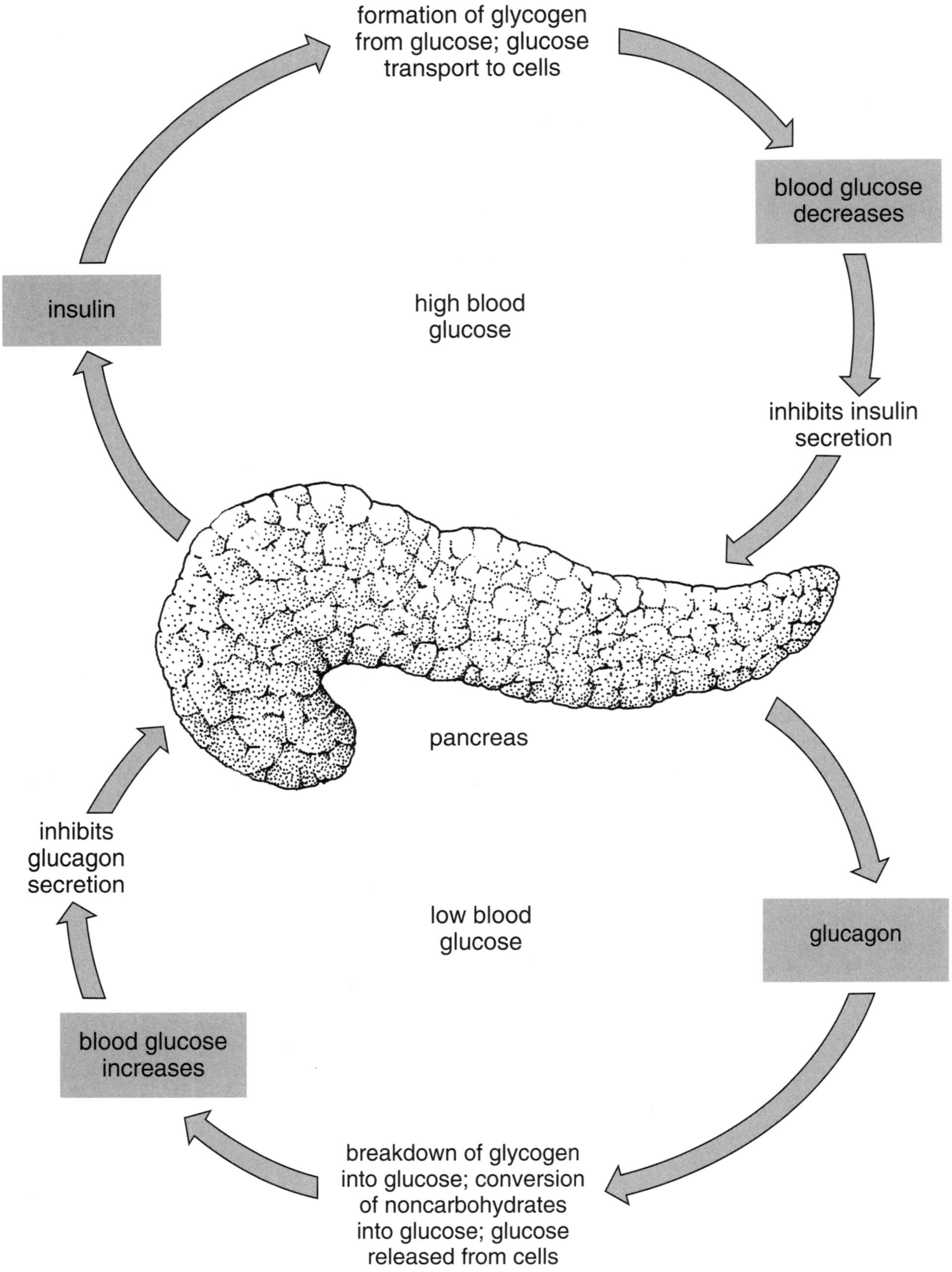
formation of glycogen from glucose; glucose transport to cells
blood glucose decreases
insulin
high blood glucose
inhibits insulin secretion
pancreas
inhibits glucagon secretion
low blood glucose
glucagon
blood glucose increases
breakdown of glycogen into glucose; conversion of noncarbohydrates into glucose; glucose released from cells

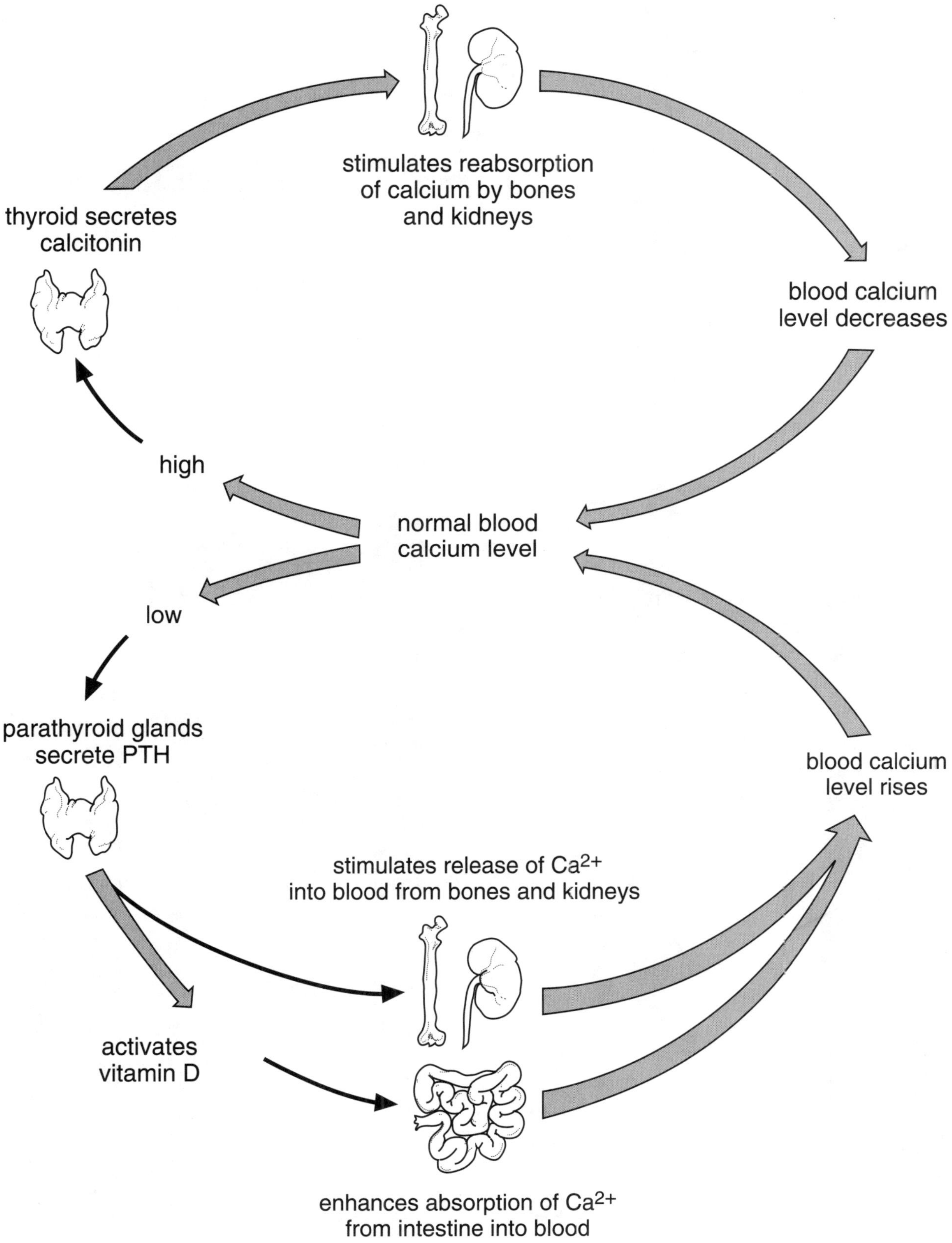
stimulates reabsorption of calcium by bones and kidneys
thyroid secretes calcitonin
blood calcium level decreases
high
normal blood calcium level
low
parathyroid glands secrete PTH
blood calcium level rises
stimulates release of Ca2+ into blood from bones and kidneys
activates vitamin D
enhances absorption of Ca2+ from intestine into blood

Figure 7.14 *Sliding Filament Model* ◆

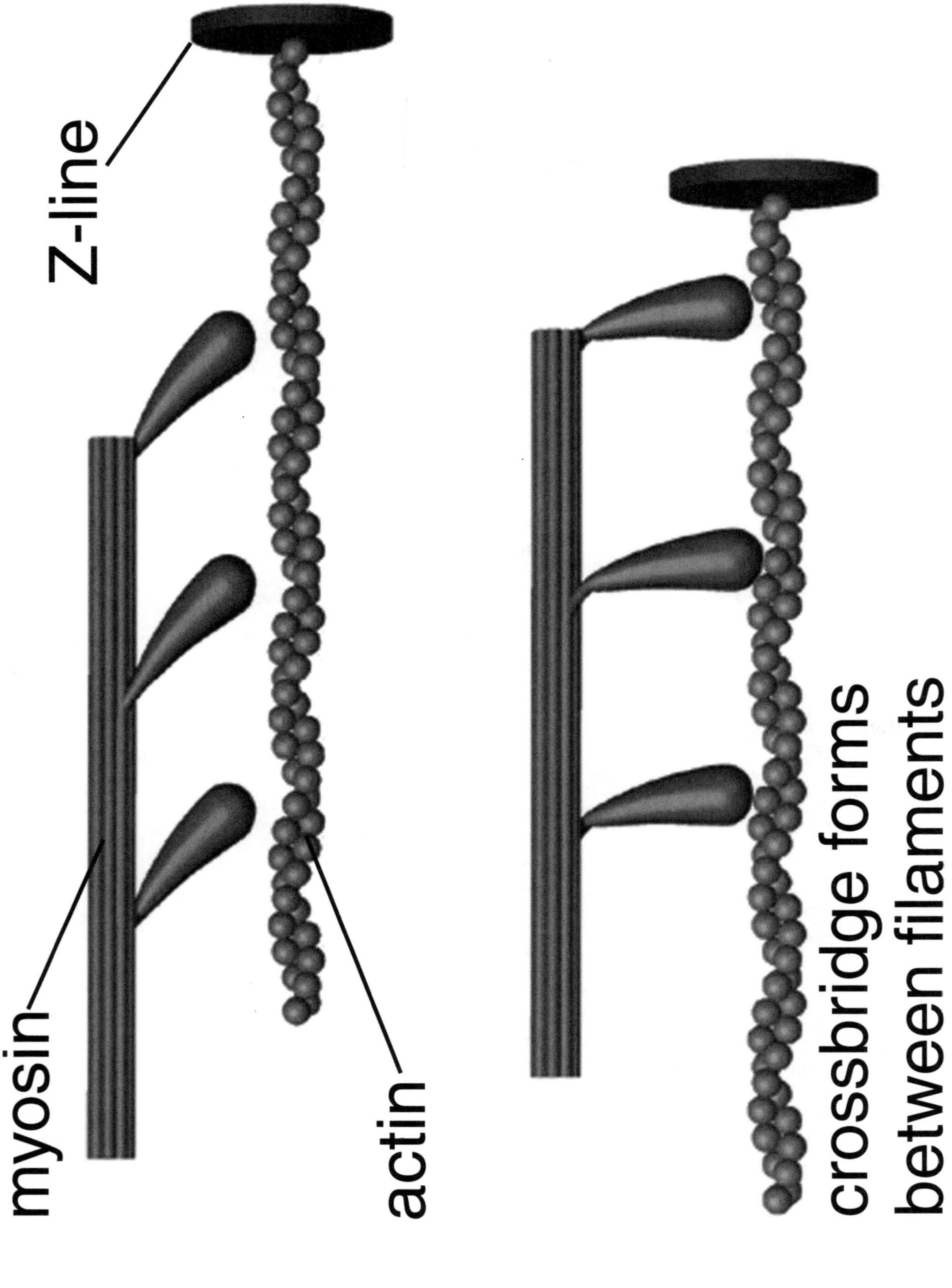


 

Figure 8.5 *Replication Fork* ◆

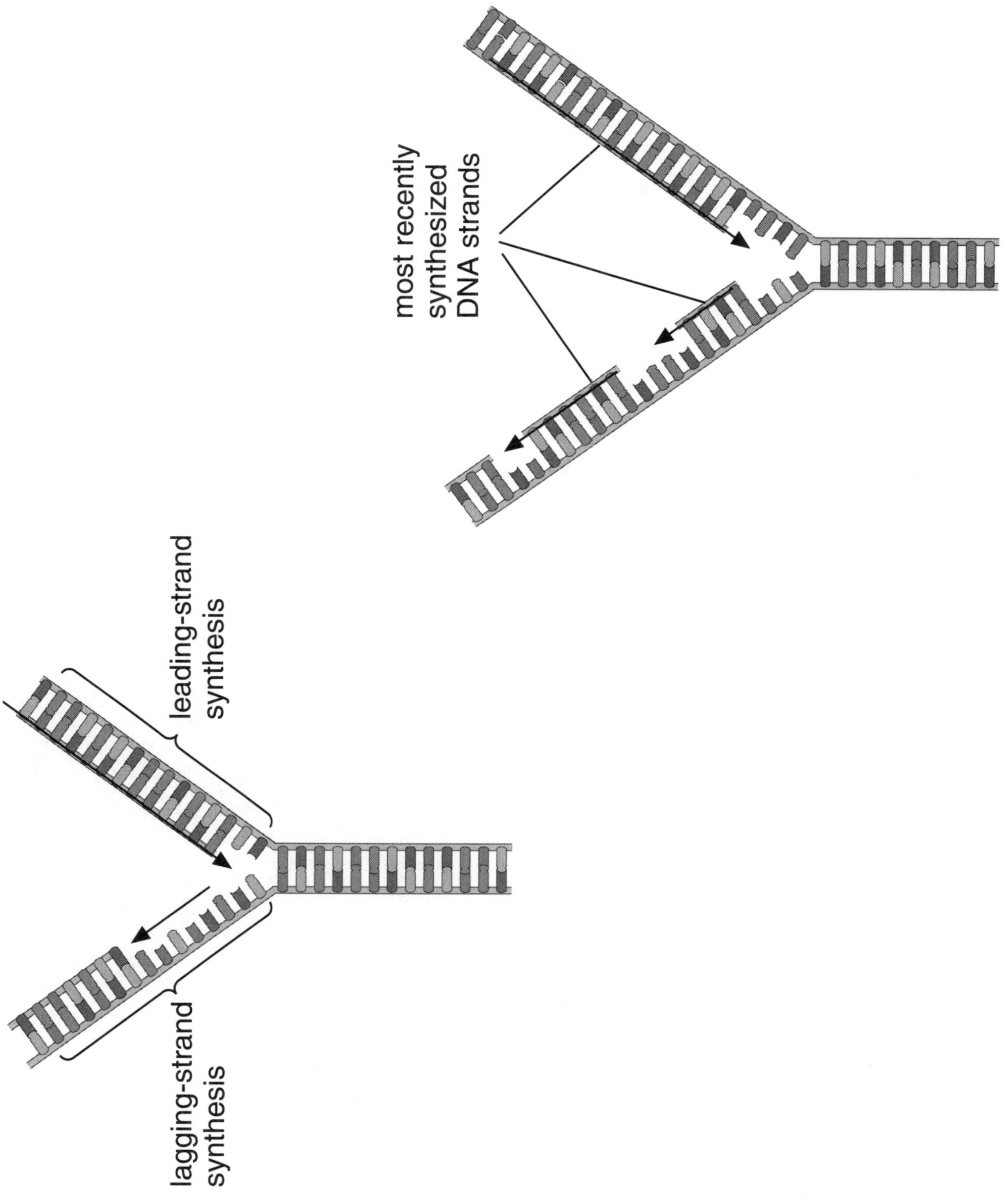

Figure 8.9 *DNA Repair* ◆

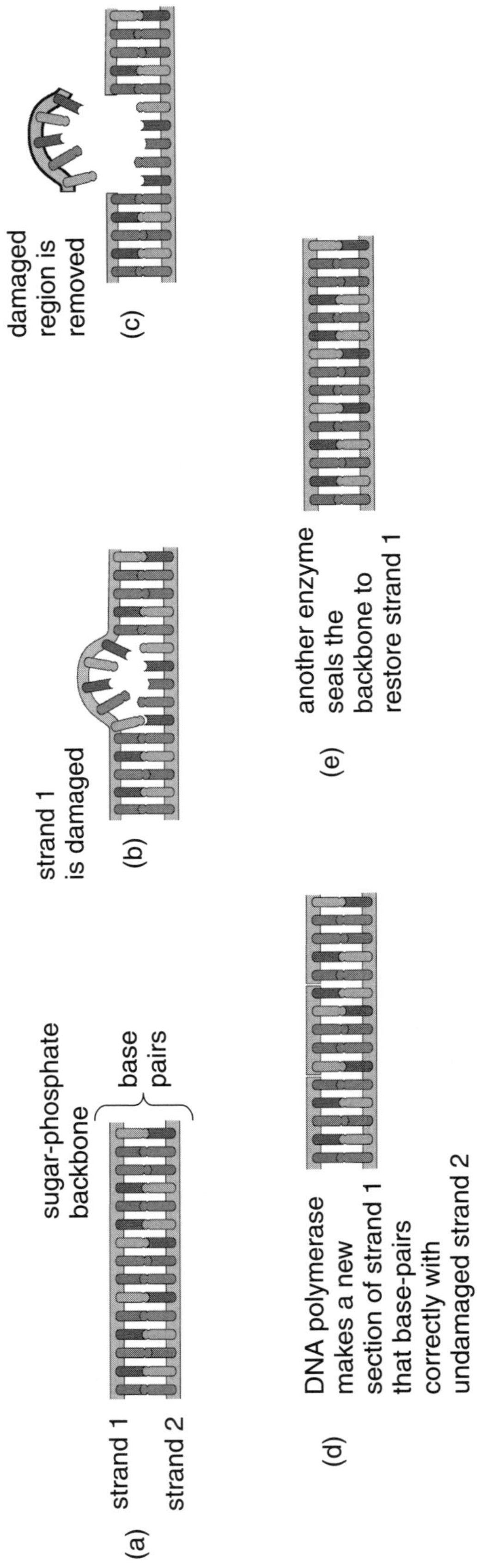

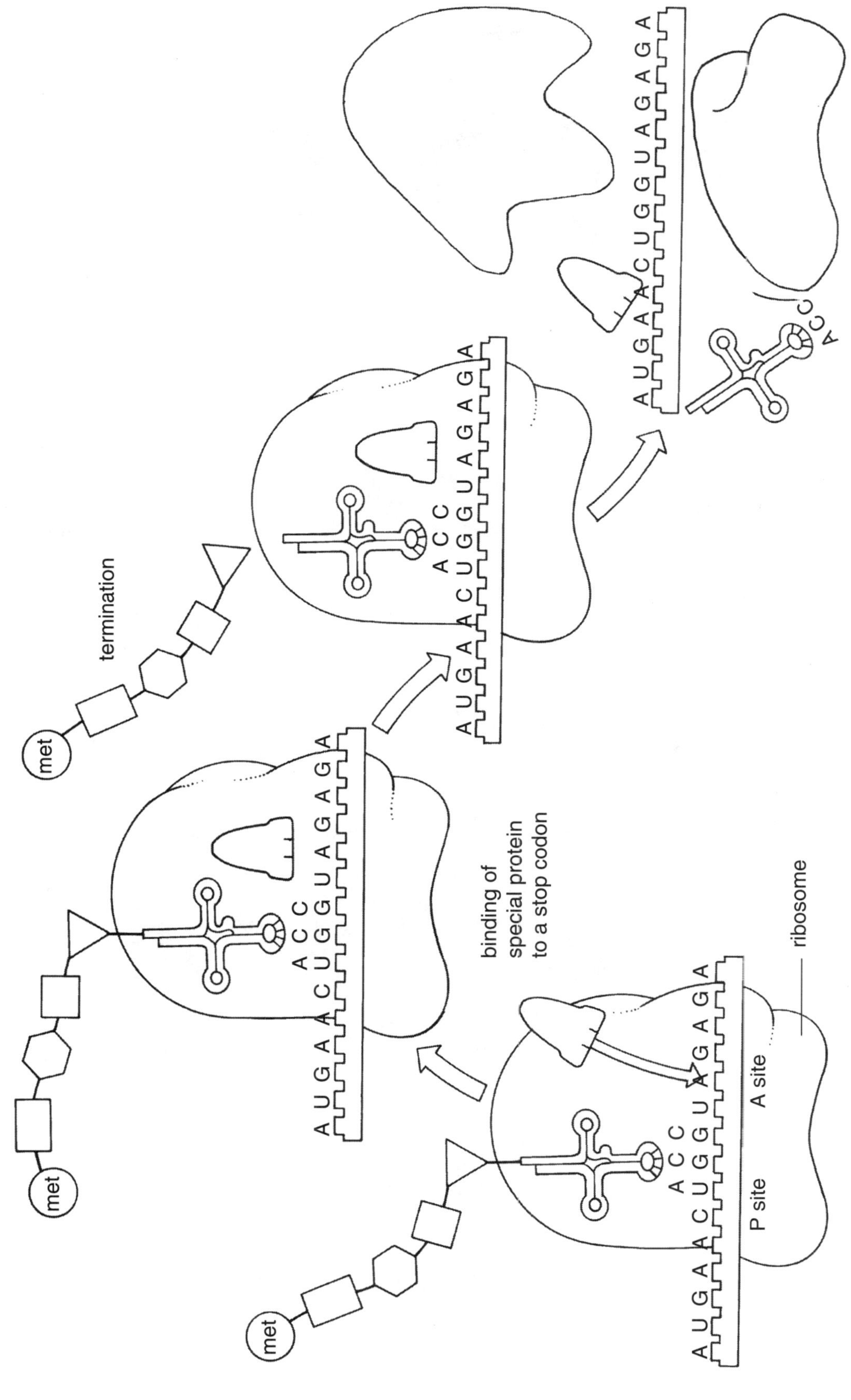
binding of
special protein
to a stop codon
termination
ribosome
P site
A site
met
A U G A A C U G G U A G A G A
A C C

# Chapter 9 Expressing Genetic Information

Figure 9.1 *Expressing Genetic Information* ◆

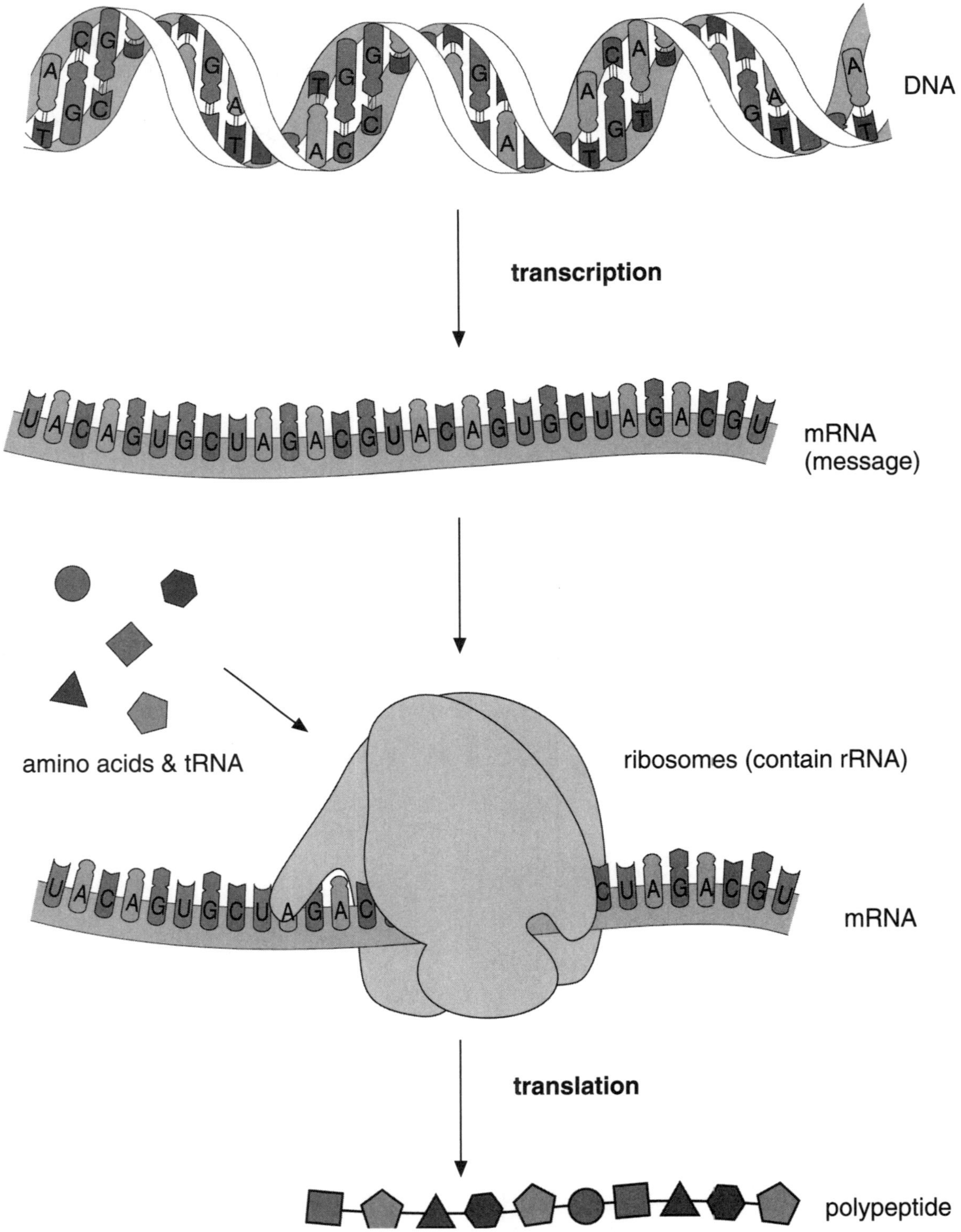

Figure 9.13 *RNA Processing* ◆

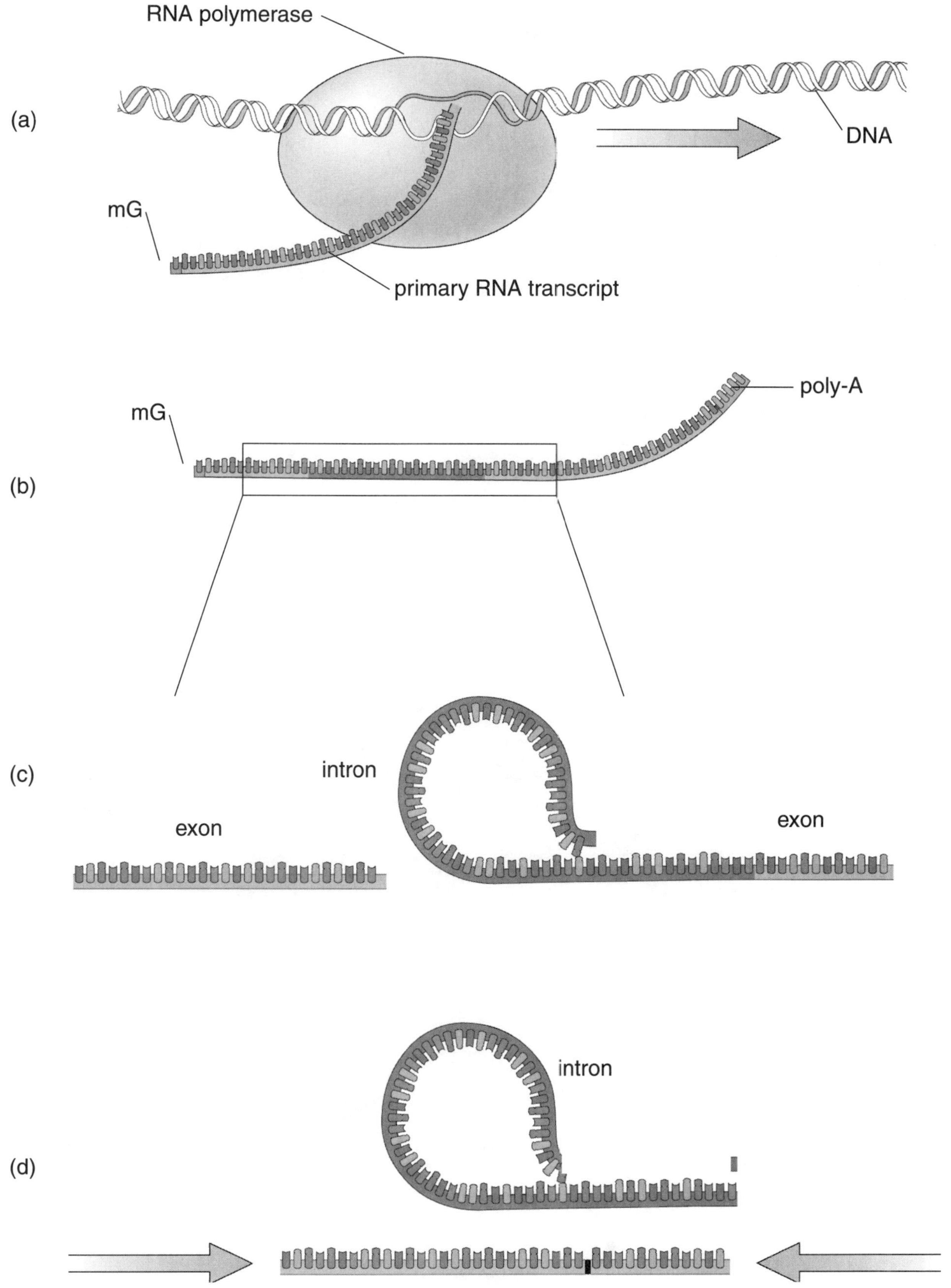

Figure 9.15 *Ribosome with Binding Sites* ◆

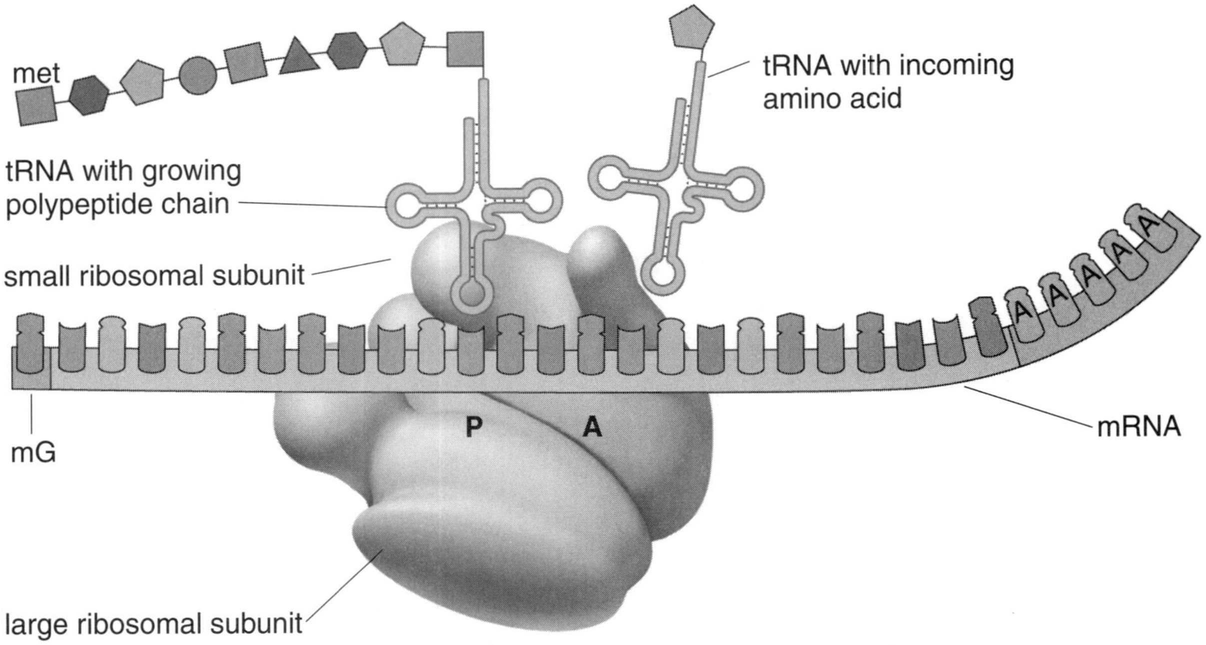


(a)

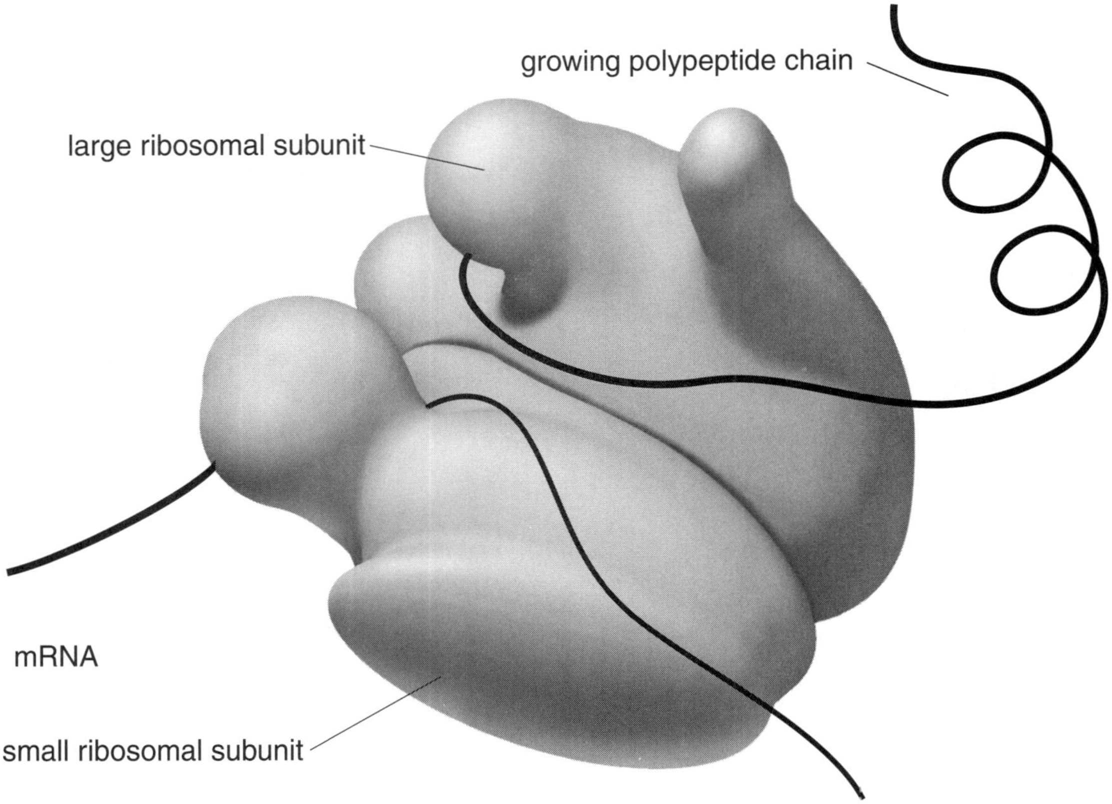


(b)

Figure 9.16 *Initiation of Translation* ◆

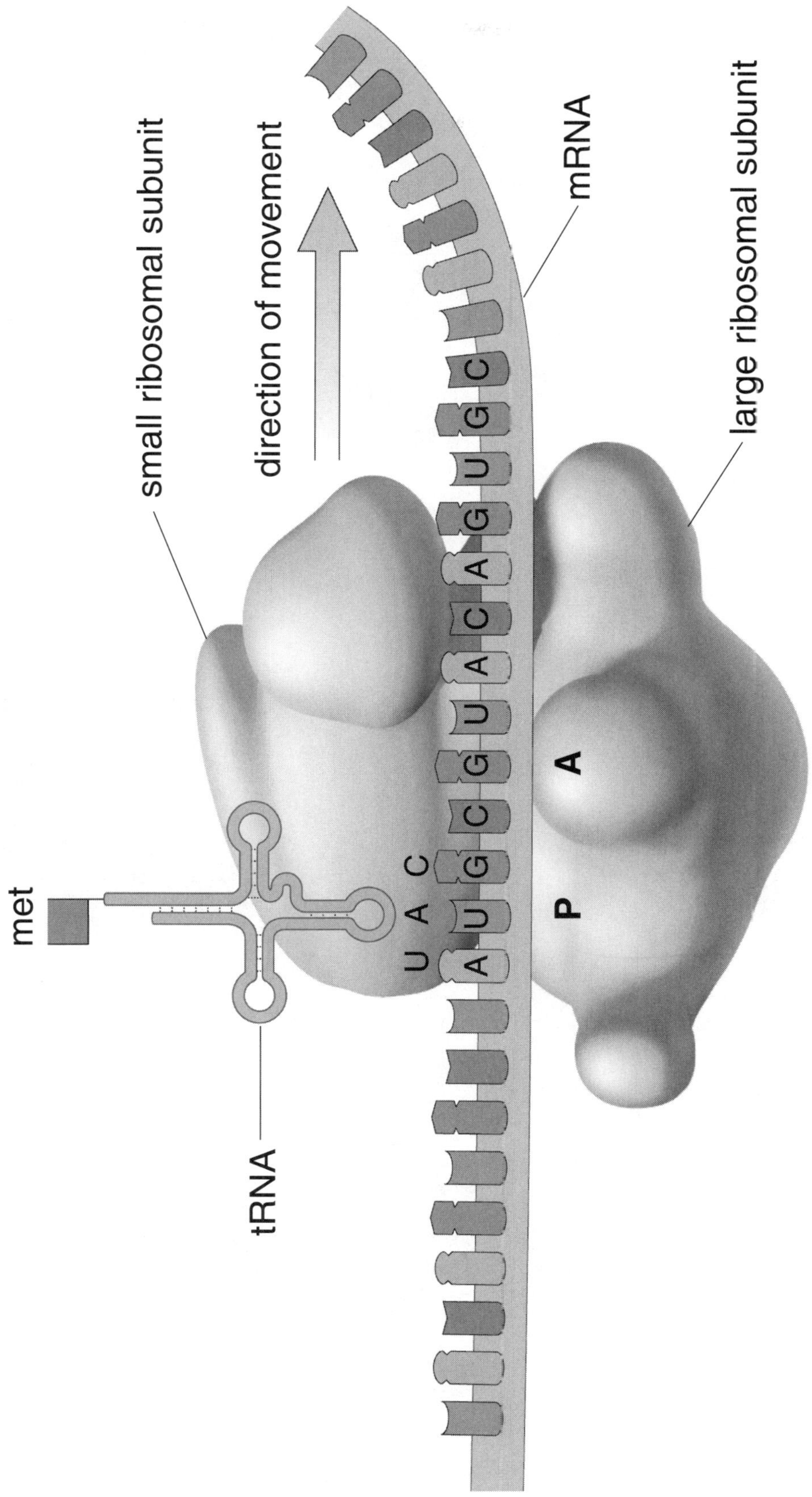

Figure 9.17 *Elongation of a Polypeptide Chain* ◆

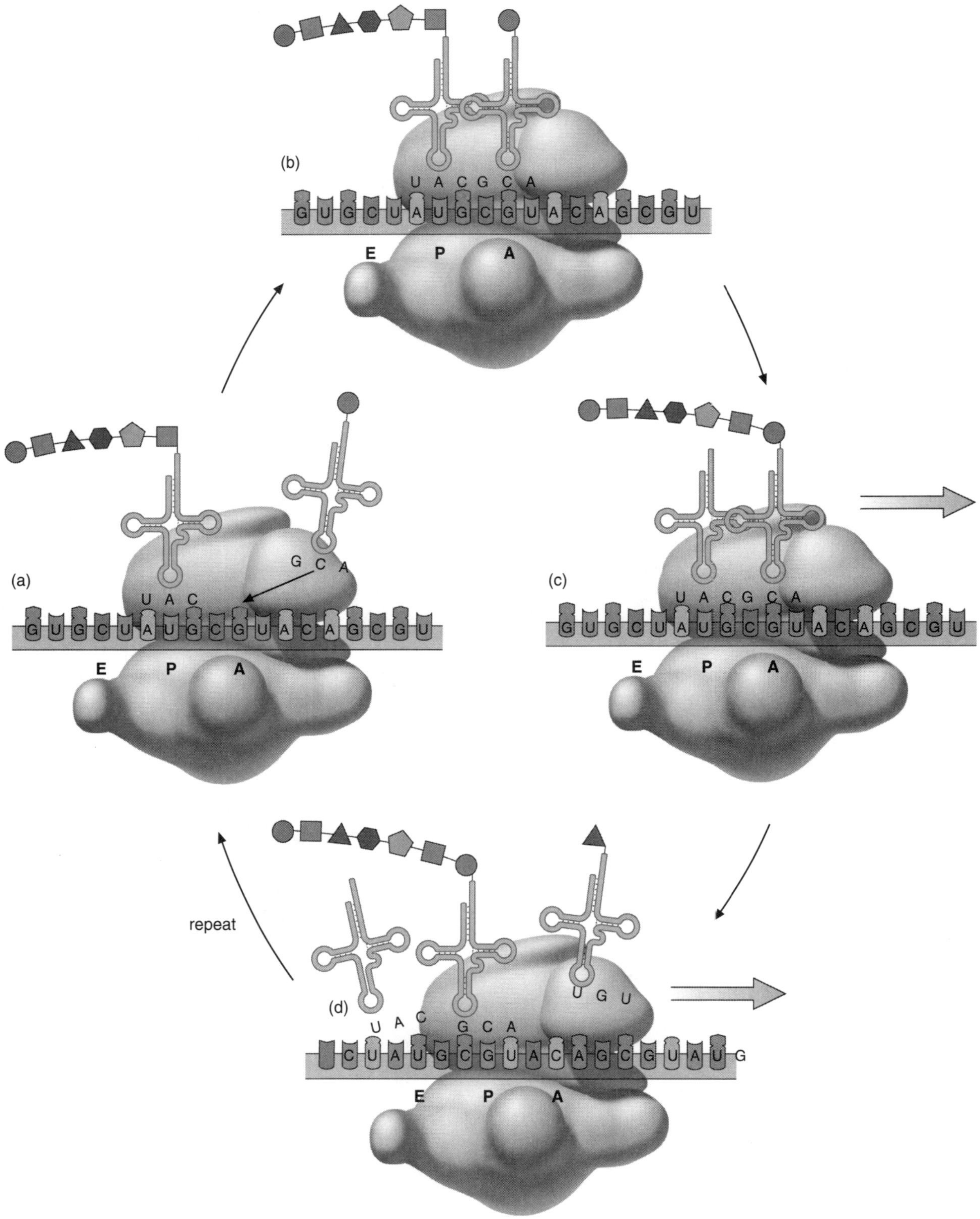


 

Figure 9.18 *Transcription and Translation* ◆

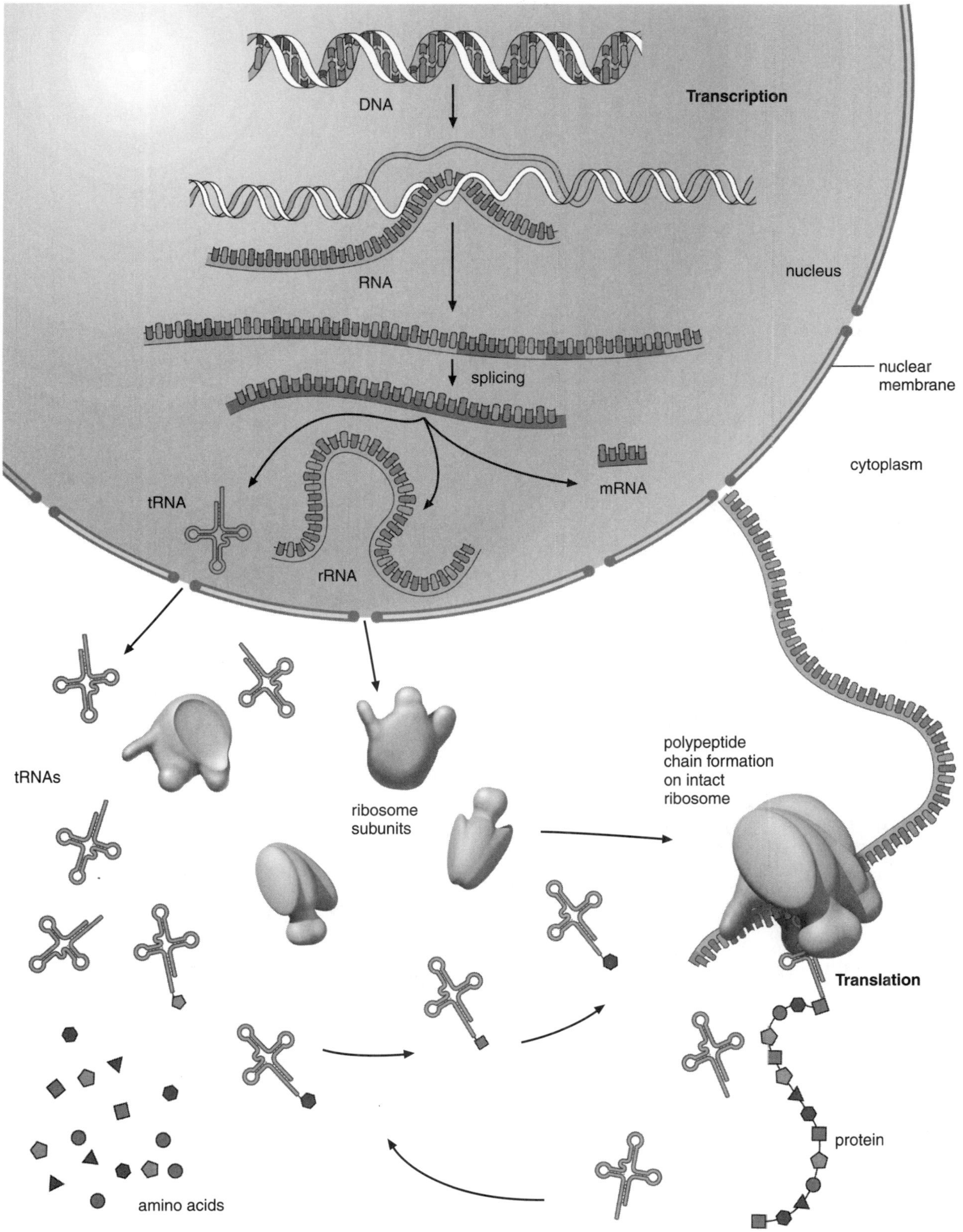

Figure 9.19 *Synthesis of Membrane Proteins*

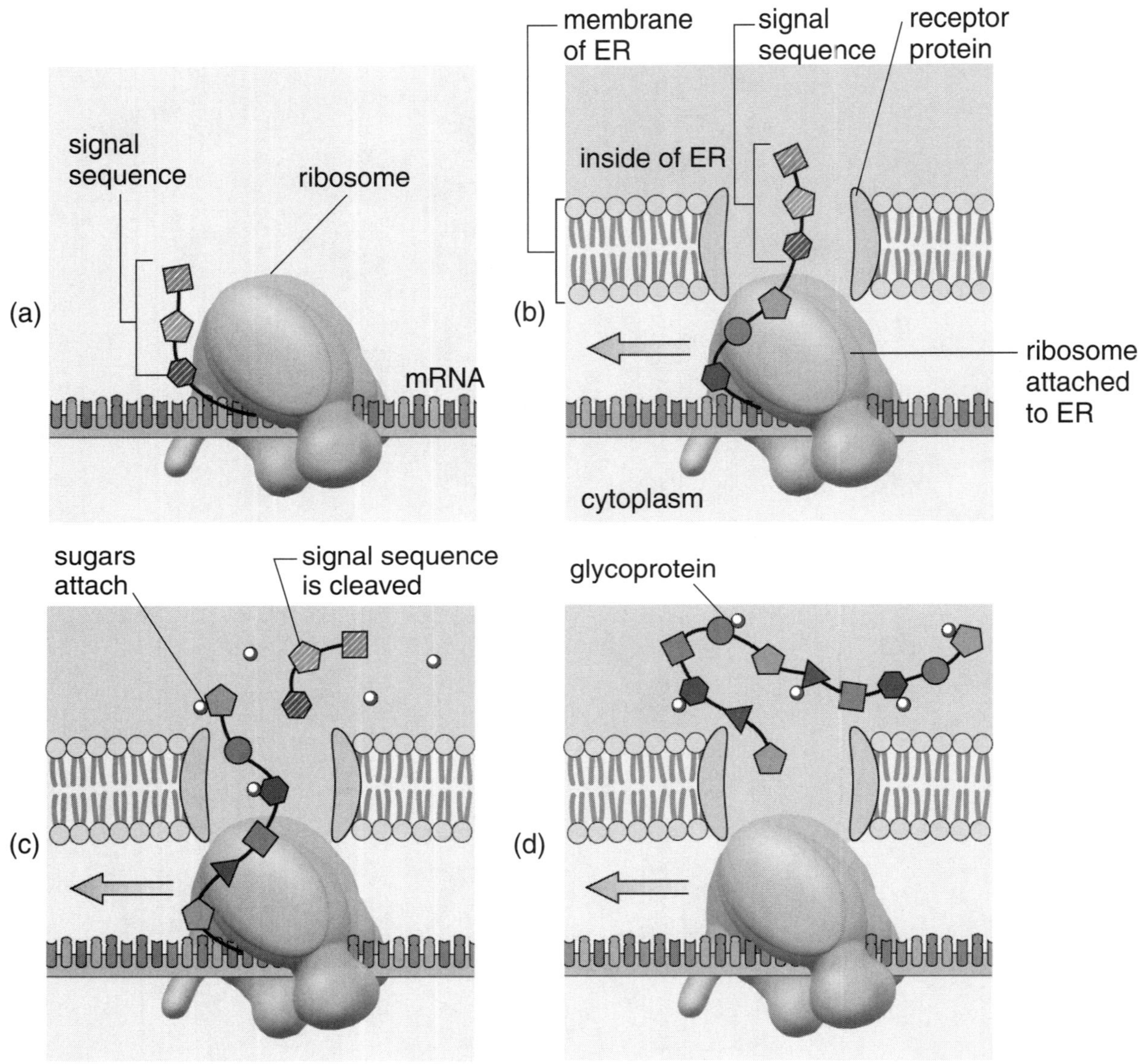


 

Figure 10.18 *Two Models of Differentiation* ◆

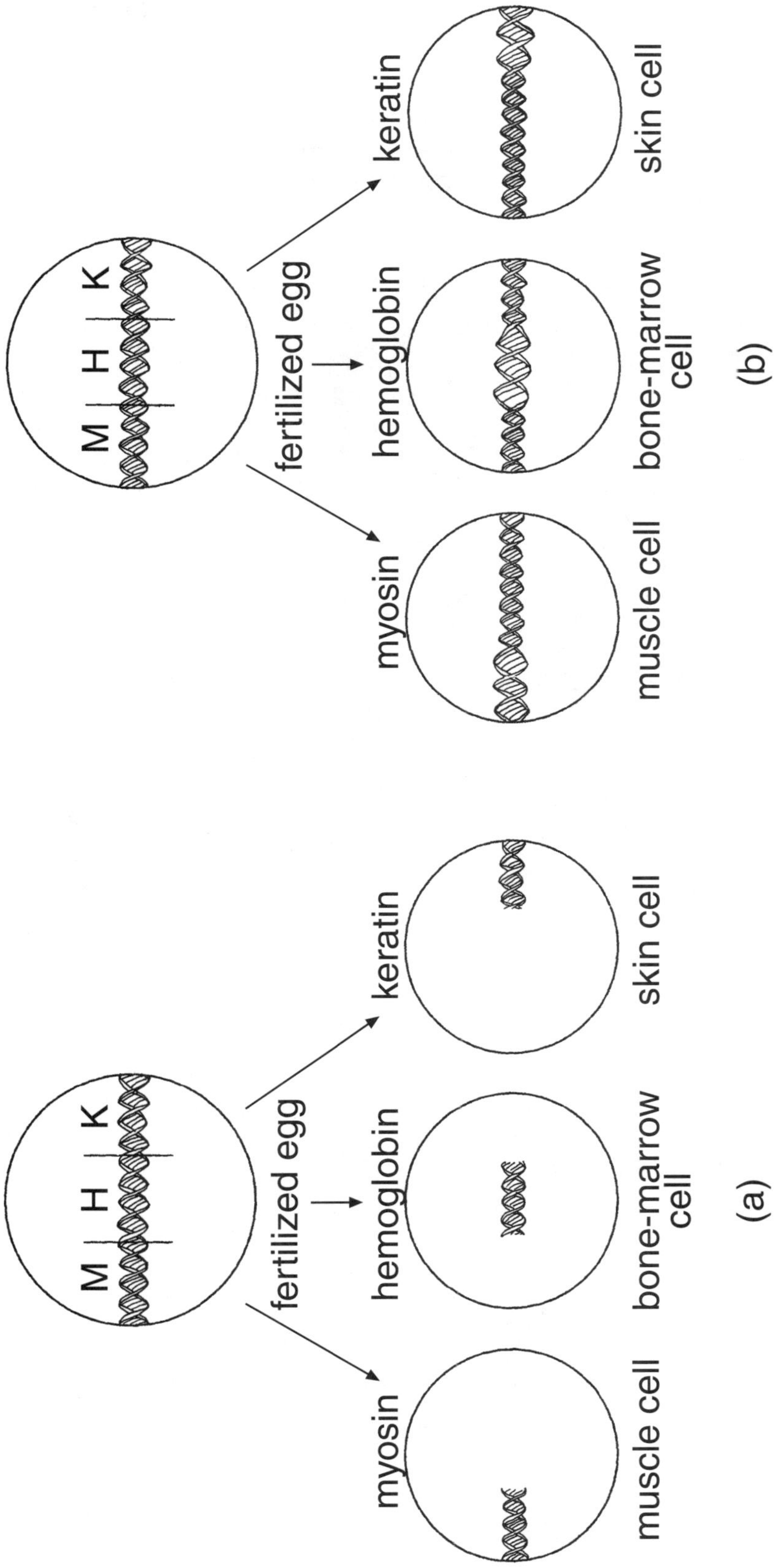

Figure 11.12 *Plant Growth and Development* ◆

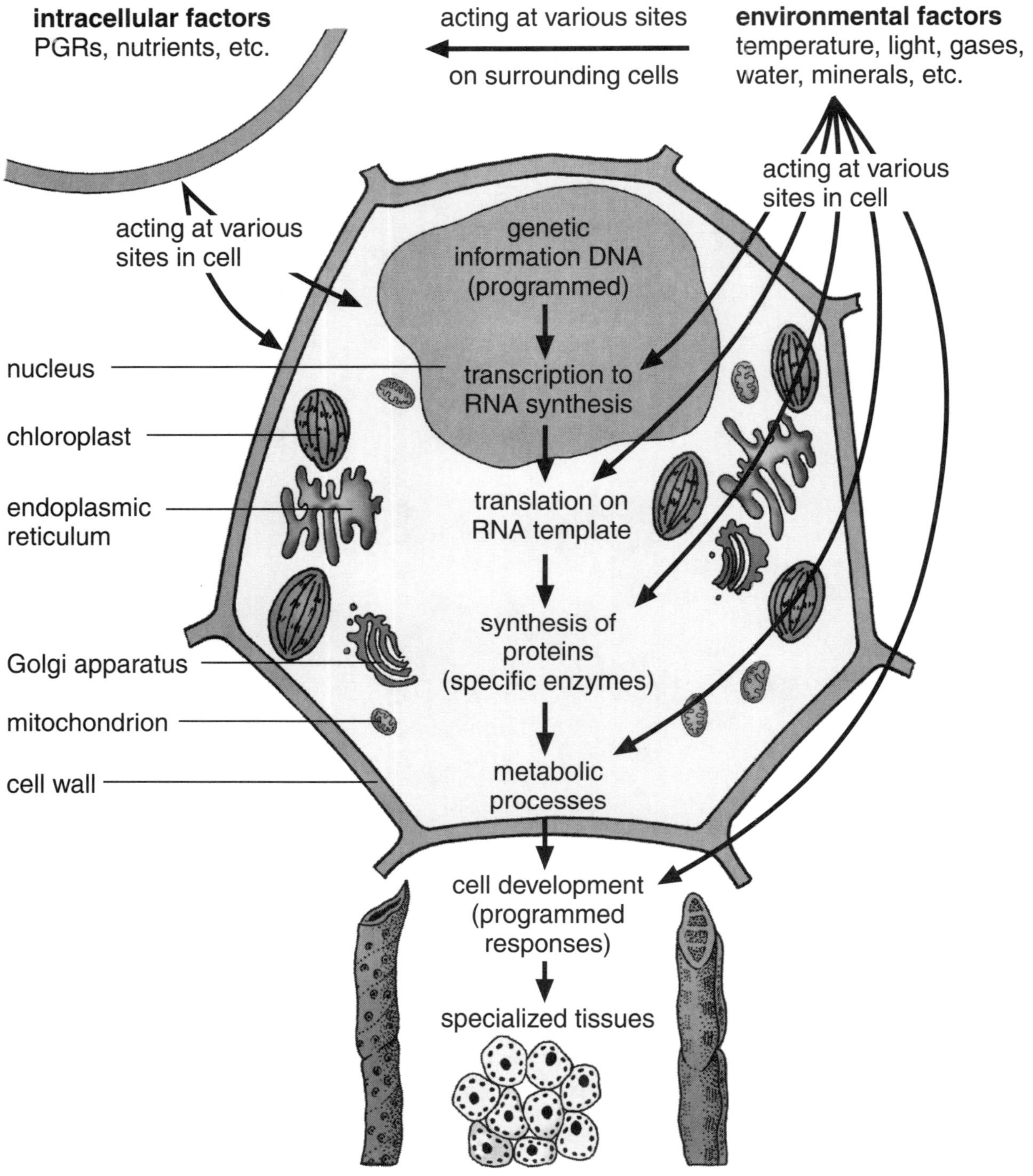

# Chapter 12 Reproduction

Figure 12.6 *The Stages of Meiosis* ◆

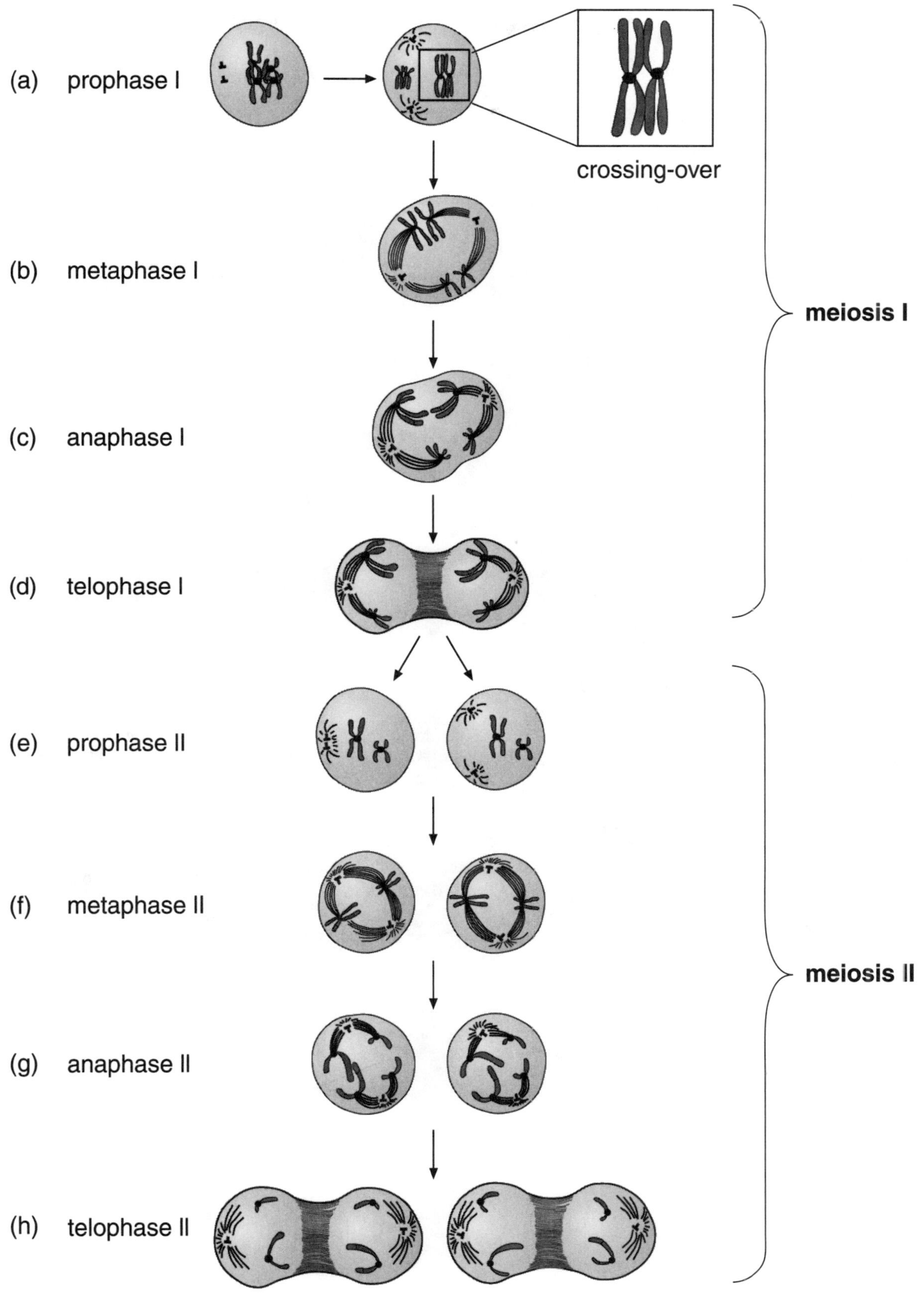

Figure 12.11 *Parts of a Flower* ◆

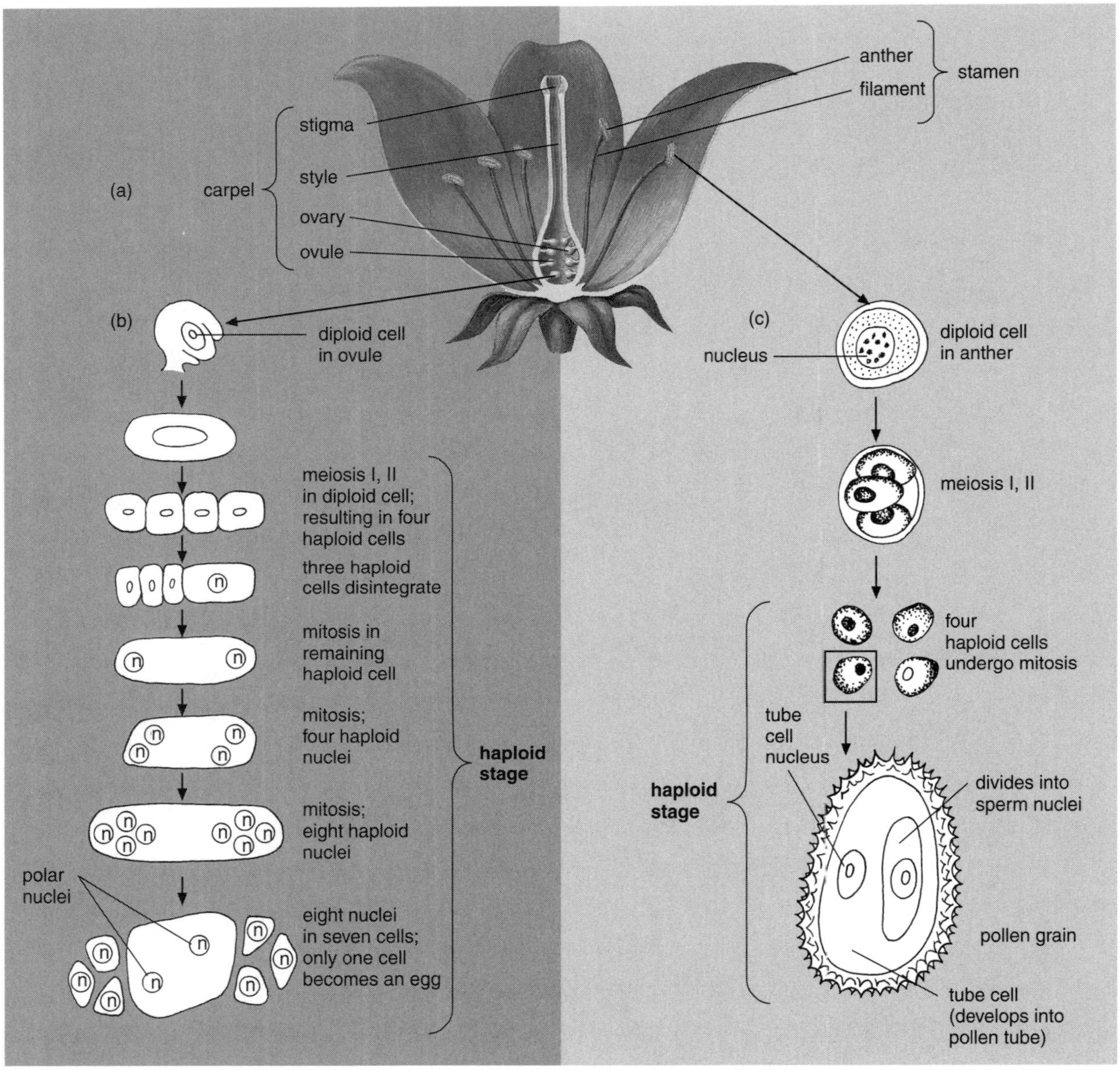


 

Figure 12.20 *The Menstrual Cycle* ◆

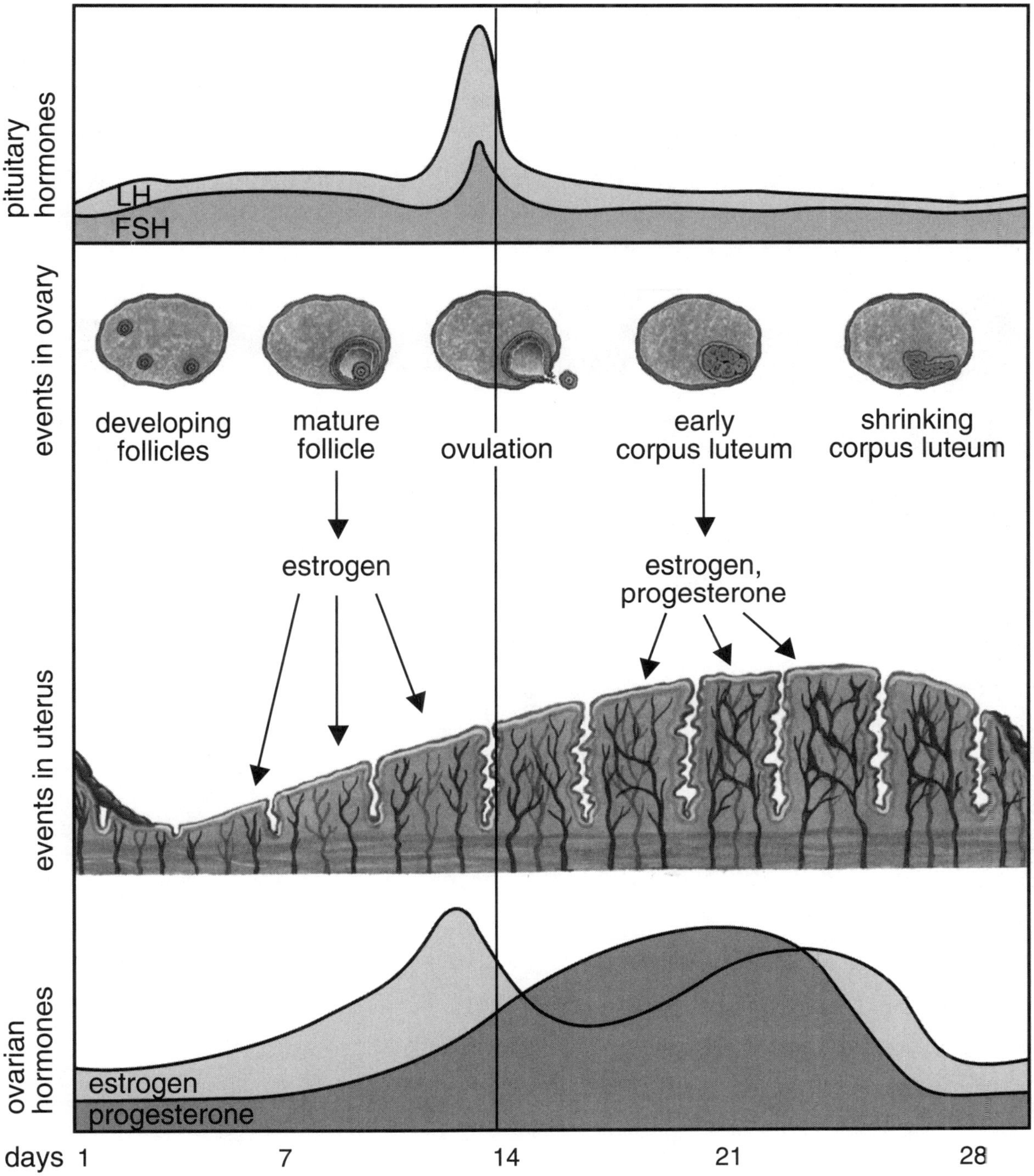

Figure 13.14 *Principle of Segregation* ◆

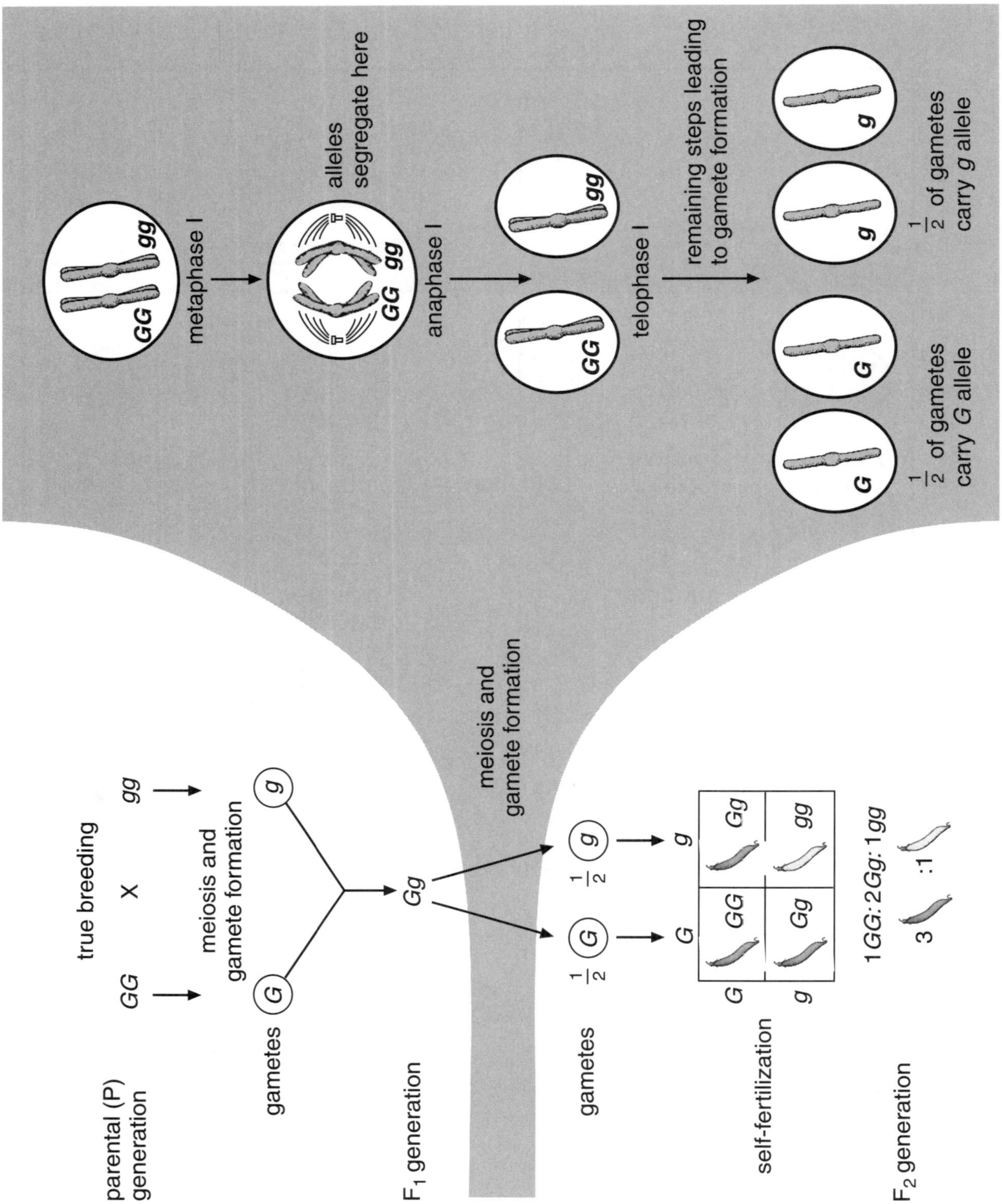

Figure 13.15 *Dihybrid Cross* ◆

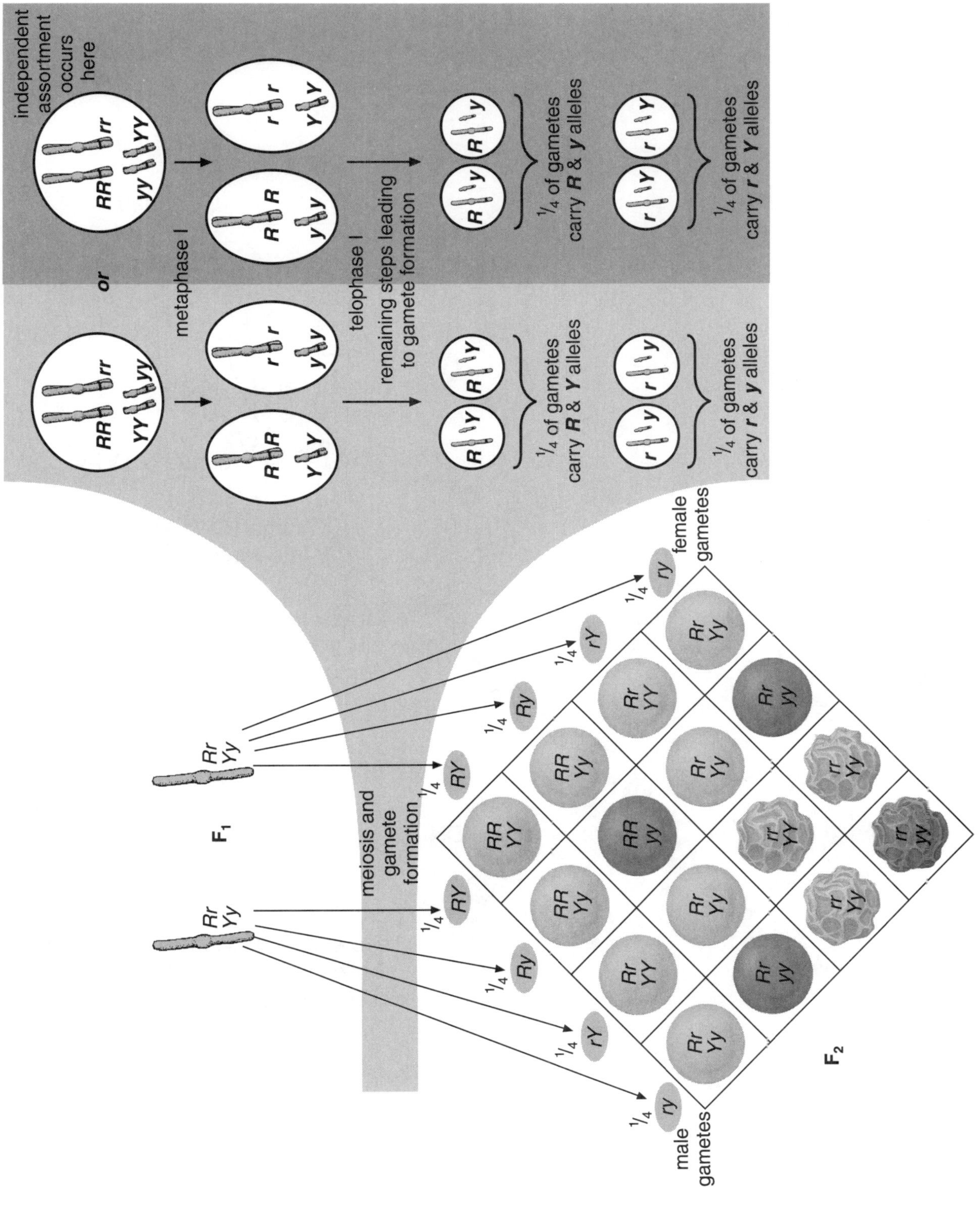

Figure 15.7 *Producing Recombinant DNA, Part 1* ◆

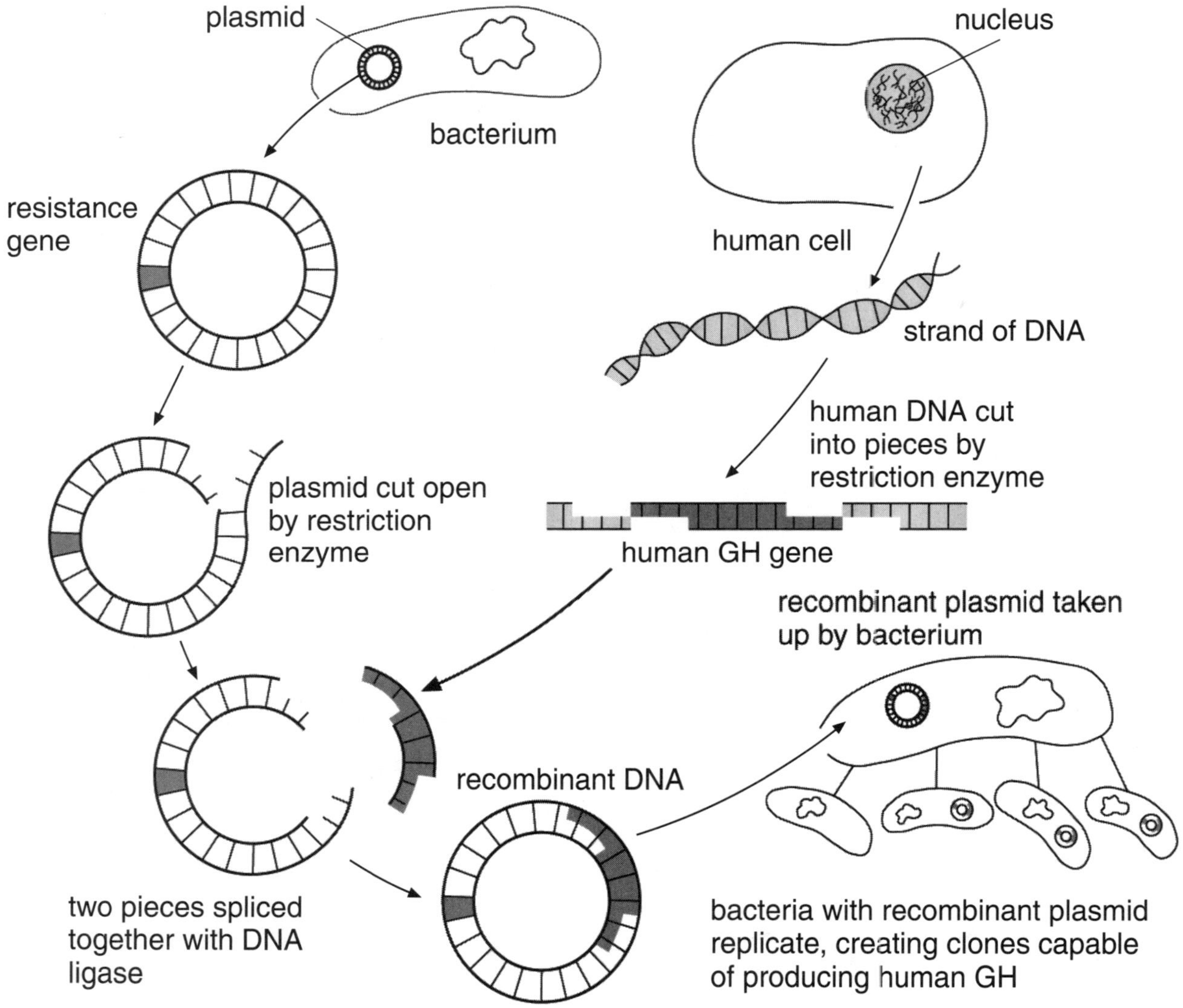

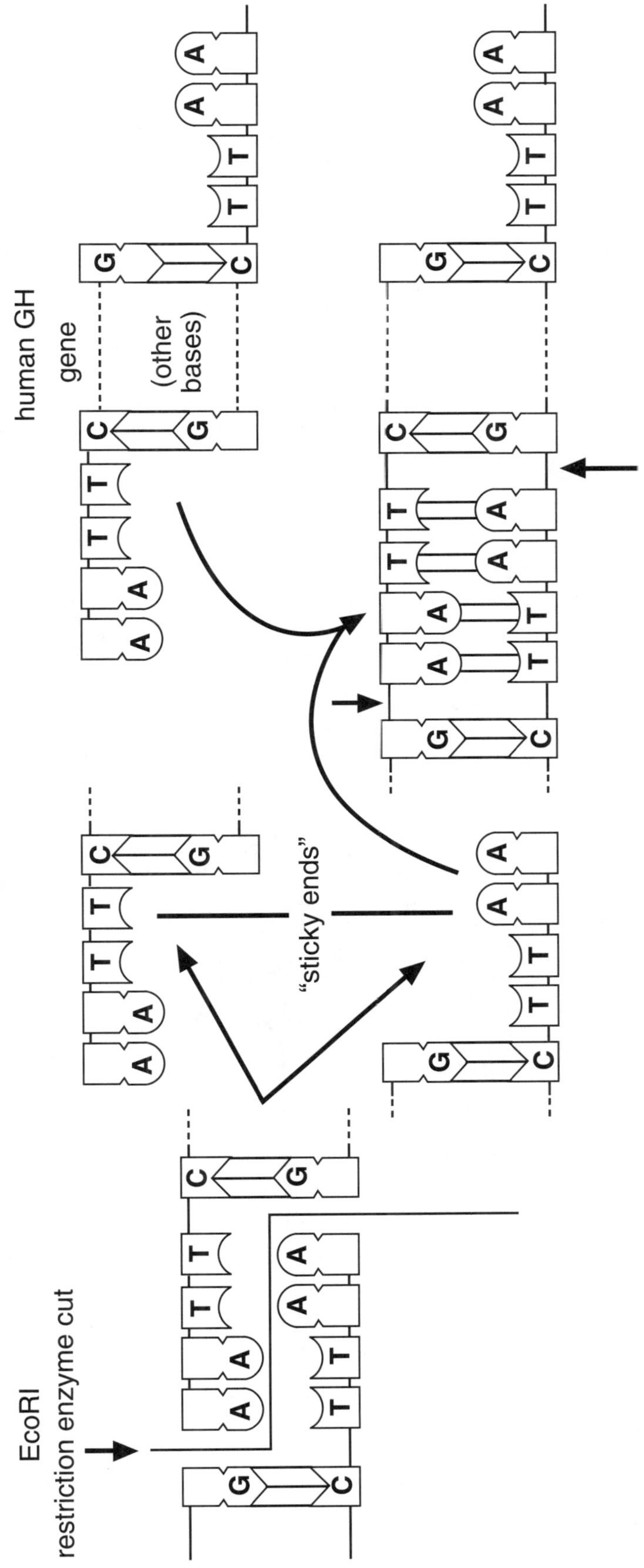

EcoRI
restriction enzyme cut
"sticky ends"
human GH
gene
(other
bases)

Figure 17.23 *Origin of Eukaryotic Cells* ◆

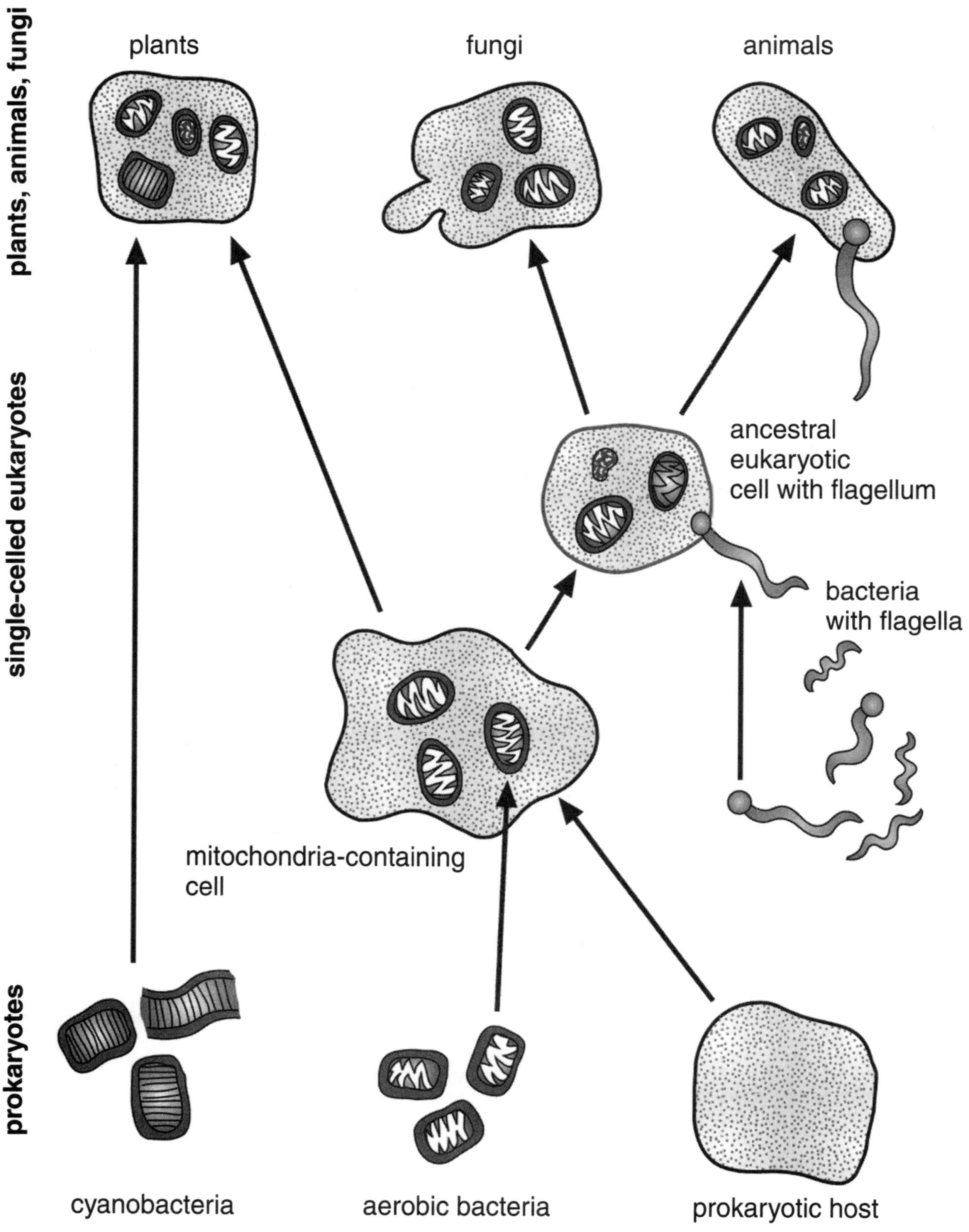


 

# The IS–Clad Method of Cladistics

## Data Table

**Title:** Traits of Plants

| Traits | tiger lily  | rose  | white pine  | sensitive fern  | polytrichum moss  |
|---|---|---|---|---|---|
| Flowers | √ | √ | X | X | X |
| Seeds | √ | √ | √ | X | X |
| Vascular tissue | √ | √ | √ | √ | X |

Note: √ = present, X = absent.

## Diagram

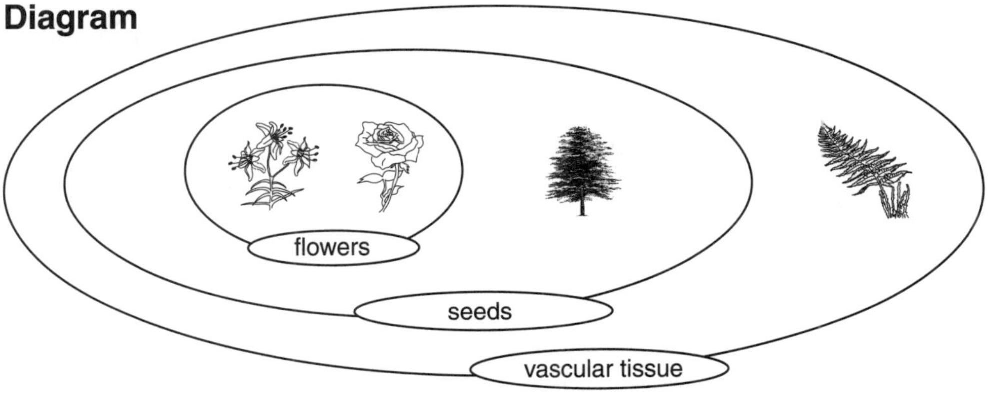


## Cladogram

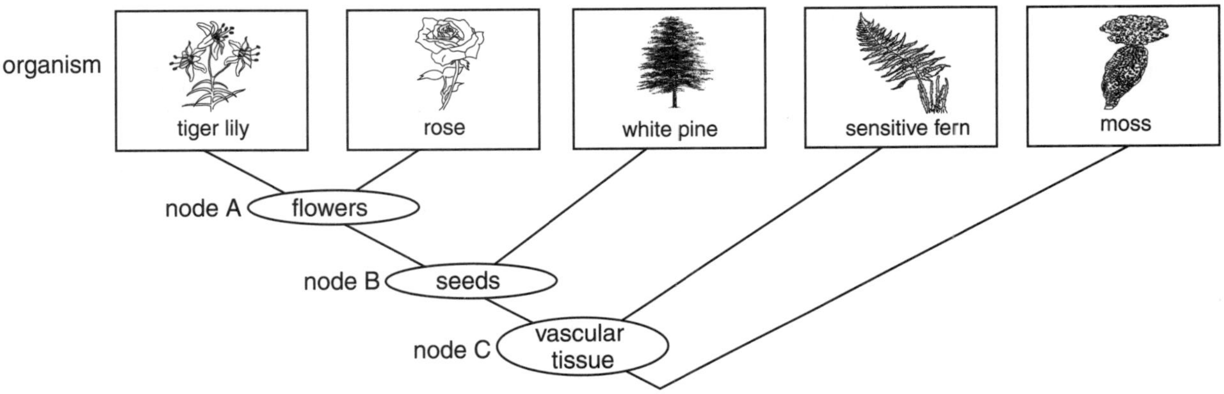

# Mammals

## Data Table

**Title:** Characteristics of Mammals for Cladistic Analysis

| Traits | Types of mammals (with specific examples) | | | | | | |
|---|---|---|---|---|---|---|---|
| | **Edentates** (armadillos) | **Sirenians** (manatees) | **Carnivores** (cats, dogs) | **Primates** (humans, chimps) | **Proboscideans** (elephants) | **South American Ungulates** (extinct) | **Perissodactyls** (horses) |
| **Middle–ear bone shaped like stirrup** | x | √ | √ | √ | √ | √ | √ |
| **Hooves** | x | √ | x | x | √ | √ | √ |
| **Eyes near the snout** | x | √ | x | x | √ | x | x |

Note: √ = present, X = absent.

## Diagram

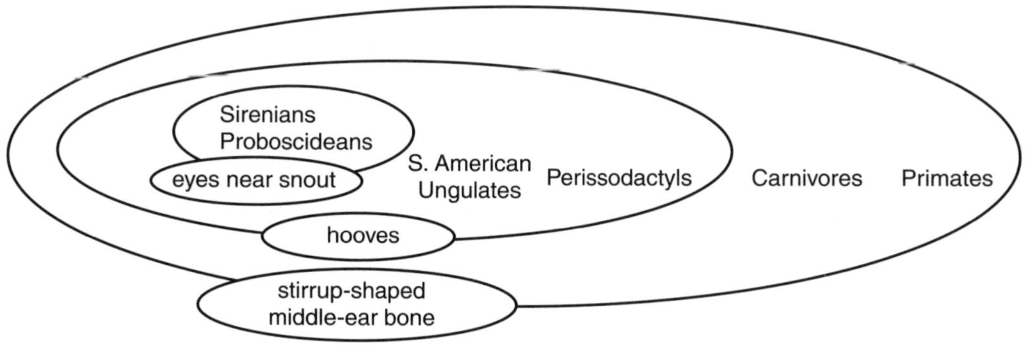

## Mammalian Cladogram

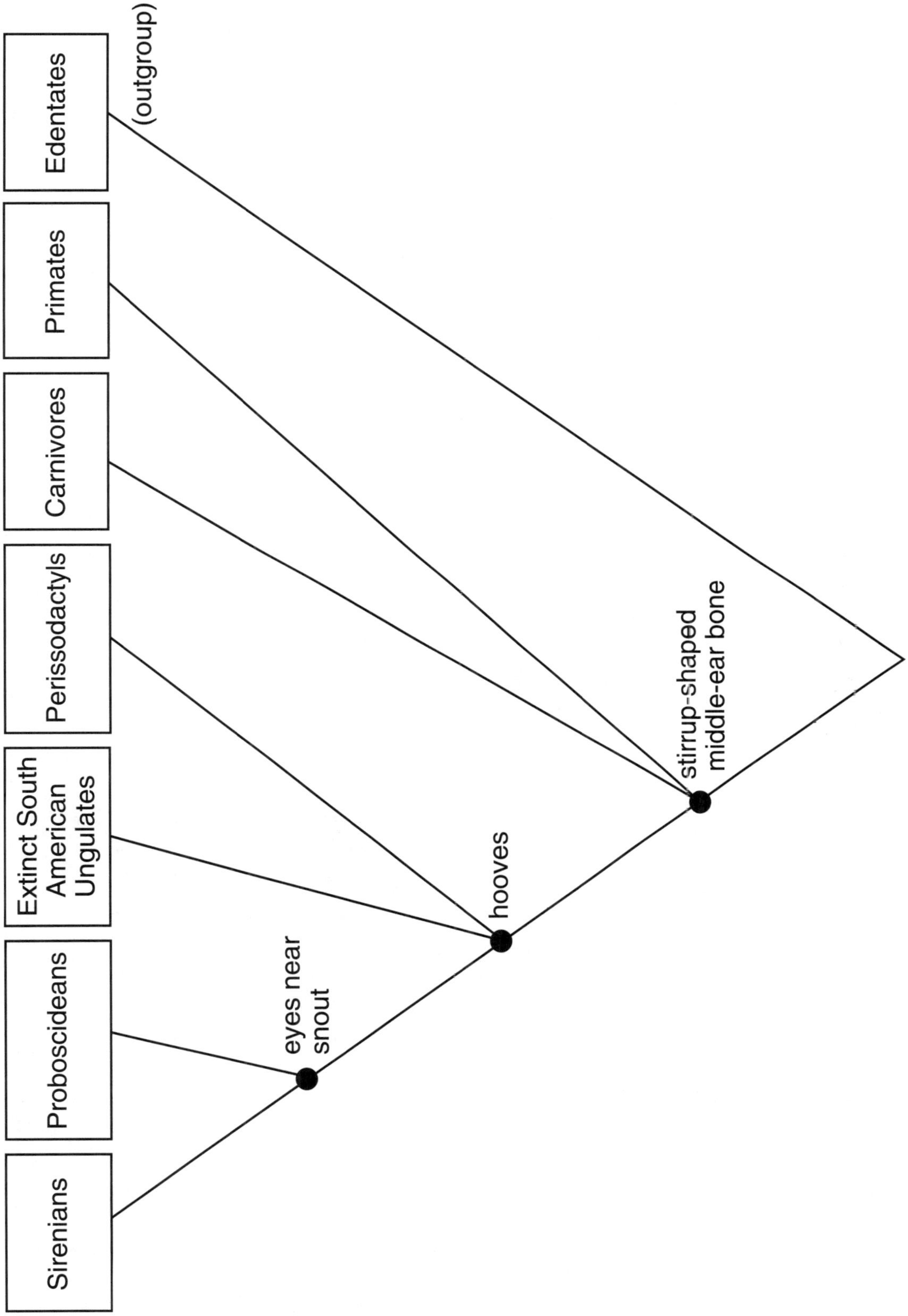

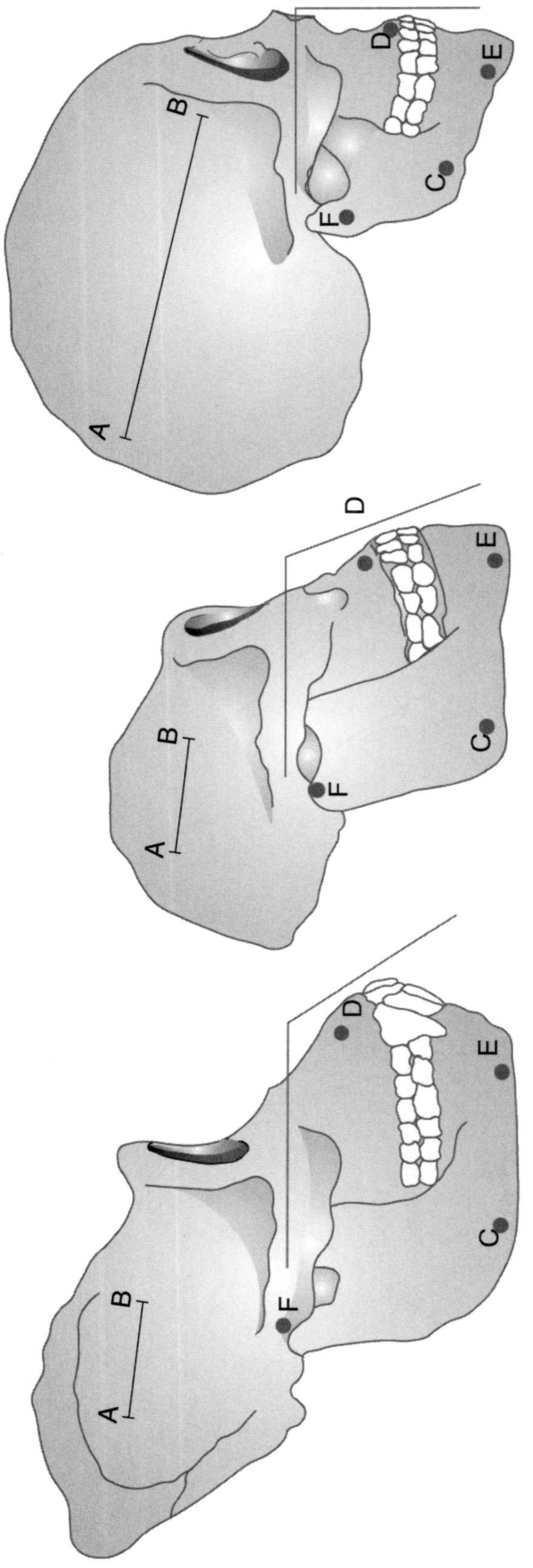


(c) (b) (a)

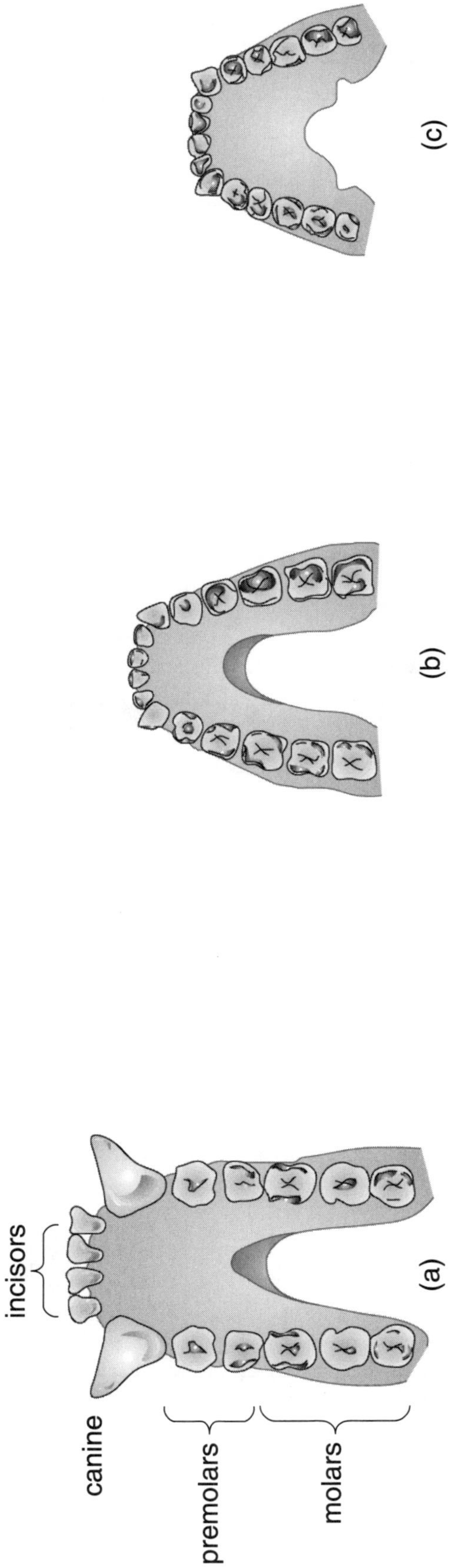

(c)
(b)
(a)
incisors
canine
premolars
molars


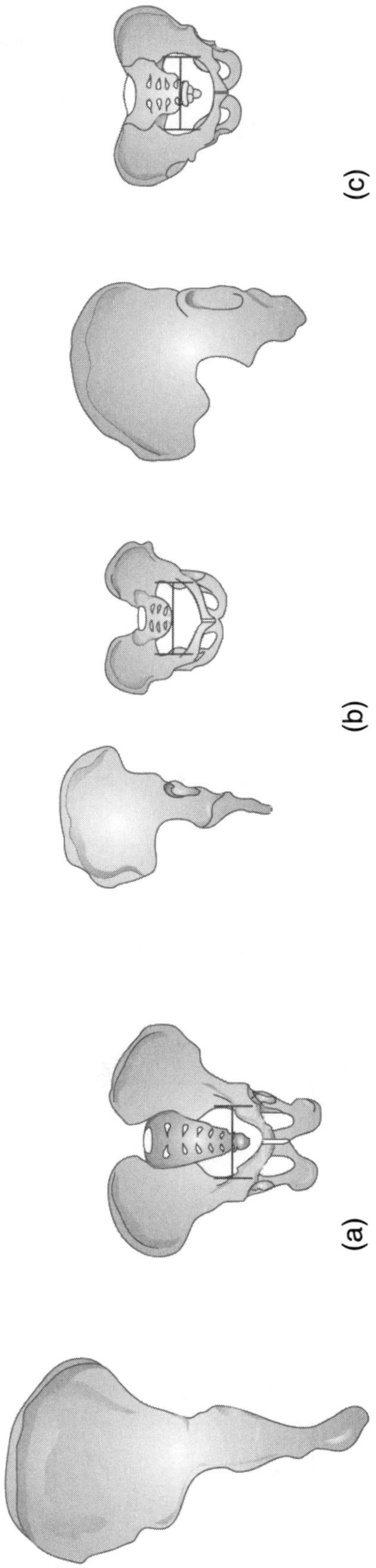

(c)
(b)
(a)

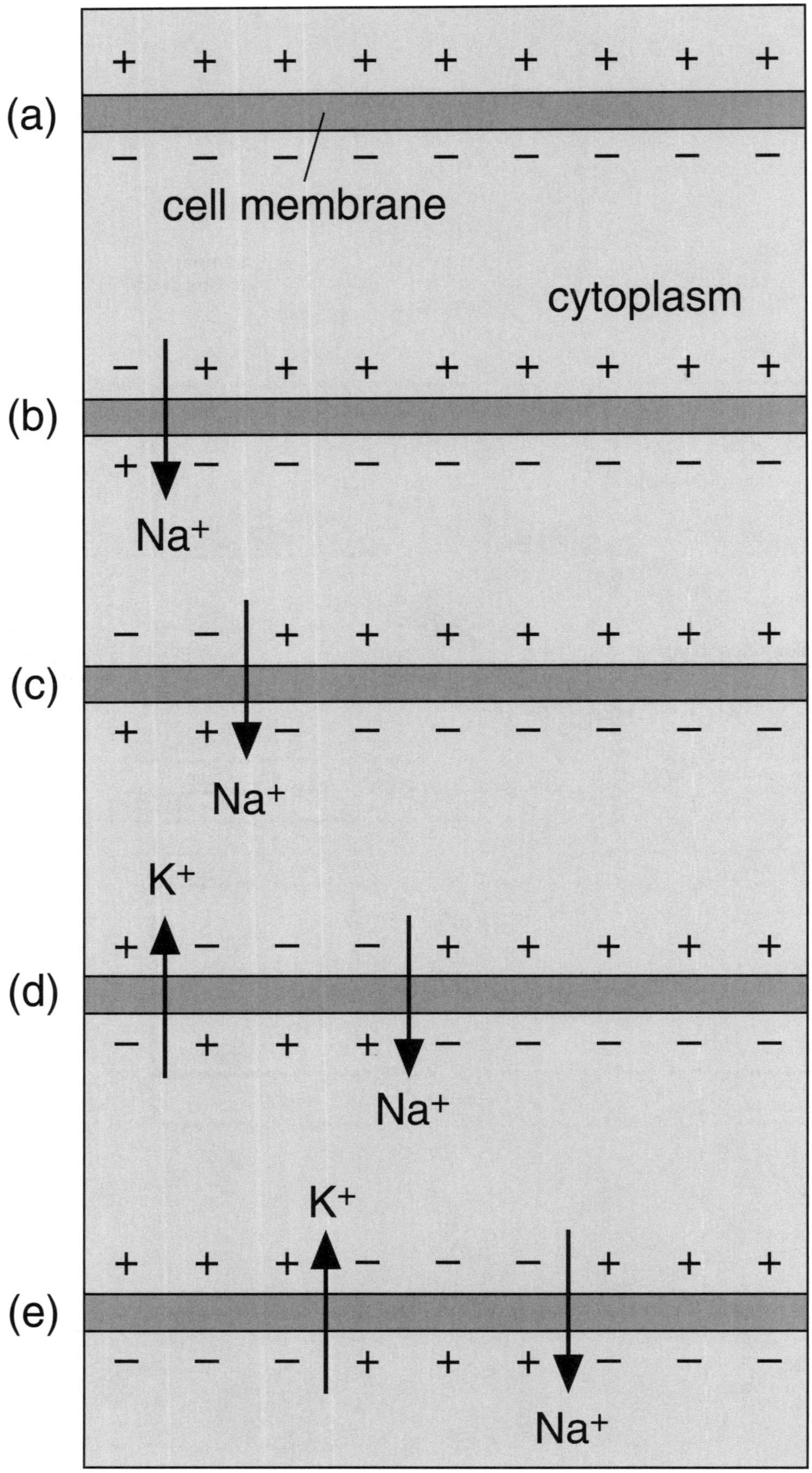
(a)
cell membrane
cytoplasm
(b)
Na+
(c)
Na+
(d)
K+
Na+
(e)
K+
Na+

 

# Chapter 21 Nervous Systems



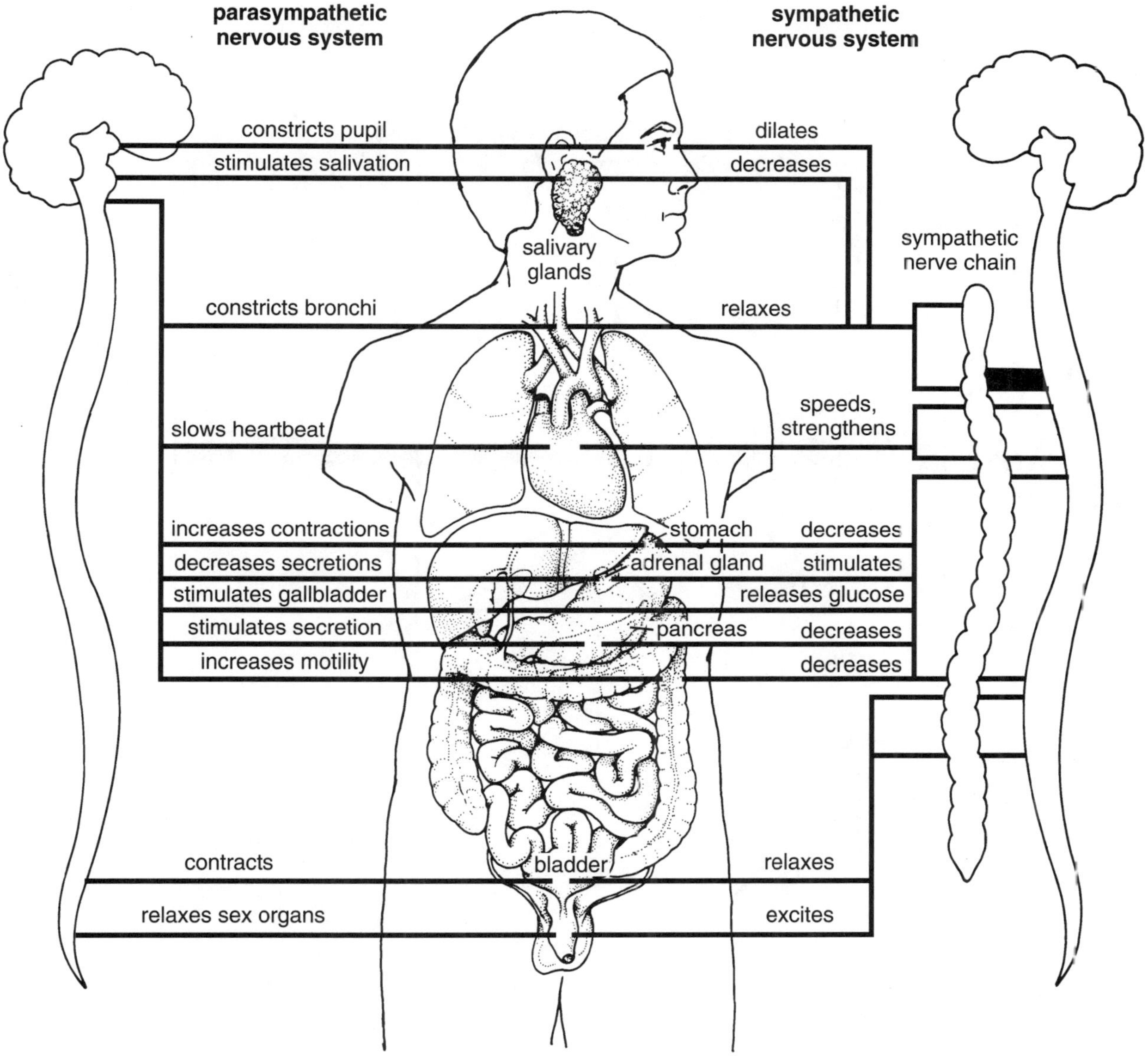

Figure 23.19 *Production of Antibodies* ◆

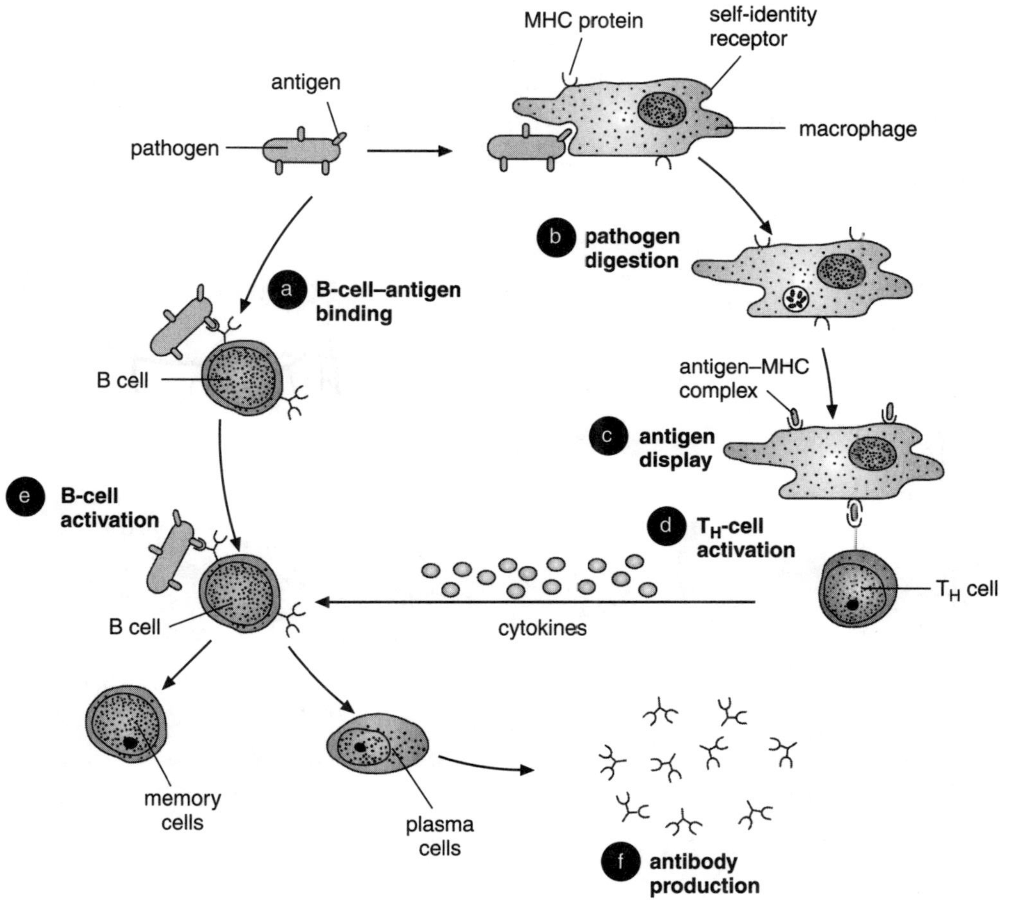

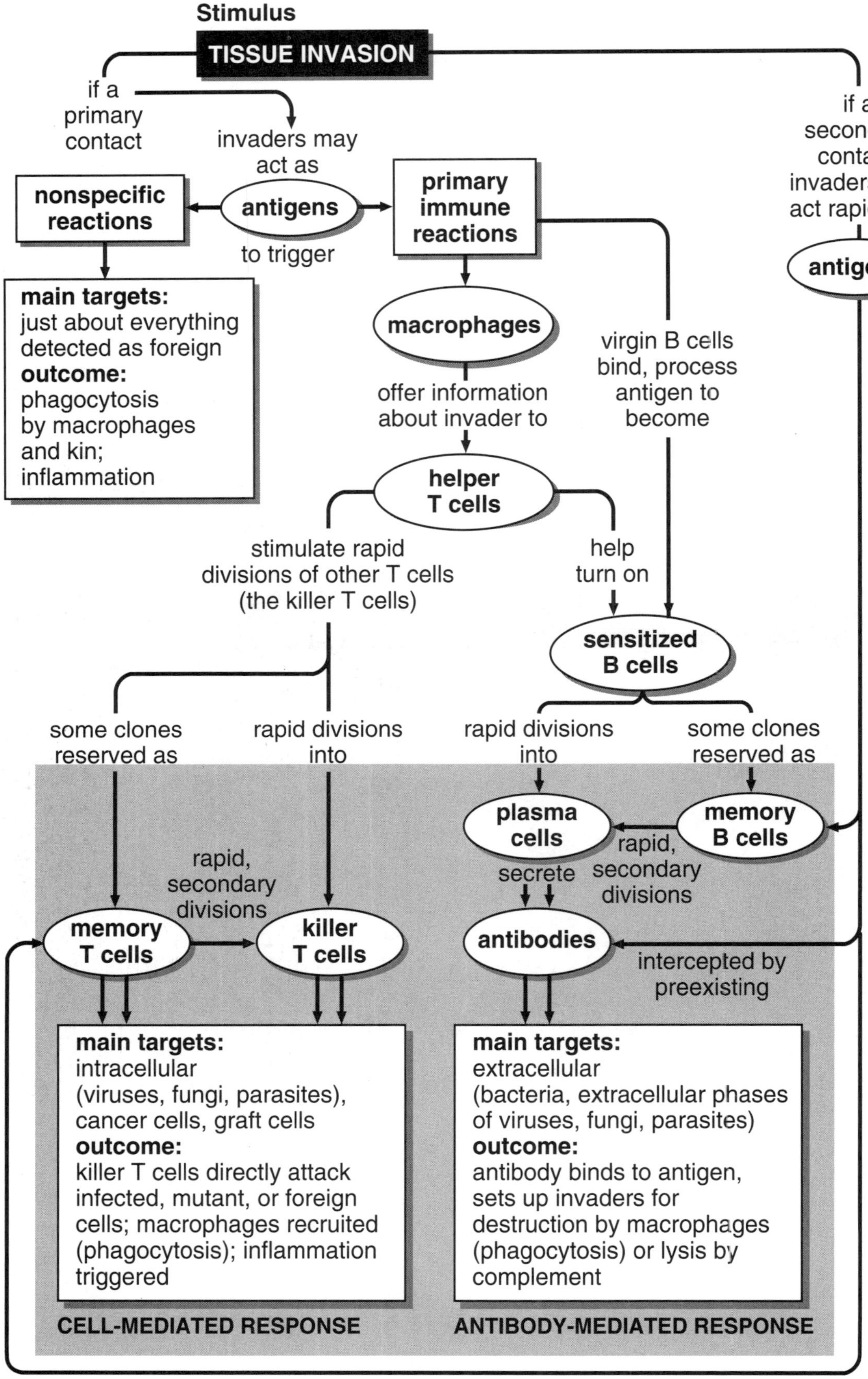
Stimulus
TISSUE INVASION
if a primary contact
invaders may act as
antigens
to trigger
nonspecific reactions
main targets: just about everything detected as foreign
outcome: phagocytosis by macrophages and kin; inflammation
primary immune reactions
macrophages
offer information about invader to
helper T cells
virgin B cells bind, process antigen to become
stimulate rapid divisions of other T cells (the killer T cells)
help turn on
sensitized B cells
if a secondary contact, invaders now act rapidly as
antigens
some clones reserved as
rapid divisions into
rapid divisions into
some clones reserved as
plasma cells
memory B cells
rapid, secondary divisions
secrete
antibodies
intercepted by preexisting
rapid, secondary divisions
memory T cells
killer T cells
main targets: intracellular (viruses, fungi, parasites), cancer cells, graft cells
outcome: killer T cells directly attack infected, mutant, or foreign cells; macrophages recruited (phagocytosis); inflammation triggered
main targets: extracellular (bacteria, extracellular phases of viruses, fungi, parasites)
outcome: antibody binds to antigen, sets up invaders for destruction by macrophages (phagocytosis) or lysis by complement
CELL-MEDIATED RESPONSE
ANTIBODY-MEDIATED RESPONSE

Figure 24.6 *Carbon Cycle* ◆

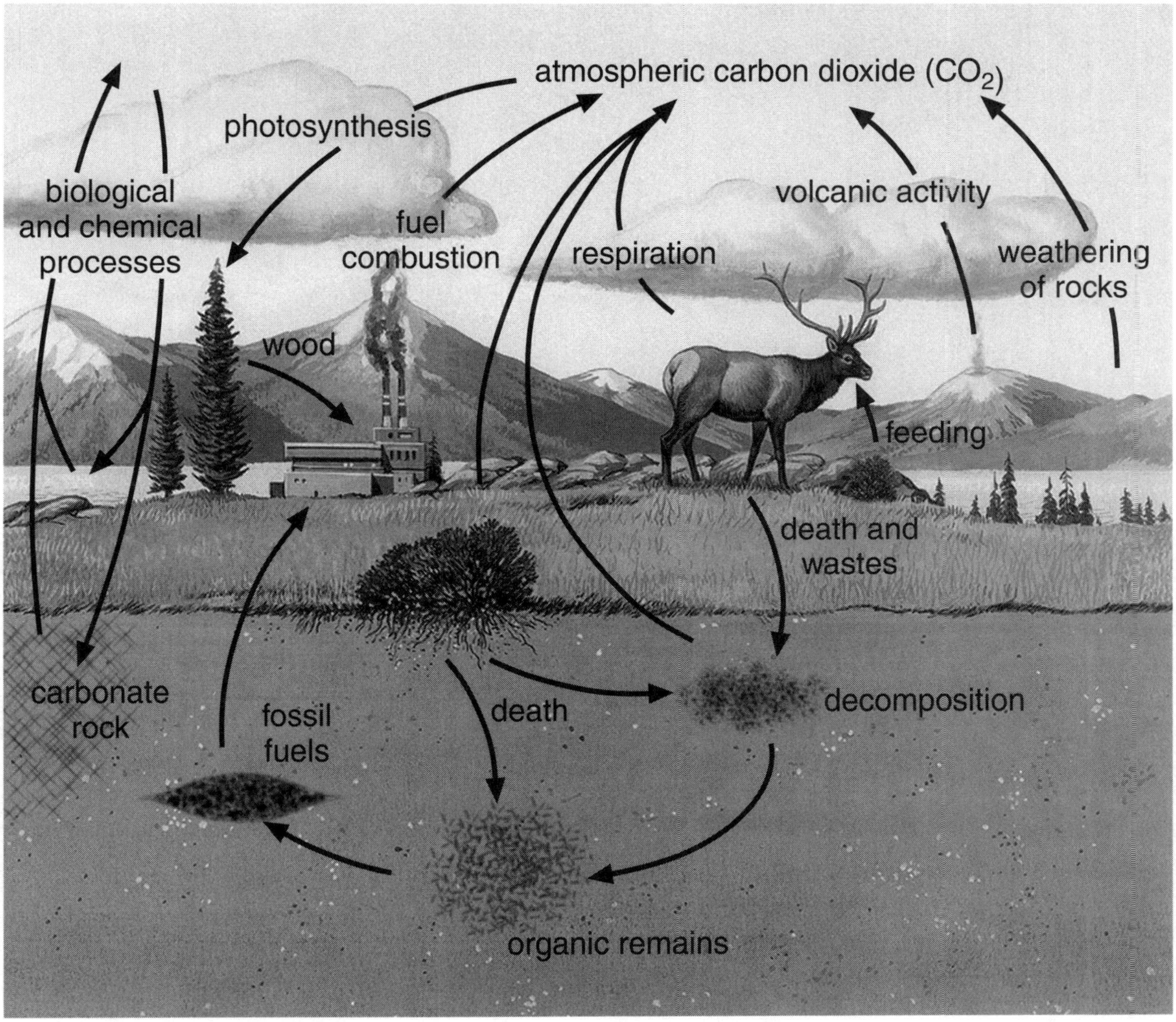

Figure 24.7 *Nitrogen Cycle* ◆

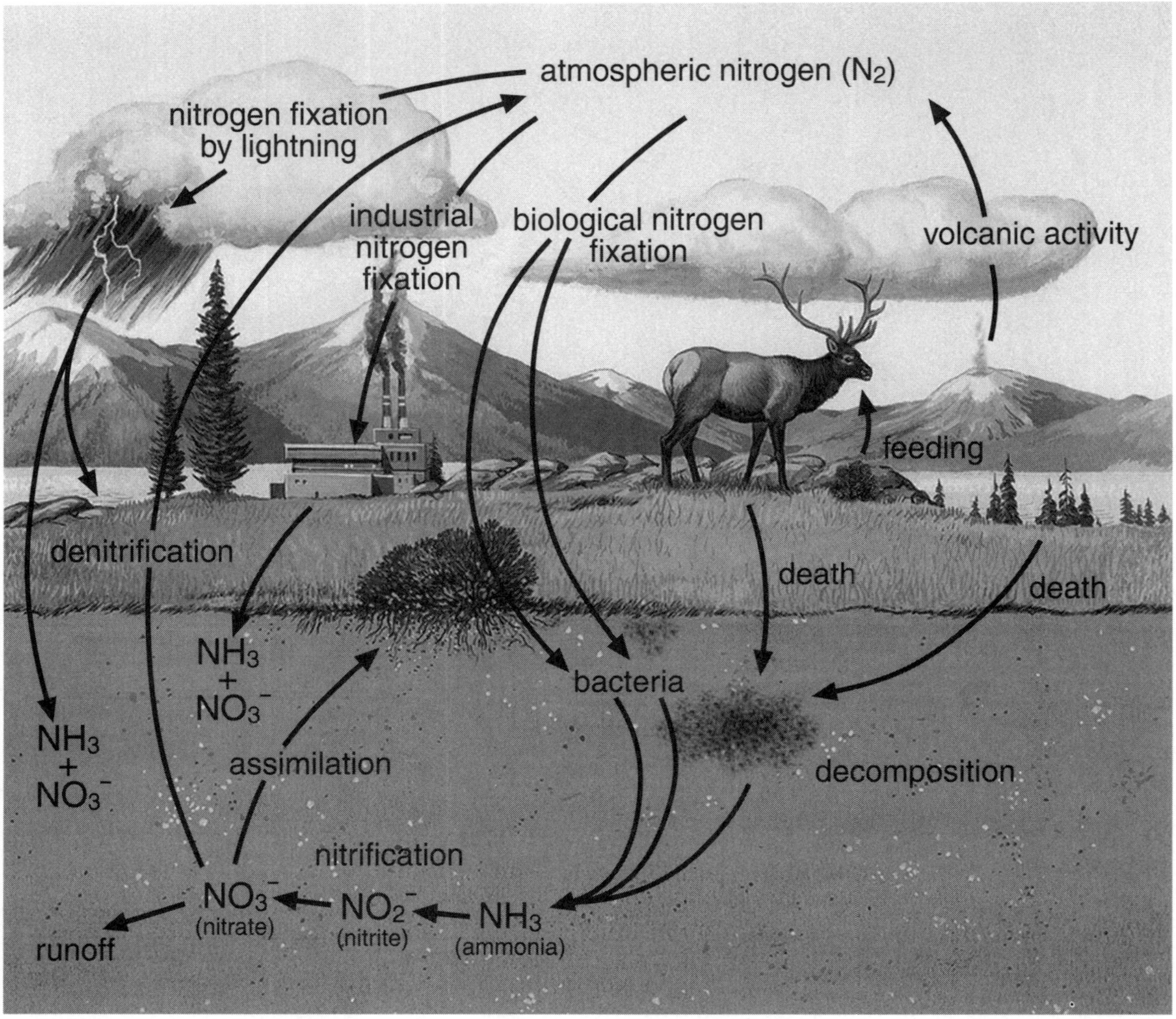